FLEXIBLE SENSORS
Materials, Devices and Applications

FLEXIBLE SENSORS
Materials, Devices and Applications

Editors

Guozhen Shen
Beijing Institute of Technology, China

Yang Li
University of Jinan, China

NEW JERSEY • LONDON • SINGAPORE • BEIJING • SHANGHAI • TAIPEI • CHENNAI

Published by

World Scientific Publishing Co. Pte. Ltd.
5 Toh Tuck Link, Singapore 596224
USA office: 27 Warren Street, Suite 401-402, Hackensack, NJ 07601
UK office: 57 Shelton Street, Covent Garden, London WC2H 9HE

Library of Congress Cataloging-in-Publication Data
Names: Shen, Guozhen (Electrical engineer) editor | Li, Yang (Professor of electrical engineering) editor
Title: Flexible sensors : materials, devices and applications / edited by Guozhen Shen, Beijing Institute of Technology, China, Yang Li. University of Jinan, China
Description: New Jersey : World Scientific, [2025] | Includes bibliographical references and index.
Identifiers: LCCN 2024056954 | ISBN 9789811266850 hardcover |
  ISBN 9789811266867 ebook for institutions | ISBN 9789811266874 ebook for individuals
Subjects: LCSH: Flexible electronics | Detectors | Wearable technology
Classification: LCC TK7872.F54 F558 2025
LC record available at https://lccn.loc.gov/2024056954

British Library Cataloguing-in-Publication Data
A catalogue record for this book is available from the British Library.

Copyright © 2025 by World Scientific Publishing Co. Pte. Ltd.

All rights reserved. This book, or parts thereof, may not be reproduced in any form or by any means, electronic or mechanical, including photocopying, recording or any information storage and retrieval system now known or to be invented, without written permission from the publisher.

For photocopying of material in this volume, please pay a copying fee through the Copyright Clearance Center, Inc., 222 Rosewood Drive, Danvers, MA 01923, USA. In this case permission to photocopy is not required from the publisher.

For any available supplementary material, please visit
https://www.worldscientific.com/worldscibooks/10.1142/13156#t=suppl

Desk Editors: Soundararajan Raghuraman/Steven Patt

Typeset by Stallion Press
Email: enquiries@stallionpress.com

About the Editors

Shen Guozhen is a professor, doctoral supervisor, and the winner of the National Outstanding Youth Science Fund. He received his PhD from the University of Science and Technology of China in 2003. He is currently working at the School of Integrated Circuits and Electronics, Beijing Institute of Technology, as the Director of the Institute of Flexible Electronic Devices and Intelligent Manufacturing. His research interests include the construction of low-dimensional semiconductor materials, flexible electronic devices, and system applications. He has published more than 300 Sci-indexed papers in *Nature Communications, Science Advances, Advanced Materials*, and other journals, and his work carries an h-index of 97. He has edited five Chinese and English monographs. He won the second prize at the Beijing Science and Technology Awards and the first prize at the Science and Technology Awards of the Chinese Materials Research Society. He was selected as a "New Century Outstanding Talent" by the Ministry of Education, a "Global Highly Cited Scientist", and Elsevier China's Highly Cited Scholar. He was also selected as an RSC Fellow, Vice President of the Nanomaterials and Devices Branch, Chinese Materials Research Society. He is also the deputy editor or editor of international journals such as *Advanced Materials Technologies, Advanced Sensor Research*, and *Science China Materials*.

 **Li Yang** is a professor, doctoral supervisor, senior member of the IEEE, and the winner of the Shandong Outstanding Youth Fund. He was selected as a Sino-Korean Young Scientist by the Ministry of Science and Technology, Shandong Taishan Scholar, leader of the Shandong University Integrated Circuit Innovation Team, candidate for the Shandong Youth Science and Technology Talent Promotion Project, and Qilu Young Scholar. He is currently working at Shandong University at school of Microelectronics. He has presided over 15 key projects at the provincial and ministerial levels. He has published more than 150 Sci-indexed papers, including 12 cover papers, 12 highly cited papers, and 4 hot-topic papers. He also had 30 authorized national invention patents and 11 Korean invention patents to his name. He is the first author or corresponding author for many articles, which have been published in top journals such as *PNAS, Chemical Society Reviews, Matter, Advanced Materials, Advanced Functional Materials, Advanced Science*, and *Nano Letters*. He is the contributing editor and editorial board member of more than 10 journals, including Journal of Semiconductors and Nano-Micro Letters.

© 2025 World Scientific Publishing Company
https://doi.org/10.1142/9789811266867_fmatter

Contents

About the Editors   v

Chapter 1   Introduction   1
Feifei Yin, Peng Wang, Li Yang, and Shen Guozhen

Chapter 2   Basic Composition and Functional Materials   7
Feifei Yin, Peng Wang, Li Yang, and Shen Guozhen

Chapter 3   Flexible Sensors   143
Yunjian Guo, Peng Wang, Li Yang, and Shen Guozhen

Chapter 4   Sensing Systems and Applications   269
Hongsen Niu, Peng Wang, Li Yang, and Shen Guozhen

Index   373

Chapter 1

Introduction

Feifei Yin[*], Peng Wang[†], Li Yang[‡,¶], and Shen Guozhen[§,]**

[*]*East China Institute of Photo-Elextro ICs,
No. 89 Longshan Road, Su Zhou 215163, China*
[†]*School of Mechanical Engineering, University of Jinan,
Jinan 250022, China*
[‡]*School of Integrated Circuits, Shandong University,
Jinan 250101, China*
[§]*School of Integrated Circuits and Electronics, Beijing Institute
of Technology, Beijing 100081, China*
[¶]*yang.li@sdu.edu.cn*
[**]*gzshen@bit.edu.cn*

1. Background

Flexible sensors have emerged as a revolutionary technology that holds immense potential in various fields, including healthcare,[1–5] robotics,[6–10] consumer electronics,[11–15] and environmental monitoring.[16–20] These sensors offer several advantages over traditional rigid sensors, such as conformability, lightweight, and the ability to monitor physiological signals and environmental parameters with enhanced sensitivity and accuracy. They have opened new avenues for the development of wearable devices, smart textiles, and flexible electronics, revolutionizing the way humans interact with technology.

The concept of flexible sensors is rooted in the growing demand for unobtrusive, comfortable, and versatile sensing solutions that can seamlessly integrate with the human body or adapt to irregular surfaces. The field of flexible sensors has evolved rapidly in recent years, driven by advancements in materials science, nanotechnology, and manufacturing techniques. This progress has paved the way for the development of sensors capable of bending, stretching, and conforming to the curvatures of different objects or the human body, while maintaining their sensing capabilities.

One of the key components in flexible sensors is the choice of flexible and stretchable materials. These materials should exhibit mechanical flexibility, durability, and electrical conductivity to ensure reliable and consistent sensor performance. Various materials have been explored, including polymers, nanomaterials, conductive textiles, and hybrid composites. Polymers such as polydimethylsiloxane (PDMS), polyimide (PI), and polyurethane (PU) have shown exceptional mechanical flexibility, making them suitable for sensor fabrication. Nanomaterials such as carbon nanotubes (CNTs), graphene, and metal nanowires offer excellent electrical conductivity and mechanical strength, enabling the development of highly sensitive and stretchable sensors.

The design and fabrication of flexible sensors involve integrating these flexible materials into sensor structures that can detect and transduce specific signals. Different types of flexible sensors have been developed, including strain sensors, pressure sensors, temperature sensors, humidity sensors, gas sensors, and biosensors. These sensors employ various transduction mechanisms, such as resistance changes, capacitance variations, piezoresistive effects, piezoelectricity, and optical or chemical changes, to convert physical or chemical signals into measurable electrical signals.

Flexible sensors have found numerous applications in healthcare and wellness monitoring. They can be used for real-time monitoring of vital signs, such as heart rate, respiration rate, blood pressure, and body temperature. By integrating these sensors into wearable devices, individuals can track their health and fitness levels, leading to personalized and preventive healthcare. Flexible sensors also have the potential to revolutionize medical diagnostics by enabling non-invasive monitoring of biomarkers, detecting diseases at an early stage, and facilitating point-of-care testing.

In the field of robotics, flexible sensors play a crucial role in enabling robots to interact with their environment and humans in a safe and intuitive manner. These sensors can provide tactile sensing capabilities, allowing robots to detect and respond to touch, pressure, and force. This opens up

Introduction 3

possibilities for applications, such as prosthetics, robotic surgery, human-robot collaboration, and soft robotics. Flexible sensors are also being integrated into smart textiles, enabling the development of interactive and responsive clothing that can monitor body posture, movement, and biometrics.

Furthermore, flexible sensors have significant implications for environmental monitoring and safety. They can be used for real-time monitoring of air quality, water quality, and hazardous gases, enabling early detection of pollution and ensuring a safer environment. These sensors can also be integrated into structural health monitoring systems to assess the condition and integrity of infrastructure, such as bridges, buildings, and pipelines, thereby enhancing safety and preventing potential disasters.

Despite the significant progress made in the field of flexible sensors, several challenges remain. The development of reliable and scalable fabrication processes as well as the integration of multiple sensing capabilities into a single device are ongoing areas of research. Improving the durability and long-term stability of flexible sensors, especially under harsh conditions, is another important consideration. Additionally, ensuring the compatibility of these sensors with existing electronics, wireless communication, and power sources is crucial for their successful integration into practical applications.

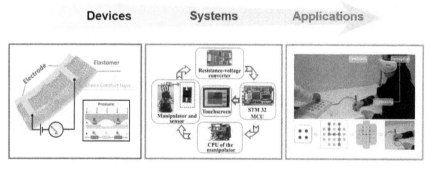

Figure 1. Flexible sensors are enabled by extensive materials for the construction of various intelligent systems and charming applications. Device: Reproduced with permission from Ref. [21]. (Copyright 2019, Elsevier Inc.), System: Reproduced with permission from Ref. [22]. (Copyright 2022, Wiley-VCH), Application: Reproduced with permission from Ref. [23]. (Copyright 2021).

In conclusion, flexible sensors have the potential to revolutionize various industries by providing unobtrusive, lightweight, and conformable sensing solutions. The integration of flexible materials, advanced fabrication techniques, and novel transduction mechanisms have paved the way for the development of highly sensitive, stretchable, and versatile sensors. With continued advancements, flexible sensors are expected to play a pivotal role in shaping the future of healthcare, robotics, consumer electronics, and environmental monitoring, leading to a more connected, personalized, and safer world (Figure 1).

2. Objective and Outline of the Book

This book aims to provide a comprehensive overview of the different types of flexible sensors, including strain sensors, pressure sensors, temperature sensors, humidity sensors, gas sensors, and biosensors. It will emphasize the significant advancements in materials, fabrication techniques, and transduction mechanisms that have contributed to the development of flexible sensors. The performance of flexible sensors will be evaluated based on metrics, such as sensitivity, accuracy, response time, durability, and stability. The factors influencing these performance characteristics will be discussed along with the strategies employed to enhance the sensing capabilities of flexible sensors. Furthermore, this book will explore the wide range of applications where flexible sensors have demonstrated promise. These applications encompass areas such as healthcare monitoring, robotics, smart textiles, environmental monitoring, and structural health monitoring. The specific requirements and challenges within each application domain will be examined, highlighting how flexible sensors effectively address these challenges. Overall, this book will serve as a valuable resource for researchers, engineers, and professionals interested in the field of flexible sensor technology. It will provide insights into the latest advancements, performance characteristics, and applications of flexible sensors, facilitating the understanding and development of innovative sensing solutions.

Chapter 2 introduces the flexible sensors that utilize extensive flexible materials with unique advantages such as thinness, softness, bendability, and stretchability. Functional material plays a vital role in affecting the functionality and reliability of flexible sensors. Therefore, this chapter provides a comprehensive overview of the advancements in flexible substrates and active layers for flexible sensors. The focus is on material selection, preparation methods, performance evaluation, and application domains. This chapter examines the performance of flexible substrate and active layers in terms of sensitivity, response time, durability, and stability. By summarizing the research progress in these areas, the chapter aims to foster a comprehensive understanding of the potential applications and future development directions of flexible substrates and active layers in the field of flexible sensors.

Chapter 3 presents flexible sensors, covering a wide range of physical and chemical sensing modalities, including pressure, strain, electrophysiology, ions, biomarkers, metabolites, gases, and more. Recent advances in the key performance, preparation strategies and operating mechanisms of flexible sensors are detailed. Due to the use of various flexible materials, flexible sensors are light in weight and thin in profile, which facilitate integration, distribution and application. Their mechanical flexibility and stretchability, shape adaptation and manufacturing scalability allow flexible sensors to measure objects with changing or dynamic shapes and large surface areas. Moreover, there are advances in integrated capabilities, such as multifunctional design, biocompatibility and even implantability. Chapter 4 introduces the advanced applications based on flexible sensors, mainly including intelligent perception systems, human–machine interaction, high-resolution sensor arrays, the Internet of Things, virtual reality/augmented reality, intelligent medical monitoring, and optoelectronic displays, and details their research progress in recent years, which will be conducive to clarify the development direction and technology expansion of flexible sensors in the future.

Chapter 2

Basic Composition and Functional Materials

Feifei Yin[*], Peng Wang[†], Li Yang[‡,¶], and Shen Guozhen[§,**]

[*]East China Institute of Photo-Elextro ICs,
No. 89 Longshan Road, Su Zhou 215163, China
[†]School of Mechanical Engineering, University of Jinan,
Jinan 250022, China
[‡]School of Integrated Circuits, Shandong University,
Jinan 250101, China
[§]School of Integrated Circuits and Electronics, Beijing Institute of
Technology, Beijing 100081, China
[¶]yang.li@sdu.edu.cn
[**]gzshen@bit.edu.cn

Flexible sensors, a new type of sensor technology based on flexible materials, have the advantages of thinness, softness, bendability and stretchability, and are suitable for applications in many fields. As an important part of flexible sensors, the development of flexible substrate layers and active layers has an important impact on the performance and application range of sensors. In this chapter, the development of flexible substrate and active layers of flexible sensors including the research progress in material selection, preparation methods, performance evaluation and application fields are concluded, aiming at a comprehensive understanding of the application potential and future development direction of flexible substrate and active layers in the field of flexible sensors.

1. Introduction

As an innovative sensor technology, flexible sensors have broad application prospects in many fields. Compared with traditional rigid sensors, flexible sensors are bendable, stretchable, and compressible, and can adapt to different shapes and surfaces, thereby achieving more flexible and precise sensing functions.

Flexible sensors can be used in medical health monitoring, smart wearable devices, human–computer interaction, virtual reality (VR), robotics and other fields, providing new possibilities for the development of these fields.

The performance of flexible sensors is mainly determined by two key components, namely the flexible substrate and the active layer. The flexible substrate layer is the underlying and key component of flexible sensors, providing strong support for other components like conductive materials and circuits. Common flexible substrate materials include polymers, rubber, fibers, etc. These materials have good flexibility and plasticity and can maintain good stability under external influences.

The choice of flexible substrate is crucial to the performance of the sensor, which determines the softness, reliability and durability of the sensor. First, the flexible substrate enables the sensor to adapt to various curved surfaces and shapes, enabling a wider range of applications. For example, in the medical field, flexible sensors can be attached to the skin to monitor physiological parameters, such as heart rate and blood oxygen concentration. Secondly, the softness and plasticity of the flexible substrate can reduce the risk of damage to the sensor during use, and improve the reliability and durability of the sensor. In addition, the flexible substrate can also realize the flexible integration of sensors, combining sensors with other functional devices to form a multifunctional flexible electronic system.

The active layer is a key part of the flexible sensor, which is responsible for sensing external stimuli and converting them into electrical signals. The design and preparation technology of the active layer directly affects the sensitivity and reliability of the sensor. Firstly, the preparation technology of the active layer has an essential impact on the performance of the sensor. High- precision, high-resolution fabrication methods enable enhanced sensor sensitivity and stability while allowing large-scale fabrication and cost-effectiveness. Additionally, the design of the active layer needs to take into account the sensitivity, selectivity, and reliability of the sensor to specific stimuli. For example, in a pressure sensor, the material and structure of the active layer should enable high-sensitivity detection of pressure and exclude other interfering signals. In recent years, with the rapid development of nanotechnology and flexible electronics, new active layer materials and preparation methods have emerged, which greatly broaden the application range and performance of flexible sensors.

In summary, flexible substrates and active layers, as the key components of flexible sensors, have an important impact on the sensitivity, reliability, and scalability of sensors. With the continuous progress and innovation of flexible electronic technology, flexible sensors will be applied in more fields, bringing more convenience and possibilities to our life and work.

2. Substrate Layers

2.1. Film substrate layers

As one of the most important material classes, polymeric materials have significant advantages, including light weight, mechanical flexibility, good compatibility, and good processability, showing their remarkable potential in realizing flexible sensors with high performance. Polydimethylsiloxane (PDMS), polyimide (PI), poly(vinylidene fluoride) (PVDF), and other insulating polymer materials have been widely explored as flexible substrate layers for these flexible sensors due to their flexibility, excellent transparency, thermal stability, and chemical resistance.[24–34] In 2013, by using two thin PDMS layers to support the two interlocked arrays of high-aspect-ratio Pt-coated polymeric nanofibers (NFs) (Figure 1(a)), Suh *et al.* presented a simple architecture for a flexible and highly sensitive wearable sensor, which successfully enables the detection of pressure, shear, and torsion (Figure 1(b)).

Xu and coworkers demonstrated an ultrathin and flexible tactile sensor based on two unconnected few-layer graphene films (GFs) deposited on a polyethylene glycol terephthalate (PET) substrate, which provides sufficient optical transparency of 80% for an aesthetically pleasing viewing.[35] Figure 2 schematically depicts the fabrication process of the flexible tactile sensor. Firstly, GFs were formed on a copper foil by a low-pressure chemical vapor deposition (CVD) method, and then polymethyl methacrylate (PMMA) film was coated on the surface of the obtained GFs. After that, an aqueous solution of $(NH_4)_2S_2O_8$ (0.5 mol/L) and $NH_3 \cdot H_2O$ (0.5 mol/L) was used to remove the copper substrate, and GFs/PMMA was fabricated. To obtain two unconnected GFs, a PI tape was pasted in the middle of the

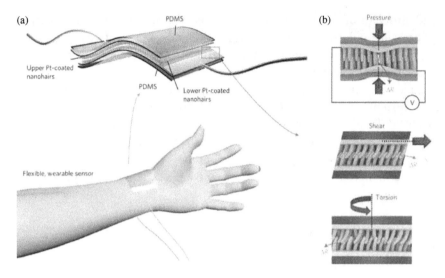

Figure 1. (a) Schematic illustration of assembly and operation of the flexible and wearable sensor that is supported by two thin PDMS layers. (b) Schematic of the sensing behaviors of the flexible and wearable sensor when loading pressure, shear, and torsion. Reproduced with permission from Ref. 24 (Copyright 2012, Macmillan Publishers Limited).

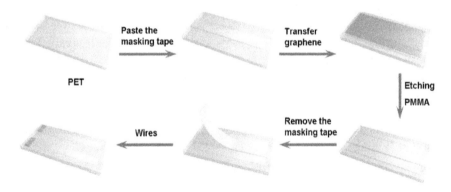

Figure 2. Schematic of the preparation process of the flexible tactile sensor composed of few-layer GFs and PET substrate. Reproduced with permission from Ref. 35 (Copyright 2018, Springer Science+Business Media, LLC, part of Springer Nature).

cleaned PET substrate as a mask, and the GFs/PMMA was transferred to the PET surface covered by the PI tape. Lastly, the PI tape was removed when a soaking procedure was completed in an acetone solution for 8 h, and a flexible tactile sensor with a channel structure was obtained.

Basic Composition and Functional Materials 11

By reproducing the bionic microstructures of natural lotus leaves, Zhang et al. prepared a uniform microwave-patterned PDMS film for supporting Au electrode materials and proposed a flexible sensor by combining it with a polystyrene (PS) microsphere dielectric layer, showing that the flexible sensor with a patterned PDMS substrate could contribute to higher sensitivity and low detection limit compared with that nonpatterned ones.[25] In addition, as shown in Figure 3(a), Pan et al. replicated a PDMS film with pyramid microstructures from the microstructured silicon mold to work as the supporting layer for bottom electrodes and developed a capacitance pressure sensor.[36] To improve the performance of the capacitance pressure sensor, the authors first fabricated four kinds of pressure sensors based on PDMS films with single-sized pyramid microstructures and different spacings (Figure 3(b)). The sensing performance of these capacitive pressure sensors was analyzed by recording the variations in capacitance under different pressures, and it was concluded that the sensitivity of the devices increased with PDMS pyramid spacing, which

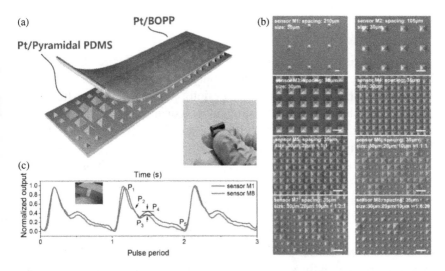

Figure 3. (a) Schematic illustration and an optical image of the flexible capacitive pressure sensor that is characterized by the bottom Pt electrodes deposited on PDMS film with pyramid microstructures. (b) SEM images of eight kinds of PDMS film with different pyramids microstructure designs for the bottom electrodes of the pressure sensor. (c) Application in human wrist pulse waveform monitoring and their comparison of pulse waveform details by using the designed pressure sensors with different hysteresis times. Reproduced with permission from Ref. 36 (Copyright 2017, IEEE).

is consistent with the simulation results of Ref. 37. However, excessive spacing between pyramids triggers the interfacial adhesion between the dielectric layer and the electrode layers, which relates the hysteresis issue to the response speed of the capacitance sensor. Therefore, they tried the hierarchical pyramid microstructure designs (Figure 3(b)) and successfully find the best strategy that inserted small pyramid arrays into large pyramid lattices for supporting the electrode, which simultaneously realized high sensitivity as well as low hysteresis. Figure 3(c) illustrates the application demonstration of real-time and accurate human pulse signal monitoring by using two pressure sensors. The blue curve is measured by the pressure sensor with high sensitivity and a high hysteresis based on the single-type pyramids film substrate of the large spacings, while the red line is recorded by a pressure sensor with a high sensitivity but a low hysteresis that adopted the hierarchical pyramid microstructure design of inserting the small pyramid arrays into the large pyramid lattices. It turns out that the hierarchically structured pressure sensor with both a low hysteresis and high sensitivity is more suitable for precise dynamic sensing.

2.2. Stretchable substrate layers

Stretchability is one of the most important properties desired for substrate layers of flexible sensors, which contributes to flexible sensors to meet the requirements that may undergo various deformations including stretching, compression, bending, and twisting in practical applications. Polymers with low Young's modulus and high stretchability, such as PDMS, Ecoflex, thermoplastic polyurethane (TPU) elastomers, styrene block copolymers, natural rubber, styrene-butadiene rubber, polyolefin elastomers, and elastic hydrogels, are some common flexible substrates that provide the stretchability of flexible sensors.[38–44] Stretchable strain sensors are highly desirable for human motion monitoring and can be used to build new forms of bionic robots. Therefore, Zhang et al. proposed a new stretchable strain sensor and designed a human-machine interface system in combination with the stretchable strain sensor to allow the remote control of a robot (Figure 4(a)–(4c)).[22] Here, as shown in Figure 4(d), an Ecoflex sleeve was used to cover a flexible magnet that was wrapped with a copper coil, which enables the flexible sensor to be stretched to a high strain of 400% due to its excellent tensile properties (Figure 4(e)). In addition, Park et al. developed a PDMS substrate with micropyramid arrays and then deposited electrode materials on that substrate, enabling a

Basic Composition and Functional Materials 13

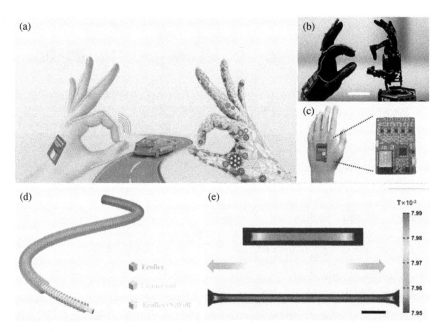

Figure 4. (a) Schematic of the wearable HMI system in combination with a new stretchable strain sensor. (b) Photograph of a hand and the manipulator, with the same gesture. (c) Schematic of a human hand wearing stretchable strain sensors and a wireless PCB. Reproduced with permission from Ref. 22. (d) Schematic of the new stretchable strain sensor. (e) Simulation results of the tensile deformation of the proposed stretchable strain sensor (Copyright 2022, Wiley-VCH).

stretchable pressure sensor.[45] Corresponding details for the preparation of those micropyramid arrays on PDMS substrate are illustrated in Figure 5. First, a silicon (Si) mold with a uniform microstructure was fabricated by combining conventional photolithography and wet etching processes. Then, the mixture of PDMS elastomer and crosslinker was cast onto the microstructured silicon mold and cured for some time. Afterward, the micropyramidal PDMS membrane was obtained by peeling it off the silicon mold. It can be seen that the Si template method involving the photolithography and wet etching processes shows considerable superiority in precisely controlling the size, spacing, and morphology of micro–nanostructures, which is exactly desired for the large-scale preparation of microstructures.

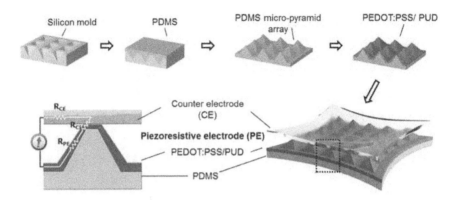

Figure 5. Si mold fabricated by photolithography and wet etching processes for the preparation of the micropyramid PDMS substrate to support electrode materials. Reproduced with permission from Ref. 45 (Copyright 2014, Wiley-VCH).

By using a laser engraving machine, Yang and coworkers reported a stretchable strain sensor based on the graphene–Ecoflex bilayer.[46] For the graphene–Ecoflex bilayer, the loading of strain will lead to a variation in resistance, which corresponds to a change in temperature of the sensor during Joule heating. As a result, the change of sensor temperature can be monitored wirelessly with an infrared (IR) camera to provide a route to realize wireless monitoring of strain. According to the author, the proposed wearable strain sensor provides the capability to wirelessly monitor small strains for intelligent robots at a high strain resolution of ≈0.1%. Therefore, the author also investigated the application of the stretchable strain sensor in wireless monitoring of the small motion of intelligent robots, and the sensor was demonstrated to be capable of wireless monitoring different bending states of the robot fingers (Figure 6(a)). The author pointed out that the research of wireless strain detection with high performance is essential for broadening the potential application of stretchable strain sensors, which will provide a promising approach to wirelessly monitor strain and even in long-range artificial intelligence fields. Liu *et al.* proposed a flexible strain sensor by using a pre-stretching process and a multi-step spin-coating method, as illustrated in Figure 6(b).[47] For this purpose, the elastomer (PDMS or VHB tape) was used to serve as the highly stretchable substrate layer and was treated with O_2 plasma to improve its hydrophilicity. Afterward, the first layer PEDOT:PSS was coated on the surface of the elastomer, baked, and stretched by 50%,

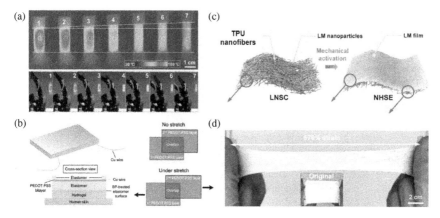

Figure 6. (a) Application demonstration of the stretchable strain sensor for monitoring the strain of a robot finger by the temperature distribution images measured under seven different bending states of the robot finger. Reproduced with permission from Ref. 46 (Copyright 2020, Wiley-VCH). (b) Schematic of the wearable strain sensor structure and its behavior under strain. Reproduced with permission from Ref. 47 (Copyright 2021, Elsevier Inc). (c) Schematic of the NHSE assembled by the electrospun elastic NF scaffolds (TPU) and electrospray liquid metal (LM) nanoparticles. (d) Photographs of the NHSE under relaxed (original state) and 570% of elongation. Reproduced with permission from Ref. 48 (Copyright 2022).

resulting in the formation of fractures and PEDOT:PSS islands. Then, the second layer PEDOT:PSS was further spin coated and baked to form the PEDOT:PSS/elastomer. After further stretching the PEDOT:PSS/elastomer to 100%, the second PEDOT:PSS layer also cracked into islands, which created overlapping areas with those from the first layer. Next, the pre-stretched elastomer was released, and a bilayer, scaly structure with overlapping microscale PEDOT:PSS islands was produced (Figure 6(b)). Lastly, a hydrogel layer was introduced below the elastomer as an interface layer between the stretchable strain sensor and the human skin, and the robustness and reliability of the elastomer/hydrogel interface were investigated by tensile as well as shear tests. It turns out that the sensor was deployed on the human body and contributed to conform sensing various skin deformations, improving the biological and mechanical compatibility of this strain sensor.

By *in situ* assembling the electrospun elastic NF scaffolds (TPU) and electrospray liquid metal (LM) nanoparticles (Figure 6(c)), Cao *et al.* proposed a highly robust stretchable electrode (NHSE), which exhibits an extremely low sheet resistance of 52 mΩ sq^{-1}.[48] As depicted in Figure 6(d), the proposed NHSE can realize a large degree of mechanical stretching (i.e., elongation up to 570%) without serious resistance change. Furthermore, the NHSE delivers superior robustness against dynamic cyclic stretching and environmental stimuli (i.e., heating, acid and alkali exposure, and submersing). Thus, the author applied the NHSE to human long-term ECG monitoring and adapted it to develop a wearable human-machine interface system for inputting computer game instruction. Most importantly, the NHSE shows superior robustness against dynamic cyclic stretching and environmental stimuli such as heating, acid and alkali exposure, and submersing, which is beneficial to the future exploration of the Internet of Things (IoT)-based wearable healthcare monitoring systems and skin-like human-machine interfaces.

2.3. Substrate layers with other desirable properties

Flexible sensors are often used in skin-contact scenarios such as wearable devices and medical monitoring. In addition to flexibility and stretchability, the ideal flexible substrate should exhibit skin-like properties, minimize thermal loading to the body, and maximize user comfort. For example, substrate materials with high breathability are crucial for the performance and comfort of flexible sensors in applications such as human contact. Choosing a breathable substrate material can improve the air permeability, moisture regulation, and environmental adaptability of the sensor, ensuring the reliability and comfort of the sensor in practical applications. These characteristics require materials with porous structures, such as fiber, paper, and aerogel.[49–54] As illustrated in Figure 7(a), Li and coworkers successfully developed a highly elastic and breathable electronic skin based on an all-fiber structure, in which three layers of NFs are included, involving the PU NFs substrate layer, the PVDF NFs sensing layer, and the carbon NFs electrode layer.[55] Benefiting from that, such an all-fiber structured electronic skin shows good sensing performance and mechanical stability even when the elastic deformation reaches up to 50%. Also, the proposed all-fiber structured electronic skin exhibits excellent breathability with a water vapor transmittance rate of 10.26 kg m^{-2} d^{-1} which is equivalent to the most advanced breathable materials and guarantees thermal-wet comfort during the wearing of this electronic skin. Gong *et al.* reported a flexible and breathable electronic device, which

Basic Composition and Functional Materials 17

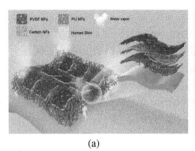
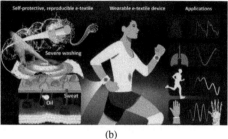

(a) (b)

Figure 7. (a) Schematic of an array composed of 3 × 3 electronic skin pixels with high elasticity and breathability. Reproduced with permission from Ref. 55 (Copyright 2020, Wiley-VCH). (b) Reproduced with permission from Ref. 57 (Copyright 2019, The Royal Society of Chemistry).

enables real-time temperature sensing and timely on-demand anti-infection therapy at wound sites are presented.[56] The electronic device is assembled by loading a crosslinked electrospun moxifloxacin (MOX) hydrochloride with a conductive pattern on a thermoresponsive polymer nanomesh film (C-PNHM). The polymer nanomesh film demonstrates excellent flexibility, reliable breathability, and robust environmental stability, and the application for real-time wireless temperature monitoring was demonstrated by coupling with a wireless transmitter. According to the author, this work is expected to pave the way for the fabrication of multifunctional flexible electronic devices with excellent flexibility, porous structure, and robust breathability for potential versatile applications in smart wound dressings, electronic skins, personal healthcare monitoring, and on-demand therapy. In addition, textile-based electronic devices have aroused considerable interest due to their excellent flexible, wearable and breathable features towards the next-generation intelligent wearable human- machine interfaces. For instance, through a hierarchical construction strategy, Zhang *et al.* designed a highly breathable e-textile (SPRET) composed of a textile substrate layer, an entangled carbon nanotube (CNT) network, and a combined polypyrrole-polydopamine-perfluorodecyltrlethoxysilane (PPy- PDA-PFDS) polymer layer, as shown in Figure 7(b).[57] Apart from the high breathability, the proposed SPRET exhibits desirable superhydrophobicity to a variety of agents, and mechanical durability under machine washing, and tape-peeling, which successfully addresses the issue of the textile-based sensors that are vulnerable to the mechanochemical attacks from sweat, oil, etc., or wear and tear. Therefore, the authors integrated the pressure sensor SPRET with the circuit and demonstrated the application in the precise and reliable detection of human motions and physiological signals either in air/wet conditions or even underwater. In view of the superior sensibility in harsh settings, as well as the soft, breathable, and comfortable features, the proposed SPRET sensor holds practical application potential in wearable health monitoring, human–machine interaction, and robot-learning fields.

Xu et al. reported the exploration of multifunctional porous on-skin electronics based on the multiscale porous SEBS substrate and spray-printed Ag NWs, and the device can not only exhibit high breathability but also simultaneously achieve several other attributes of outstanding passive-cooling performance, waterproofing, and recyclability.[58] According to the author, the combination of these fascinating attributes is owing to the multiscale porous PS-block-poly(ethylene-ran-butylene)-block-PS (SEBS) substrates and has not been demonstrated in state-of-the-art on-skin electronics yet. As depicted in Figure 8(a), the preparation of the multifunctional porous on-skin electronics was achieved by a solution-printing–based fabrication, which offers a simple, cost-effective, high-throughput way for patterning devices on flexible supporting substrates. Here, the multiscale porous SEBS was first pre-stretched by 150%, and the serpentine-like Ag NWs were spray-coated on it with the help of a mask. Figures 8(b) and 8(c) illustrate the photograph and SEM image of the serpentine-like Ag NWs on the multiscale porous SEBS substrate. Due to the outstanding mechanical compliance and high electrical conductivity, the Ag NWs on the multiscale porous SEBS substrate serve as the basic building blocks for a variety of bioelectronic devices and the multifunctional porous on-skin electronics were obtained (Figure 8(d)). Lastly, the author demonstrated the capability of the multifunctional porous on-skin electronics in the application of human pulse waveforms and ECG signals monitoring even though the volunteer fully submerged the forearm with the wearable device in the water (Figure 8(e)).

Flexible sensors are commonly employed in applications that involve repeated bending and stretching, which can lead to wear and damage to the sensor's substrate. To address this issue, the utilization of self-healing substrate materials becomes crucial. These materials can autonomously repair certain types of damage, such as scratches, tears, or small cracks.[59–63] By incorporating self-healing substrate materials, the lifespan of the sensor can be extended, while enhancing its reliability and stability.

Basic Composition and Functional Materials 19

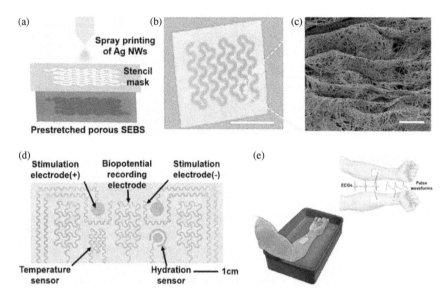

Figure 8. (a) Pre-stretched multiscale porous SEBS film working as the substrate to support the spray-printed Ag NW. (b) A photograph of a serpentine-like Ag NW conductive trace, made of spray-printed Ag NWs, on the multiscale porous SEBS film. (Scale bar: 1 cm.) (C) SEM image, showing the formation of microscale wavy structures of Ag NWs networks with underneath porous SEBS. (Scale bar: 5 μm.) (d) A photograph of multimodal on-skin electronics based on the multiscale porous SEBS substrate and spray-printed Ag NWs. (e) Demonstrations of waterproofing of the multimodal on-skin electronics based on the multiscale porous SEBS substrate and spray-printed Ag NWs and the recorded human physical signals by the device. Reproduced with permission from Ref. 58 (Copyright 2020).

Considering the excellent reversible properties of synthetically controllable hydrogels, including high stretchability, rapid self-healing, and significant swelling in water, Park et al. presented a highly deformable and self-healable device based on the hydrogel, as shown in Figure 9(a).[64] Here, the interactions between numerous hydroxyl groups (–OH) in a hydrogel associated with water molecules and the native gallium oxide (Ga$_2$O$_3$) layer of EGaIn allows EGaIn droplets to spread on hydrogels, thus it is possible to direct pattern-transfer printing of EGaIn and facilitating patterning of a variety of EGaIn electrodes on hydrogels. Figures 9(b) and 9(c) present the photograph and optical microscopy photographs of the fabricated hydrogel with the surface transferred EGaIn micropatterned

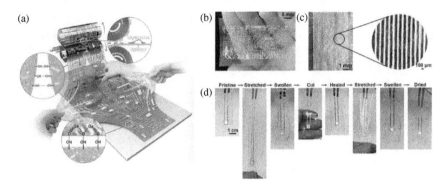

Figure 9. (a) Schematic illustration of a deformable and self- healable hydrogel with micropatterned liquid-metal (EGaIn) electrode. (b) Photographs showing the fabricated hydrogel with the surface transferred EGaIn micropatterned electrode. (c) Optical microscopy photographs of periodically patterned lines with a width of 100 μm of EGaIn on the hydrogel. (d) Photographs of an EGaIn electrode on hydrogel under consecutive mechanical conditions of stretching, swelling in water, cutting, and self-healing. Reproduced with permission from Ref. 64 (Copyright 2020, Wiley-VCH).

electrode. EGaIn patterned on a hydrogel autonomously reconciliations its surface to form a firm hydrogel interface upon mechanical deformation of the hydrogel, contributing to a maximum stretchability of 1500% with a low resistance of ≈2 Ω and low variation in conductance (variation factor is ≈2.3 with the resistance of ≈4.6 Ω at the strain of 1500%). Furthermore, the author confirmed that an EGaIn electrode can even successfully maintain its properties under consecutive events of stretching, swelling in water, cutting, and healing, as shown in the photographs in Figure 9(d). And, this autonomous surface reconciliation of EGaIn on hydrogels allows researchers to reap the benefits of chemically modified hydrogels, such as reversible stretching, self-healing, and water-swelling capability, thereby facilitating the fabrication of super stretchable, self-healable, and water-swellable liquid-metal electrodes with very high conductance tolerance upon deformation. Such electrodes are suitable for a variety of deformable microelectronic applications.

Conventional non-biodegradable electronic devices have potential pollution problems and may contain a large number of environmental pollutants, such as toxic heavy metals and hazardous chemicals. These pollutants would migrate into the drinking water or food chain, posing a serious threat to our health.[65–68] To reduce the impact on the environment, biodegradable

Basic Composition and Functional Materials 21

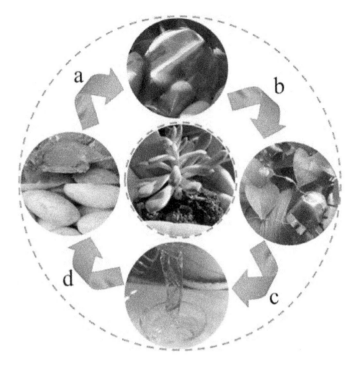

Figure 10. Cyclic diagram of biodegradable conductive nanocomposites from material to device. Reproduced with permission from Ref. 71 (Copyright 2018, American Chemical Society).

materials have received more and more attention, and have demonstrated their potential in the applications of transient e-skin.[69,70] Starch and chitosan (CS) extracted from natural potato and crab shells, respectively, are cheap, abundant, and biodegradable. For example, by incorporating the three-dimensional (3D) interconnected conductive nanocomposites into an edible starch–CS substrate, Zhang *et al.* introduced a biodegradable and flexible transparent electrode,[71] as shown in Figure 10. The construction of a 3D interconnected network of single-walled carbon nanotubes (SWCNTs), pristine graphene, and PEDOT, results in a conducting network with a remarkably low sheet resistance, which allows for natural adaptation to the morphology of the skin or any other surface it comes into contact with. Most importantly, due to the edible starch–CS substrate layer, the transparent electrode demonstrated rapid biodegradability when exposed to lysozyme solution at room temperature, and there were no any toxic residues occurred in the degradation process. The same with the former research group, Huang et al. utilized starch material as the substrate layer to create a versatile and biodegradable flexible sensor with triple-modal functionality.[72] Firstly, through a laser, the PI film was carbonized in the air with a designed pattern. Then, the starch dispersion was coated on the laser-carbonized PI film and dried for a while at 60°C for the starch film. Lastly, when peeling off the starch film from the PI film, the laser-induced carbonized layer was successfully transferred onto the starch film with the patterns, contributing to a degradable multimodal flexible sensor.

Because of their softness, cost-effectiveness, abundance, and biodegradability, composite cellulose-based material shows considerable potential in the realm of large-scale wearable and biodegradable electronic devices.[73–78] For instance, Xu et al. presented a biodegradable and diverse array of on-skin electronic devices by using pencil and paper, as depicted in Figure 11, which involve biophysical sensors for measuring temperature and biopotentials, as well as sweat biochemical sensors capable of detecting pH levels, uric acid, and glucose.[79] Additionally, the author developed thermal stimulators and humidity energy harvesters, in which the conductive traces and sensing electrodes in these devices are created using pencil-drawn graphite patterns or in combination with other compounds, and the flexible supporting substrates are composed of office-copy papers. As the author discussed, they have achieved real-time, continuous, and accurate monitoring of various vital biophysical and biochemical signals from the

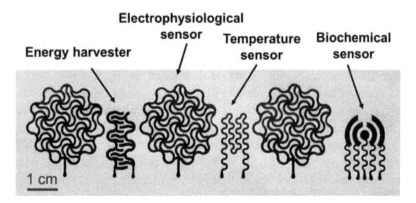

Figure 11. A photograph showing the pencil–paper based on-skin electronics. Reproduced with permission from Ref. 79 (Copyright 2020, Published under the PNAS license).

human body, such as skin temperature, electrocardiograms (ECGs), electromyograms (EMGs), alpha, beta, and theta rhythms, instantaneous heart rates, respiratory rates, and sweat pH, uric acid, and glucose levels.

Furthermore, commonly used biodegradable polymers like polyvinyl alcohol (PVA),[80–88] polycaprolactone (PCL),[89–92] polylactic acid (PLA),[93–97] and poly(lactic-co-glycolic acid)[98–103] are frequently employed as substrate layers in biodegradable electronic products. As illustrated in Figure 12, Guo et al. introduced a sensitive and biodegradable flexible pressure sensor by sandwiching a porous $Ti_3C_2T_x$ MXene-infused tissue paper between a biodegradable PLA film and a PLA substrate coated with the interdigitated electrodes.[93] The pressure sensor exhibits remarkable attributes such as high sensitivity, a wide detection range, fast response time, low power consumption, and excellent degradability, which is capable of effectively performing various medical monitoring tasks, ranging from detecting small deformations to tracking significant movements. The author believe that the innovative approach offers new possibilities for the development of flexible wearable transient pressure sensors that exhibit high sensitivity, reproducibility, and wireless capability.

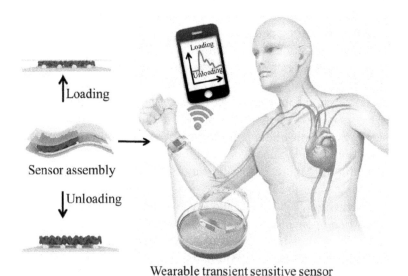

Wearable transient sensitive sensor

Figure 12. Schematic illustration of the wearable pressure sensors composed of PLA substrate layer coated with an interdigitated electrode and the sandwiched porous MXene-impregnated tissue paper. Reproduced with permission from Ref. 93 (Copyright 2019, American Chemical Society).

Furthermore, biocompatible flexible substrates are of great significance in developing flexible sensors. Such substrates can be compatible with biological tissues without causing adverse reactions or damage, thus offering great potential for medical treatment and health monitoring.[70,104–110] The biocompatible flexible substrate also has good adjustability and deformability and can adapt to different biological environments and motion states, allowing it to fit body curves and surfaces, providing a comfortable wearing experience, and can be used in physiological monitoring, health areas such as management and rehabilitation play an important role. For example, by introducing the Zein into the PVA NFs substrate layer, the flexible sensor developed by Zhang et al. demonstrates exceptional characteristics including high sensitivity for pressure detection across a wide range, impressive self-powered capability, and outstanding biocompatibility and degradability.[111] By culturing rat mesenchymal stem cells (rMSCs) on the proposed sensor, the author demonstrated that the proposed sensor supports high cell viability of rMSCs, meeting the criteria for non-cytotoxic materials. Therefore, the author believes that the proposed sensor holds promise as a biocompatible wearable device that can be safely applied directly to human skin without any detrimental effects on human health. In addition, Kim et al. present a novel approach for fabricating patterns of laser-induced graphene (LIG) on various natural woods and leaves in ambient conditions.[112] Here, a technique, known as one-step femtosecond laser direct writing (FsLDW), utilizes high-repetition-rate ultraviolet (UV) femtosecond (FS) laser pulses to maximize the photoconversion process, resulting in the formation of LIG patterns. As shown in Figure 13(a), the utilization of an FS laser with a UV emission wavelength offers significant advantages due to its ultrashort pulse duration and high photon energy, leading to enhanced photoabsorption capabilities. Figure 13(b) illustrates the mechanism based on the interaction between the laser and the materials make the process of one-step LIG formation from woods and leaves in ambient conditions clear. As a result, all components of woods and leaves, including not only lignin with phenolic groups but also cellulose and hemicellulose composed of sugar monomers, can be successfully transformed into LIG. According to the author, this approach involves the utilization of a UV FS laser for direct photoconversion to LIG, which can be attributed to two consecutive conversion stages. Firstly, the high- repetition-rate laser pulses generate heat accumulation, leading to the carbonization of wood and the formation of an intermediate char. Secondly, the high photon energy and ultrashort

Basic Composition and Functional Materials 25

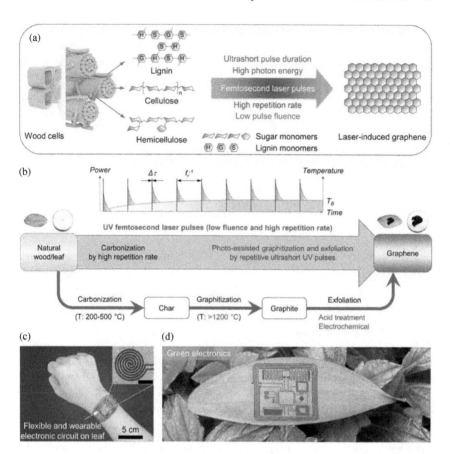

Figure 13. (a) Schematic illustration of femtosecond laser direct writing (FsLDW) for one-step laser-induced graphene (LIG) synthesis on wood. All the chemical compositions (cellulose, hemicellulose, and lignin) in wood can be transformed into graphene by the irradiation of UV femtosecond pulses in ambient conditions. (b) Schematic illustration of LIG formation mechanism from natural wood/leaf. (c) Photo of a LIG electronic circuit on a thin leaf for flexible and wearable devices. The inset shows an enlarged optical image of the temperature sensor (scale bar: 1 mm). (d) Photo of LIG electronics on a leaf for green electronics. Reproduced with permission from Ref. 112 (Copyright 2019, Wiley-VCH).

pulse duration of the laser induce the photoassisted graphitization and exfoliation of the char, resulting in the formation of LIG. In conventional thermal processing (as depicted in Figure 12(b)), wood can undergo thermal decomposition (200–500°C) to produce char through carbonization. The char can then be transformed into graphitic materials via annealing at high temperatures (1200–3000°C) in a reducing gas environment. The formation of graphene requires the subsequent exfoliation of the graphitic materials through mechanical, thermal, chemical, or electrochemical treatments. By this way, the synthesis of LIG directly on the surface of natural substrates was realized. As presented in Figures 13(c) and 13(d), the leaf exhibited remarkable flexibility and could be easily wrapped around a human wrist, showcasing the potential applications of graphene induced by FsLDW in wearable and biocompatible electronics.

3. Active Layer

3.1. Roles in flexible sensor

Flexible sensors rely on active layers, which consist of a sensing layer and an electrode layer. The primary function of these layers is to convert external stimuli into electrical signals and transmit them. By incorporating functional materials and innovative structures into the active layer, it becomes possible to enhance the flexibility, stretchability, and sensing performance of the sensor. This includes improving sensitivity, which refers to the sensor's ability to detect and respond to changes in external stimuli and is calculated as the ratio between the output electrical signal and the rate of change of the stimulus. Additionally, optimizing the active layer can expand the dynamic range of the sensor, which represents the range of external stimuli within which the sensor can operate accurately and reliably. The response/relaxation time, which is the time required for the sensor to generate a stable signal output after applying stimuli, can also be improved. Lastly, the limit of detection, which defines the minimum and maximum values of the external stimuli that the sensor can accurately measure, can be influenced by the active layer's properties. The subsequent sections will further elaborate on the impact of these factors on flexible sensors.

3.2. Working mechanisms

As presented in Figure 14, flexible sensors can be divided into the following categories based on the different types of detection electrical signals:

- **Resistive sensor:** This type of sensor detects a physical quantity based on a change in electrical resistance. Common resistive sensors include pressure sensors, stretch sensors, and bend sensors.

Basic Composition and Functional Materials 27

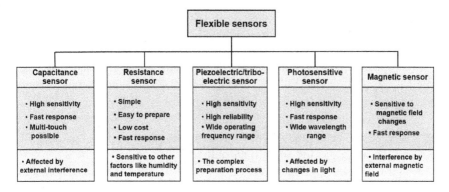

Figure 14. Pros and cons of common flexible sensor types.

They determine the size or location of an external stimulus by measuring a change in electrical resistance.
- **Capacitive sensors:** Capacitive sensors use changes in capacitance to sense changes in physical quantities. Flexible touch sensors and flexible humidity sensors are common capacitive sensors. When the flexible sensor is stimulated externally, the capacitance value will change, which will be converted into a corresponding electrical signal.
- **Piezoelectric sensors:** Piezoelectric sensors convert mechanical stress into electrical charges or voltage signals through the piezoelectric effect. These sensors can be used to measure physical quantities such as pressure, vibration, acceleration, etc.
- **Magnetic sensor:** Magnetic sensor senses physical quantity according to the change of external magnetic field. Flexible magnetic sensors can be used in applications such as measuring magnetic field strength, direction, and proximity to objects.
- **Photosensitive sensor:** The photosensitive sensor uses the sensitivity of photosensitive materials to light signals to sense parameters such as light intensity and light frequency. Flexible photosensitive sensors have a wide range of applications in the fields of optoelectronics, optical communication, and photobiosensing.

3.2.1. Capacitance sensor

The capacitance sensor, commonly utilizing a "sandwich" configuration or planar structure, demonstrates excellent capability in detecting external information like pressure or strain by relying on the change in capacitance.

The capacitance (C) of the sensor is influenced by the effective dielectric constant (ε) of the dielectric layer, as well as the overlapping area (A) and distance (d) between the two electrodes. Due to the notable advantages of their simple device structure, low power consumption, high sensitivity, and good repeatability, flexible capacitance sensors have gained widespread use in replicating the tactile perception behavior of human skin.[67,113–120]

Bao et al. presented a flexible capacitance sensor based on a microstructure of an elastomer PDMS dielectric layer and demonstrated its potential to enhance the sensitivity and response speed of the sensor.[121] Through the investigation of the sensitivities of three capacitive sensors using different PDMS film structures (pyramidal, linear, and unstructured) as dielectric layers, the authors found that the sensor employing the unstructured PDMS dielectric film exhibited the lowest pressure sensitivity. The linear-structured and pyramid-structured PDMS dielectric films showed pressure sensitivities that were 5 and 30 times higher than the unstructured film, respectively. The authors attributed the significantly improved sensing properties of the structured PDMS dielectric films to two main factors. Firstly, the presence of voids in the structured PDMS reduced the elastic modulus of the dielectric layer, allowing for greater deformation when subjected to equivalent pressures, thereby achieving higher sensitivity. Secondly, under compression caused by applied pressure, the voids were replaced by the PDMS matrix with a higher dielectric constant, leading to an increased effective dielectric constant and capacitance variation of the sensor. Furthermore, the linear- and pyramid-structured PDMS dielectric films exhibited low hysteresis and rapid response speed. This can be explained by the fact that the introduction of microstructures into the PDMS dielectric layer ultimately reduced the viscoelasticity and adhesion between the polymer dielectric and electrode layers during the pressure-loading and releasing processes.

Xiong et al. developed a flexible and highly sensitive capacitance pressure sensor by using flexible electrodes with convex microarrays and an ultrathin dielectric layer.[122] Figure 15(a) illustrates the preparation of the flexible electrodes with convex microarrays, and Figure 15(b) shows the assembled capacitance pressure sensor. To manifest the working mechanism of the pressure sensor, Figure 15(c) shows the details of the variation of distance and contact area when loading pressures on the capacitance sensor. The sensor was easy to deform under pressure because external pressure was focused on the microstructure, resulting in patterned microarrays that were easier to deform. The distance between the two

Basic Composition and Functional Materials 29

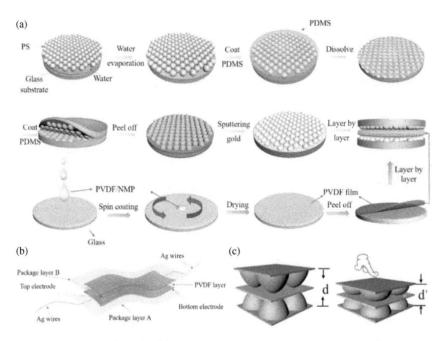

Figure 15. (a) Schematic illustration of the preparation process of the flexible capacitance pressure sensor with convex microarrays. (b) The capacitance pressure sensor packaging with a sequential sandwiching process, where a top PDMS-Au electrode, a middle PVDF dielectric layer, and a bottom PDMS- Au electrode are aligned and placed face to face. (c) Schematic of the variation of distance and contact area when loading pressures on the capacitance sensor. Reproduced with permission from Ref. 122 (Copyright 2020, Elsevier Ltd).

electrodes but also the contact area was dramatically changed under applied pressure, leading to observable improvement of its sensitivity. Therefore, the proposed capacitance sensor exhibits remarkable performance, including an ultra-high sensitivity of 30.2 kPa^{-1} within the pressure range of 0–130 Pa, a fast response time of 25 ms, a low detection limit of 0.7 Pa, and exceptional stability with 100,000 fatigue-free cycles. In addition, various microstructures such as micropyramids,[123,124] micropillars,[125,126] and micro wrinkles[127] have been introduced in the electrode and dielectric layers to prove that it is an effective way of boosting the sensing properties of the flexible capacitance sensor.

In addition to incorporating interfacial microstructures into dielectric and electrode layers, a sponge or foam-like polymer dielectric layer was

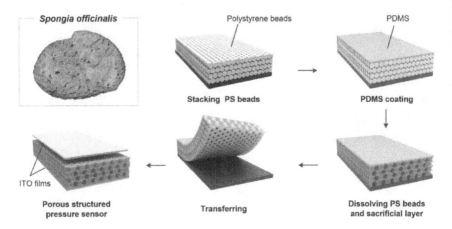

Figure 16. Schematic illustration of the preparation procedures for a capacitance pressure sensor based on a porous dielectric layer that was inspired by spongia officinalis. Reproduced with permission from Ref. 134 (Copyright 2016, Wiley-VCH).

also applied to adjust the sensing behaviors of the capacitive tactile e-skin devices due to its high porosity, good mechanical flexibility, and low compression modulus.[128–133] For example, inspired by the natural multi-layered porous structures seen in mushrooms, diatoms, and spongia officinalis, Kang et al. developed a porous structure of PDMS thin film and adopted it to propose a highly sensitive capacitive pressure sensor.[134] As shown in Figure 16, there are three primary steps in the fabrication process of the spongia officinalis-inspired porous elastomeric film for a high-performance capacitance sensor:

(i) **Stacking PS beads:** PS are stacked on a silicon substrate in a layered arrangement.
(ii) **Coating PDMS film:** A PDMS film is applied to cover the stacked PS on the substrate.
(iii) **Dissolving PS and transferring porous PDMS film:** The PS is dissolved, leaving behind a porous PDMS film. This porous PDMS film is then transferred onto the electrodes, completing the fabrication of the porous elastomeric film for the high-performance pressure sensor.

Additionally, Park et al. recommended another droplet-based and microfluidic-assisted emulsion self-assembly (DMESA) method to prepare a 3D microstructured PDMS dielectric layer and proposed a high-performance capacitive sensor for tactile e-skin applications.[135]

Basic Composition and Functional Materials 31

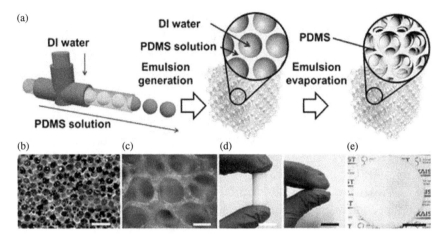

Figure 17. (a) Schematic of the DMESA process for fabrication of 3D microstructured PDMS. (b) SEM and (c) optical images of the microstructured PDMS when evaporating the solvent. (d) Compression test of a cylindrical PDMS with microstructures when loading and unloading the pressure. (e) Optical image of the wafer-scale microstructured PDMS fabricated by such a DMESA process. Reproduced with permission from Ref. 135 (Copyright 2018, American Chemical Society).

Figure 17(a) depicts the preparation details of the microstructured PDMS layer via such a DMESA method. An aqueous solution of deionized water and an oil solution containing PDMS and surfactant were prepared respectively and then flowed through a perpendicularly arranged T-tube to meet at the junction. As a result, deionized water droplets were injected into the PDMS mixture and stabilized by the surfactant, forming a closely packed stacked lattice. The deionized water droplets were evaporated by a heating process, and a 3D uniformly microporous PDMS layer was left behind, as presented in Figures 17(b) and 17(c). Figure 17(d) exhibits that a 3 cm tall micro structured PDMS can be easily compressed by hand, which contributes to the construction of a highly sensitive capacitance pressure sensor. Lastly, the authors pointed out that it is possible to realize the fabrication of a large-area (9 cm in diameter) microstructured PDMS, which can easily shorten the processing time by implementing more microfluidic channels.

3.2.2. Resistance sensor

Resistance sensors have gained significant importance in the development of high-performance flexible sensors due to their ability to detect changes in device resistance under external stimuli. They offer several distinct advantages, including a straightforward readout circuit and excellent performance. The resistance of a resistance sensor is influenced by factors such as the electrical resistivity, length, and cross-sectional area of its functional layer. In terms of electrode configuration, there are two dominant types of resistance devices: sensors with planar electrodes and sensors with top-bottom electrodes. These configurations play a crucial role in determining the sensing capabilities and overall performance of the resistance sensors.[7,51,136–141]

By stacking a functional layer on a pair of in-plane electrodes, a resistance sensor with planar electrodes is assembled, which serves as an important component of the flexible sensors. Polymer materials with flexible, stretchable, or even other excellent properties tend to be designed with microstructures as the supporting layer or combined with other conductive materials like CNTs and graphene as the functional layer of the resistance flexible sensors.[29,142–150] For example, Chung et al. proposed a piezoresistive tactile sensor by incorporating a hierarchical structured PDMS with the deposition of monolayer graphene, which achieves a high sensitivity of 8.5 kPa^{-1} and a good linear response over a wide pressure range of 0–12 kPa.[143] Figure 18(a) illustrates the SEM image of the hierarchical structured PDMS functional layer coated with conducting graphene for such a resistance sensor. As presented in Figure 18(b), the proposed resistance tactile sensor consisted of two main components, including the two bottom graphene interdigitated electrodes fabricated from CVD and the top hierarchical structured PDMS/graphene functional layer. Here, the top hierarchical structured PDMS/graphene worked as a conductive bridge between the separated coplanar electrodes at the bottom. The design of this device was such that when normal pressure was applied, the graphene/PDMS structures would deform, leading to an increase in the contact area between the top graphene/PDMS structures and the bottom electrodes. As a result, the total resistance would decrease, causing an increase in the electrical current at a given voltage, as depicted in Figure 18(b).

For the resistance sensors that adopt the top-bottom electrode configuration, surface microstructures are also intentionally introduced on both top and bottom electrode layers to achieve the construction of flexible sensors with outstanding behaviors.[24,77,146–148,150–157] For instance, Suh et al. proposed a simple architecture of two interlocked arrays of high-aspect-ratio Pt-coated ultraviolet-curable polyurethane acrylate (PUA)

Basic Composition and Functional Materials 33

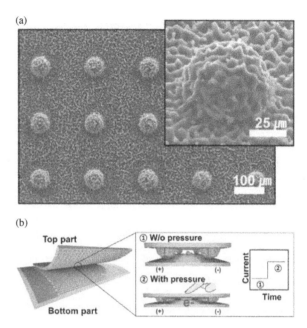

Figure 18. (a) SEM image of the fabricated hierarchically structured PDMS with the deposition of monolayer graphene. (b) Schematic illustration of the assembled resistance tactile sensor based on the hierarchically structured PDMS/graphene layer and the working principle of the resistance tactile sensor. Reproduced with permission from Ref. 143 (Copyright 2016, Wiley-VCH).

NFs for building a flexible and highly sensitive tactile sensor.[24] It is the key to obtaining excellent tactile perception that numerous tiny contacts are ensured between the conductive Pt layers on both the top and bottom PUA NFs. Ko et al. interlocked the top and bottom conductive layers of the CNTs/PDMS composite with a microdome array by simulating the epidermal-dermal layers in the human skin (Figures 19(a)–19(d)).[146] Similar to the stress-concentrating function of interlocked epidermal-dermal ridges, which magnify the tactile stimuli, the arrays could induce exclusive stress concentration at the contact spot and thus deformation of the micro-domes, resulting in enhanced sensitivity of the piezoresistive response to stress. In particular, the unique geometry of the interlocked micro-dome arrays led to different deformation patterns that depend on the type and direction of mechanical stress, enabling the detection and differentiation of various mechanical stimuli including normal, shear, stretch, bending, and twisting forces.

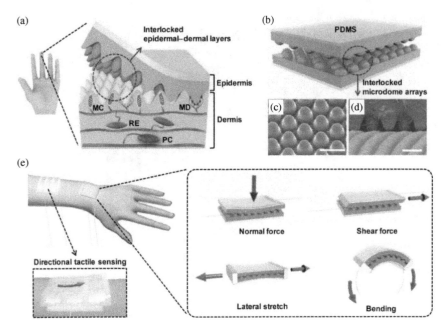

Figure 19. (a) Schematic illustration of human skin structure that is composed of the interlocked epidermal-dermal layers and mechanoreceptors (MD: Merkel disk; MC: Meissner corpuscle; PC: Pacinian corpuscle; RE: Ruffini ending). (b) Schematic of a highly sensitive resistance tactile sensor involving a sensing film with the interlocked micro-dome array. (c) and (d) SEM images of the sensing film with the interlocked microdome array. (e) The diagram illustrating a resistance sensor that is capable of detecting and distinguishing different types of mechanical stimuli, such as normal, shear, stretching, bending, and twisting forces. Reproduced with permission from Ref. 146 (Copyright 2014, American Chemical Society).

Another effective approach for building a highly improved resistance sensor is to introduce microporous features in the elastic and conductive resistance sensing layers. Such porous resistance sensing layers are more easily compressed due to the presence of the microporous features, causing the increase of the changes in the contact area and enhancing the sensitivity of the tactile e-skin.[158–162] For instance, Bao et al. fabricated an elastic microstructured conductive polymer (EMCP) and invented a resistive tactile sensor by sandwiching the EMCP layer between a copper electrode and a PET film deposited with an indium tin oxide (ITO) electrode.[158] Here, a multiphase synthesis technique was used to fabricate the EMCP layer composed of interconnected PPy hollow-sphere structures. First, an oxidative reagent solution was mixed with a mixture of isopropanol, pyrrole monomer, and phytic acid in a petri dish. As a result, an emulsion was obtained because of the phase separation effect between the aqueous and organic bi-component. With the help of the dopant and crosslinker of PPy, PPy polymerization was realized and gelated within ~3 s. Finally, a conductive sensing layer of PPy hollow-sphere structures with 3D porous foam morphology was formed after that the impurities were exchanged with the deionized water and then underwent a drying process. The authors claimed that the hollow-sphere structures formed by such a multiphase reaction endowed the conductive sensing layer of PPy with an effectively decreased elastic modulus and excellent elasticity, and demonstrated that such a sensing layer enabled tactile e-skin could achieve the detection of tiny pressures less than 1 Pa.

3.2.3. Piezoelectric/triboelectric sensor

Benefiting from the piezoelectric effect of the piezoelectric materials, the piezoelectric sensor does not need an external power supply to respond to the pressure information and has been widely used to build high-performance self-powered tactile e-skins. Common piezoelectric polymer materials such as PVDF and P(VDF-TrFE) present fascinating piezoelectric properties and high flexibility, which have been paid increasing attention in the field of fabricating flexible piezoelectric tactile e-skins.[63,151,163–172] For example, P(VDF-TrFE) piezoelectric layers with nanowire and micropillar structures were fabricated by the nanoimprinting method and were developed to construct tactile sensors for satisfying the requirements of e-skins (Figure 20(a)).[167] Figure 20(b) presents the steps of such a nanoimprinting method for the preparation of a flexible piezoelectric tactile sensor assembled by the P(VDF-TrFE) nanowire structures. First, a spin-coating process was performed to deposit a thickness of 4–5 μm P(VDF-TrFE) film on the surface of the Au-coated Kapton substrate and then pressed with an AAO porous template under a temperature of 170°C. After that, free-standing P(VDF-TrFE) nanowire structures were fabricated by successively immersing the AAO template in $CuCl_2$ and HCl mixed solution and NaOH solution to remove it. Lastly, a PMMA layer for avoiding short circuits and a top flexible PEDOT:PSS conducting electrode were sequentially deposited on top of the P(VDF-TrFE) nanowires by using spin-coating processes, and the P(VDF-TrFE) nanowires-based piezoelectric tactile sensor was successfully obtained.

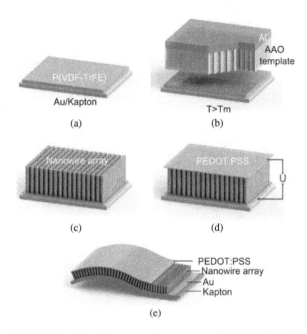

Figure 20. (a)–(d) Schematic of the preparation process of the P(VDF-TrFE) piezoelectric layers with nanowire and micropillar structures. (e) Schematic of the assembled flexible piezoelectric sensor based on the P(VDF-TrFE) piezoelectric layers. Reproduced with permission from Ref. 167 (Copyright 2014, IEEE).

In addition, considerable researchers have developed nanoscale inorganic piezoelectric materials such as ultrathin membranes, nanorods, nanowires and NFs to be aligned with a flexible polymer matrix.[173–175] For instance, Rogers et al. proposed a thin conformable piezoelectric tactile sensor by connecting an ultrathin (400 nm) PZT piezoelectric membrane supported by a flexible PI substrate to the gate electrode of a MOSFET that was composed of silicon nanomembranes (SiNMs).[174] Here, the coupling of the SiNMs-based MOSFET and the PZT piezoelectric membrane could effectively amplify the voltage response of this pressure sensor, enabling low hysteresis and highly sensitive measurements for human arterial pulse signals. Alternatively, the strategy of the piezoelectric composite materials that combine the advantages of both inorganic piezoelectric materials and polymer materials has been adopted by researchers to enable piezoelectric sensors with enlarged output performance, high sensitivity, and excellent flexibility.[176–179] Shao et al. fabricated a piezoelectrically enhanced nanocomposite layer by embedding inorganic BTO

Basic Composition and Functional Materials 37

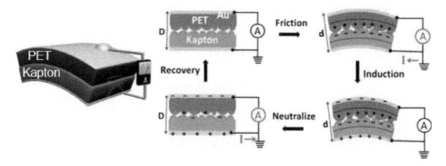

Figure 21. Schematic of a TENG sensor in vertical contact-separation mode and its working principle. Reproduced with permission from Ref. 183 (Copyright 2012, Elsevier Ltd).

nanoparticles in a highly flexible P(VDF-TrFE) micropillar array polymer and demonstrated a high-performance piezoelectric pressure sensor for the energy harvesting and self-powered sensing.[176]

Since Wang et al. first reported a seminal work on TENG in 2012,[180] TENG utilizing the coupling effects between triboelectrification and electrostatic induction has attracted considerable attention in the construction of self-powered tactile sensors and wearable e-skins.[181–184] According to the different operation modes, vertical contact-separation mode, contact-sliding mode, and single-electrode mode three different TENGs are invented.[185]

Vertical contact-separation mode: Figure 21 presents a typical example of an all-polymer-based TENG sensor in vertical contact-separation mode,[183] in which two films (PET and Kapton) with different triboelectric properties were stacked and then metal electrodes on the top and bottom of these two polymer films are deposited. Also, Figure 21 schematically illustrates the working mechanism of this TENG device. When loading pressure to the TENG sensor, the PET and Kapton films are contacted and rubbed with each other, resulting in the distribution of opposite electrical charges on the surfaces of these two films because of their different abilities to capture electrons. After releasing the pressure, the PET and Kapton films are separated, and the compensating charges are induced on the two electrodes, which gives rise to a potential difference between the top and bottom electrodes. Based on the working principle of the above TENG sensor, Wang *et al.* studied the effect of the surface structure and morphology of the functional layer on nanogenerator efficiency by fabricating three kinds of TENG sensors using differently patterned PDMS films (line, cube, and pyramid).[185] The pyramid microstructured PDMS film-enabled TENG tactile sensor exhibited the largest power generation and far surpassed the sensors made of non-microstructured PDMS films, which can be interpreted that the triboelectric effect and the generated surface charges during the friction progress are significantly improved for the pyramid microstructured PDMS film.

Contact-sliding mode: Figure 22(a) shows the configuration schematic of a TENG sensor based on the contact-sliding mode.[186] Here, Nylon and PTFE are respectively utilized as the positive and negative friction layers. As depicted in Figure 22(a-i), the contact triboelectrification process of such a TENG sensor is similar to that in the vertical contact-separation mode. Once the Nylon and the PTFE relatively slide to each other (Figure 22(a-ii)), the contact surface area between the Nylon and PTFE decreases. Then, the opposite charges are electrostatically induced on

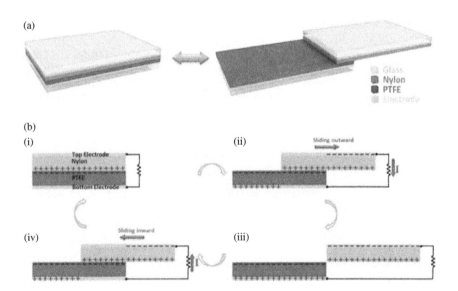

Figure 22. (a) A typical example of a TENG sensor based on the contact-sliding mode and (b) the schematic of its working mechanism. Reproduced with permission from Ref. 186 (Copyright 2013, American Chemical Society).

the surfaces of the top and bottom electrodes, generating a potential difference between electrodes and driving a current flow from the top electrode to the bottom electrode. And, the generated potential difference constantly increases until the Nylon layer slides out of the PTFE layer (Figure 22(a-iii)). When the Nylon and PTFE layers are in contact again, positive charges on the Nylon surface and negative charges on the PTFE surface cancel each other (Figure 22(a-iv)), and the redundant induced charges on the electrodes flow away through the external load to keep the electrostatic equilibrium. In fact, due to the necessary condition of the relative sliding between two different friction material layers, only a few research works on the construction of tactile e-skins adopted such a TENG sensor based on the contact-sliding mode.[186–191]

Single-electrode mode: Different from the above two situations, the TENG sensor based on a single-electrode mode is straightforward, where only one functional layer and electrode layer connected to the ground are needed.[192–194] Figure 23(a) shows the working principle of a typical single-electrode TENG sensor.[195] Here, the Al film deposited on a flexible acrylic substrate layer is simultaneously used as the electrification layer and electrode layer. A counter friction layer, polyamide (PA) film, was utilized to contact the Al foil for analyzing the contact sensing behaviors and electrical output performance of such a single-electrode TENG sensor. When applying pressures, the PA film contacts the Al layer, and charge transfer between these two layers is triggered because of the different triboelectric properties. As a result, the negative charges are injected from PA film into the Al layer, leaving positive charges on the PA film. Once the pressure is

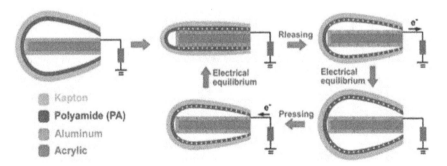

Figure 23. Schematic of the working mechanism of a single- electrode TENG sensor. Reproduced with permission from Ref. 195 (Copyright 2013, Wiley-VCH).

released, with the help of the elasticity of the Kapton supporting layer, the PA film immediately separates from the Al layer. Negative charges would flow from the Al layer to the ground through an external load, and an electrical current and an output voltage could be observed on the load. Subsequently, when the sensor is pressed and the PA film contacts the Al layer again, negative charges tend to flow from the ground to the surface of the Al layer to balance the positive charges on the PA film, which induces an electrical current and output voltage in the opposite direction. In light of the simple device layout and fabrication process, increasingly works based on such single-electrode TENG tactile sensors have been explored to build the advanced, self-powered, high-performance tactile e-skins in the past years.[55,195–200]

3.2.4. Photosensitive sensor

A photosensitive sensor, also known as a light sensor or a photodetector, is a device that detects and measures light levels in its surroundings. Photosensitive sensors are designed to convert light energy into an electrical signal that can be processed or used in various applications, which have been widely used in fields such as photography, robotics, industrial automation, security systems, and many other areas where light detection is necessary. There are different types of photosensitive sensors, including photodiode, phototransistor, photoresistor, and photomultiplier tube. As a non-contact and non-destructive measurement strategy, flexible photosensitive sensors based on extensive materials have also obtained great advances and progress in terms of structure design, the working mechanism, and state-of-the-art approaches, exhibiting their potential application prospects in electronic skin, wearable devices, and soft robotics.[201–211]

For example, Zirkl reported a flexible optothermal sensor that integrates a transistor device and pyroelectric polymer sensor (Figures 24(a) and 24(b)), which can serve as a laser pointer-activated switch and a sensitive infrared sensor.[212] As shown in Figure 24(c), the bottom electrode of the pyroelectric polymer sensor serves as the gate electrode of the transistor device, and then two components are integrated. Due to the pyroelectric effect of the P(VDF-TrFE) copolymer film, the transistor device could be switched on by a 40 on-off ratio when an infrared laser with a modulation frequency of 0.01 Hz is applied on the top electrode of the pyroelectric sensor.

Basic Composition and Functional Materials 41

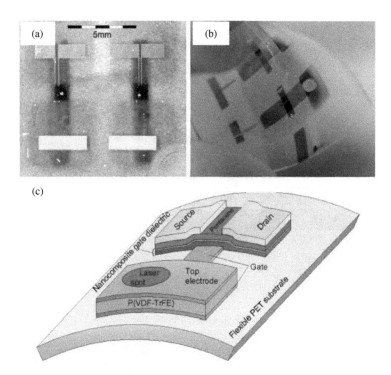

Figure 24. (a) Microscope image and (b) optical image of the proposed optothermal sensor serving as a light-activated switch. (c) Schematic illustration of the configuration of the optothermal sensor integrated by a transistor device with a pyroelectric polymer sensor. Reproduced with permission from Ref. 212 (Copyright 2021, Wiley-VCH).

Zhang et al. assembled the polymer microwire arrays via an efficient solution-processing method combined with the strictly regulated dewetting of liquid. They proposed a high-performance UV photodetector with a high on/off ratio of 137 and responsivity of 19.1 mA W^{-1}.[213] An array composed of 14 × 18 polymer array-based photodetectors was fabricated to demonstrate the application in image sensing. Figure 25(a) presents the image- sensing system that involves a 365-nm UV irradiation, a shadow photomask with the pattern of Arabic numerals, a photodetectors array, and a quartz substrate. As a result, the photocurrent response mappings of the photodetectors array consistent with the Arabic numerals on the shadow photomask were obtained with UV irradiation. More importantly,

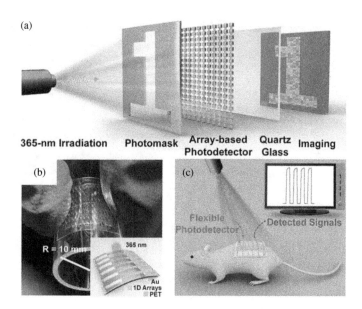

Figure 25. (a) An array composed of 14 × 18 photodetectors for application in image sensing. (b) Photograph and schematic illustration of the sensor array with excellent mechanical flexibility. (c) Schematic of the demonstration of the sensor array working as flexible photodetectors. Reproduced with permission from Ref. 213 (Copyright 2022, American Chemical Society).

such a photodetector array with good flexibility (Figure 25(b)) shows an outstanding fatigue characteristic even after 4000 cycles of bending processes and demonstrates its promising potential in wearable electronics by attaching it to the mouse skin (Figure 25(c)).

The heterojunction structures of polymer materials and other materials are widely used to design and create the high-performance photodetector.[214–216] Li *et al.* reported a phototransistor based on the heterostructure of In_2O_3/poly5,5'-*bis*-3,5-*bis*(thienyl)phenyl-2,2'-bithiophene-3-ethylesterthiophene (PTPBT-ET), which combines the merits of the fast electron transport of In_2O_3 and the high photoresponse of PTPBT-ET, achieving a large current on/off ratio (>10⁷).[214] Considering the superiorities of the inorganic perovskite materials, Liu *et al.* presented a photodetector enabled by lead-free perovskite ($FASnI_3$)/polymer material (PEDOT:PSS) vertical heterostructure, proving its excellent photoresponse over a broad range from UV to near-infrared (NIR).[215] According to the authors, the sensitivity and the photoresponse speed of the $FASnI_3$/

PEDOT:PSS photodetector can be adjusted by altering the thickness of the PEDOT:PSS polymer layer. Also, many research works have focused on the investigation of polymer composites and adopted them to develop more outstanding photodetectors than those based on pure polymer materials.[217-219] For instance, Zhao et al. fabricated a new composite (RE–SCP) consisting of a semiconducting polymer (SCP) and rare earth-doped nanoparticles (RENPs) and assembled a flexible shortwave infrared (SWIR) photodetector. Due to the introduction of RENPs with distinctive SWIR-responsive characteristics, the photon-to-electron conversion of such a photodetector was efficiently improved, which enhanced the SCP's detection performance at multiple wavelengths including 808, 975, and 1532 nm.[218] Most recently, by dispersing 2D materials of graphene and black phosphorus (BP) in a polymer matrix with good self-healing capability, An et al. explored a self-healing photodetector that does not present electrical and optoelectrical behavior degradation even after cutting-and-healing for 30 cycles.[219] With the help of graphene and BP, such a self-healing photodetector shows enhanced broad light spectrum absorption and empowers prominent photocurrent under visible and NIR light illumination, demonstrating its feasibility for wearable electronics in various application circumstances.

3.2.5. Magnetic sensor

A magnetic sensor is a device that can detect and measure the strength or changes in a magnetic field. It operates based on principles such as the magnetoelectric effect, Hall effect, or magnetoresistance effect.

Magnetoresistive sensors are based on the magnetoresistance effect, which utilizes the change in material resistance in a magnetic field to detect the magnetic field. The most common types are magnetoresistive sensors, which can be further classified into magnetoresistive stripe sensors and magnetoresistive rotation sensors.

Hall effect sensors are based on the Hall effect, which involves measuring the Hall voltage generated when a current passes through a conductor in a perpendicular magnetic field. Hall effect sensors are known for their fast response and high accuracy, and they are widely used in applications such as position sensing, speed measurement, and proximity switches.

Magneto-Resistance Sensors: Magneto-resistance sensors utilize the magnetoresistance effect, measuring the change in material resistance

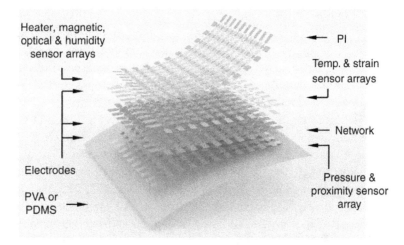

Figure 26. Schematic layout of a multi-functional sensor array involving typical GMR sensors. Reproduced with permission from Ref. 220 (Copyright 2018).

caused by a magnetic field. Two common types of magneto-resistance sensors are giant magnetoresistance (GMR) and anisotropic magnetoresistance (AMR) sensors. For example, Wang *et al.* deposited the Co/Cu multilayers on highly stretchable and conformable polymer substrates, and a giant GMR-sensing e- skin was obtained to achieve the detection of the static or dynamic magnetic field (Figure 26). Relying on the GMR effect, such a magnetoreceptive e-skin was capable of detecting the distance between the e-skin device and the magnetic field, which will likely attract increasing attention for the application in magnetic field proximity detection, navigation, and noncontact control.[220] Magnetic induction sensors operate based on the principle of magnetic induction. They detect the influence of a magnetic field on the induced electromotive force in a sensor coil to measure the magnetic field. This type of sensor is often used in applications such as non-contact distance measurement, metal detection, and magnetic field imaging.

4. Functional Materials in Flexible Sensors

4.1. Conductive polymers

Conductive polymer materials have become most promising candidates for flexible electrode materials compared with the previous rigid materials, playing a dominant role in the emerging flexible and wearable e-skins.

Basic Composition and Functional Materials   45

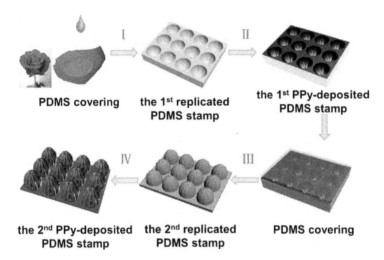

Figure 27. Schematic of the fabrication of the sensing layer composed of a microstructured PDMS supporting layer and the nanostructured PPy conductive layer. Reproduced with permission from Ref. 30 (Copyright 2020, Wiley-VCH).

For example, Lu et al. deposited an active polypyrrole (PPy) conductive layer with surface wrinkling nanopatterns on PDMS stamps that were replicated from rose petals and reported a highly sensitive pressure sensor.[31]

In addition, an approach combining the rose petal–templating method and surface wrinkling process was proposed by Lu et al. to fabricate a sensing layer with a multiscale hierarchical structure of a microstructured PDMS supporting layer and the nanostructured PPy conductive layer.[30] As illustrated in Figure 27, there are four steps involved in the preparation process of such a multiscale hierarchical structured sensing layer: (I) replicating the 1st PDMS stamp from a rose petal template, (II) depositing and in situ self-wrinkling of the 1st PPy layer on the surface of the 1st PDMS stamp (the preparation of the 1st PPy-deposited PDMS stamp), (III) fabricating the 2nd replicated PDMS stamp by using the 1st PPy-deposited PDMS stamp as the new template, (IV) preparing the final multiscale hierarchical structured sensing layer through the deposition and in situ self-wrinkling process of the 2nd PPy layer on the surface of the 2nd replicated PDMS stamp. Finally, two multiscale hierarchical structured sensing layers were assembled face-to-face to construct a sensor with excellent pressure-sensing performance. The authors pointed out that the sensitivity of such a sensor could be highly boosted under the condition of light illumination owing to photocurrent response and the photothermal effect of the PPy layer. According to the variation of the contact resistance, these sensors have significantly optimized the device sensing performance, which offers opportunities for the construction of high-performance tactile e-skins.

Another one of the most classic conductive polymers, PEDOT:PSS, because of its optical transparency at the visible range, tunable electrical conductivity and work function, high flexibility, stretchability, etc., has attracted extensive attention in fabricating the existing photovoltaics, displays, and transistors and various sensing electronics including strain, pressure, temperature, humid, and biosensors.[24–29,55,221–230] For example, Sun et al. reported a stretchable wrinkled PEDOT:PSS electrode and fabricated a triboelectric nanogenerator (TENG) to achieve high-performance active tactile sensing.[196] Here, the stretchable wrinkled PEDOT:PSS was prepared by blade-coating PEDOT:PSS solution on a pre-stretched PDMS plate (Figure 28). First, the PDMS mixture was poured into a groove model. After a curing process, the PDMS film was obtained and then pre-stretched by 100% onto a glass plate. Next, PEDOT:PSS solution was coated onto the pre-stretched PDMS film by using the blade technique and dried for a period. Lastly, the stretchable wrinkled PEDOT:PSS electrode was obtained after the PDMS was released.

Inspired by the imbricate scales on the snakeskin that are conducive to improving stretchability, a novel strain-sensing approach was achieved

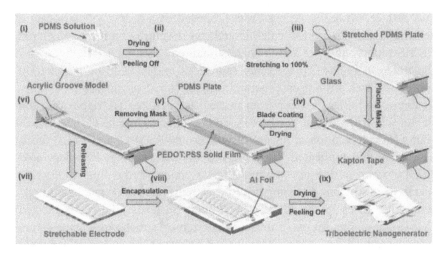

Figure 28. Schematic of the preparation process of the stretchable wrinkled PEDOT:PSS fim. Reproduced with permission from Ref. 196 (Copyright 2018, Wiley-VCH).

by involving the development of a bilayer of PEDOT:PSS.[231] This film exhibited high stretchability due to the presence of micro-sized PEDOT:PSS islands that were interconnected in a percolative manner. In addition, the strain sensor based on such a PEDOT:PSS bilayer, demonstrated the capability to detect small strains by monitoring the relative displacement of overlapping PEDOT:PSS micro-islands, accompanied by remarkable durability and functionality when subjected to large strains. According to the author, this structural robustness allowed the sensor to reliably operate across a wide range of strain magnitudes.

Lee et al. presented a high-performance flexible sensor, as shown in Figure 29(a), which realizes the simultaneous regulation of tactile stress transfer to an active sensing area and the corresponding electrical current because of the gradient structures.[232] Figures 29(b) and 29(c) present the multilayered PEDOT:PSS/PUD on the top PDMS micro-dome layer and the interdigitated electrode on the bottom TPU micro-dome layer. The conductivity-gradient PEDOT:PSS multilayer on the stiffness-gradient

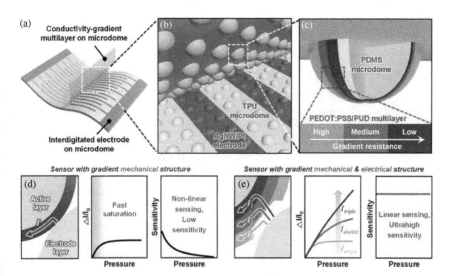

Figure 29. (a) Schematic of a flexible pressure sensor based on multilayered PEDOT:PSS. (b) Enlarged picture from dashed box in (a), showing the multilayered PEDOT:PSS/PUD on the top PDMS micro-dome layer and the interdigitated electrode on the bottom TPU micro-dome layer. (c) Schematic of the multilayered PEDOT:PSS/PUD on the top PDMS micro-dome layer. Schematic of the sensor with (d) gradient mechanical structure and with (e) gradient mechanical & electrical structure. Reproduced with permission from Ref. 232 (Copyright 2020, American Chemical Society).

interlocked microtome geometry endows the flexible sensor with extraordinarily high piezoresistive sensitivity at low power and good linearity over a broad pressure range. In specific, Figure 29(d) shows the working mechanism of the sensor with a gradient mechanical structure, observing that the stiffness-gradient interlocked microtome structures allow the efficient transfer and localization of applied stress to the sensing area. Furthermore, the multilayered structure with gradient conductivity (Figures 29(e)) enables the efficient regulation of piezoresistance in response to applied pressure by gradual activation of current pathways from outer to inner layers, resulting in a pressure sensitivity of 3.8 × 105 kPa^{-1} with linear response over a wide range of up to 100 kPa.

Apart from that, the PEDOT:PSS has been utilized as a conductive filler and embedded into an insulating elastomeric matrix such as polyurethane (PU) or PDMS to form a stretchable conductor.[233,234] Further, to develop robust and stretchable PEDOT:PSS conducting polymer films that meet the deformability requirements of e-skins, Jeong et al. presented a simple strategy of incorporating an excessive amount of a nonvolatile surfactant plasticizer.[235] In this way, the plasticizing effect of the plasticizer could contribute to improving the mechanical properties and stretchability of the PEDOT:PSS conducting polymer film without degradation in conductivity. Kee et al. proposed a stretchable PEDOT:PSS transparent electrode by mixing it with the ionic liquid compound, in which the ionic liquid not only works as a secondary dopant for enhancing the polymer conductivity but also acts as a plasticizer for prolonging polymer elongations.[236] Wang et al. used serpentine PEDOT:PSS/PVA/G sensing layer for the temperature sensor with excellent stretchability and stability.[237]

Some other extensively studied conductive polymers are polyaniline (PANI),[238–240] and poly(3-hexylthiophene).[241–243] Park et al.[244] developed a highly sensitive flexible piezoresistive sensor based on PANI-nanospines deposited *in situ* on hybrid hierarchical NFs. Herein, as the pressure increases, more contact points are created between the gradient PANI-nanospine arrays, which enables an ultra-high sensitivity of 179.1 kPa^{-1} over a broad linear range of up to 50 kPa. Wang et al.[245] reported a new self-powered ammonia (NH_3) nanosensor based on a TENG with PANI NFs with good flexibility, portability, selectivity, sensitivity, conductivity, and high specific surface area. This gas nanosensor displays response signals for NH_3 detection, and provides power for the sensor measurement process by converting ambient mechanical energy to electric energy without any external power sources.

4.2. *Hydrogel*

Hydrogels are crucial components in flexible sensors, offering remarkable benefits. These highly absorbent materials possess exceptional softness and stretchability. Their flexibility allows for seamless adaptation to surfaces of varying shapes and curvatures, ensuring optimal contact and adaptability to human skin or soft robots. Moreover, hydrogels exhibit excellent biocompatibility and biosafety, making them extensively utilized in medical treatments, health monitoring, and other related domains.

In the context of flexible sensors, hydrogels serve as ideal candidates for constructing sensing layers or filling materials. This enables them to facilitate the sensitive detection of various physical parameters, including pressure, strain, and temperature. The absorbent nature of hydrogels allows them to respond effectively to external stimuli, translating these physical changes into measurable signals. Conductive hydrogels have gained significant attention in various applications due to their unique properties, such as high electrical conductivity, high specific surface energy, and the ability to exhibit optical, magnetic, and catalytic characteristics. Among pure metals, silver (Ag) is known for its exceptional electrical conductivity, making Ag nanomaterials highly desirable for electron transport.

To prevent the agglomeration of Ag nanoparticles, Xiang and Chen utilized the deprotonated carboxylic acid groups within the hydrogel network to anchor Ag^+ ions, which were subsequently reduced to Ag nanoparticles.[246] This approach resulted in a pH-responsive conductive hydrogel with excellent electron transport capabilities. Similarly, Devaki and their colleagues reported the synthesis of an electron-conductive hydrogel by combining the in situ polymerization of acrylic acid with the in situ reduction of Ag^+ ions, resulting in a homogeneous network structure.[247] These studies highlight the potential of incorporating metallic nanomaterials, specifically Ag nanoparticles, into hydrogel networks to achieve desirable electrical properties and expand their range of applications. Besides, through the reduction of Cu^{2+} within a poly(vinyl alcohol) (PVA) network grafted with poly(acrylamide) (PAM) branches, Wei *et al.* fabricated Cu nanoparticles by an in situ method.[248]

The resulting conductive hydrogel demonstrated excellent vapor sensing capabilities. Typically, the direct incorporation of metallic nanomaterials into hydrogel networks has minimal impact on their mechanical properties. However, these conductive hydrogels find extensive application as sensors that can detect changes in the environment, such as stress, temperature, pH, and oxidation. Their versatility in sensing various environmental factors makes them highly valuable for a wide range of applications.[249–254]

In addition, Xuan et al. developed a highly stretchable and self-healing $Ti_3C_2T_x$ MXene/PVA hydrogel electrode by mixing MXene solution with a PVA aqueous solution and subjecting the mixture to a gelatinization process with the help of transparent sodium tetraborate solution (Figures 30(a) and 30(b)).[255] As a demonstration, a very high bond (VHB) tape was sandwiched as a dielectric layer between two pieces of MXene/PVA hydrogel

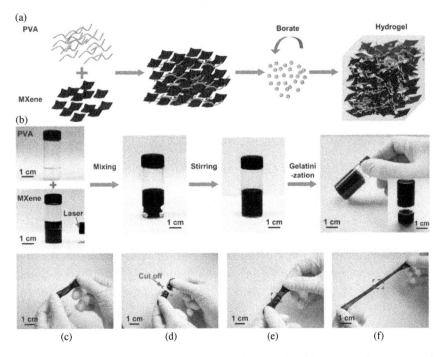

Figure 30. (a) Schematic illustration and (b) photographs of the preparation process of the stretchable and self-healing MXene/PVA hydrogel electrode. (c)–(f) Photographs illustrating the self-healability and stretchability of the MXene/PVA hydrogel electrode. Reproduced with permission from Ref. 255 (Copyright 2019, Wiley-VCH).

electrodes and used to construct a capacitive strain sensor, showing its sensing capability in the subtle and large-strain ranges was proved by mounting the sensor directly on human skin.

The mechanical properties of PVA hydrogels can be easily adjusted through a cyclic freeze-thaw process, making PVA a widely explored material in various applications. For example, Rong *et al.* and Gotovtsev *et al.* have incorporated conductive components, such as PEDOT:PSS, into PVA to create conductive hydrogels for constructing flexible strain sensors.[255,256] To enhance intermolecular interactions, ethylene glycol and iota-carrageenan were utilized in the respective studies. In addition to simple blending methods, Mawad and coworkers functionalized PEDOT with carboxyl groups and double bond-bearing pendant groups.[257] By copolymerizing the modified PEDOT with acrylic acid and poly(ethylene glycol) diacrylate, they developed conductive hydrogels suitable for cell adhesion, proliferation, and differentiation. Most recently, Lu *et al.* fabricated pure PEDOT:PSS hydrogels with interconnected networks of PEDOT:PSS nanofibrils.[258] This was achieved by adding dimethyl sulfoxide (DMSO) and subsequently undergoing controlled dry-annealing and rehydration processes (Figure 31). The resulting hydrogel demonstrated exceptional electrochemical stability during charge storage and injection processes in wet physiological environments, and it could be patterned into complex geometries, as depicted in Figure 31.

However, traditional hydrogels have limitations in terms of retaining water and functioning effectively under cold or hot conditions. To overcome these challenges, inspired by natural antifreeze and anti-heating

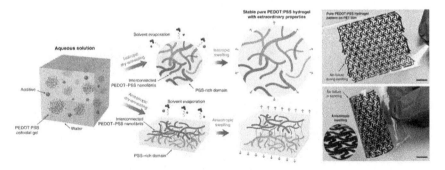

Figure 31. Dry-annealing and swelling processes of PEDOT:PSS by using the DMSO as the additive and patterning of the fabricated PEDOT:PSS hydrogels. Reproduced with permission from Ref. 258 (Copyright 2019).

mechanisms, Han *et al.* developed a novel adhesive and conductive hydrogel by incorporating mussel chemistry principles.[259] Firstly, polydopamine (PDA)-decorated CNTs (PDA-CNTs) (Figure 32(a)) were selected to work with the hydrogel to confer conductivity to the hydrogel and improve its mechanical properties. For comparison, the authors have copolymerized acrylamide (AM) and acrylic acid (AA) monomers in pure water and

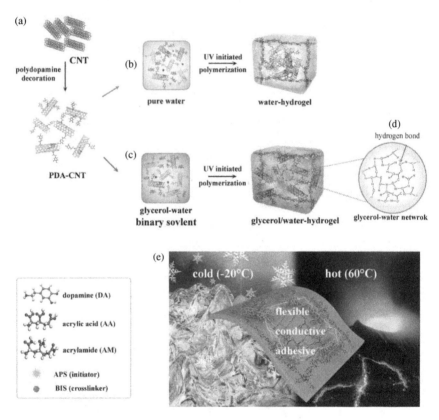

Figure 32. (a) Schematic of the PDA-CNTs dispersed in an aqueous solution. Uniformly dissolving process of AM and AA monomers in PDA-CNT dispersion and the UV-initiated polymerization processes of AA and AM to form (b) water-based (W) hydrogel in pure water and (c) glycerol–water (GW) in glycerol–water binary solvent. (d) Formed strong hydrogen bonds between glycerol and water for securing water molecules from evaporation in the GW hydrogel. (e) Schematic of the GW hydrogel, showing antifreezing and anti-heating performance to maintain good flexibility, conductivity, and adhesive behaviors even under cold (−20°C) and hot (60°C) environments. Reproduced with permission from Ref. 259 (Copyright 2017, Wiley-VCH).

glycerol–water binary-solvent systems to form the water-based (W) and glycerol–water (GW) hydrogels, respectively shown in Figures 32(b) and 32(c). Due to the synergetic interactions among glycerol, PDA-CNTs, and poly(acrylic acid)-poly(acrylamide) (PAA-PAM) covalent network, the developed GW-hydrogels exhibited the properties of high toughness and excellent recoverability. In addition, benefiting from the existence of the strong cooperative hydrogen bonding between glycerol and water (Figure 32(d)), the water molecule was firmly anchored in the hydrogel network, Resulting in the GW-hydrogel with good anti-freezing, anti-heating, and long-term stability (Figure 32(e)), which cannot be achieved using water-based hydrogels.

Considering that the flexible sensor plays a crucial role in the development of human–machine interfaces, functional prostheses, and health monitoring devices, it is significantly necessary to achieve the extensional design of flexible sensing devices from a single function to a multifunction by exploiting the hydrogel materials. Charaya et al. presented a flexible sensor that is capable of responding to both pressure and temperature information through electrical and optical approaches, respectively.[260] Figure 33 shows the configuration of the flexible sensor composed of two conductive PAm hydrogel electrodes and a sandwiched dielectric layer of

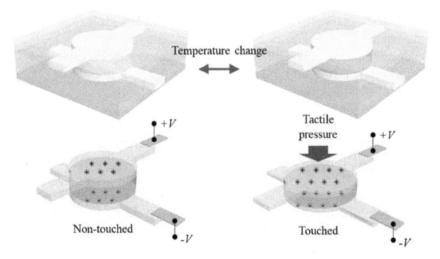

Figure 33. Schematic of temperature colorimetric responsibility and pressure-sensing behavior of the proposed flexible sensor. Reproduced with permission from Ref. 260 (Copyright 2019, Wiley-VCH).

the PDMS (Sylgard 184)/thermochromic liquid crystal (PDMS/TLC) composite.

As the author discussed, the functions of thermochromic response and capacitive pressure detection were realized by sandwiching a composite material containing a thermochromic dye elastomer into tough and transparent stretchable electrodes of electrolyte hydrogel. In this way, the pressure and temperature did not affect each other.

4.3. *Ionic liquid*

Ionic liquid, also known as low-temperature molten salt, refers to a salt that appears liquid at or near room temperature and is completely composed of organic cations and organic or inorganic anions with melting points below 100°C.[261–264] The common cations are imidazolium, pyrrolidinium, and quaternary ammonium salts, and their derivatives. The common anions are *bis*(trifluoromethanesulphonyl)imide, methanesulfonate, tricyanomethanide, and hexafluorophosphate. As a new polar solvent, ionic liquid is the so-called "green" chemical solvent that can replace the traditional volatile toxic solvent,[265,266] and has many valuable characteristics, such as low vapor pressure, high ionic conductivity, nonvolatility, high chemical and thermal stability, recyclability, and environmental friendliness. These characteristics have made ionic liquids highly sought-after functional materials for constructing flexible sensors.

In addition, ionic liquid-based capacitive tactile sensors (supercapacitive iontronic tactile sensors)[267–273] and triboelectric tactile sensors[274–276] are the most widely studied.

The sensing mechanism of supercapacitive iontronic tactile sensors derives from the electric double layer (EDL), which is a well-known interface effect in electrochemistry. Compared to traditional parallel plate capacitors, EDL-based tactile sensors can achieve ultrahigh capacitance per unit area of up to several μF cm^{-2}, and their capacitance value can increase by more than 1000 times. This phenomenon is attributed to the emergence of EDL, which enables the electrons on electrodes and the ions in the sensing layer to gather and attract each other at nanometer distances, thereby forming numerous paralleled small capacitors and substantially improving the specific capacitance of the device.[277–280] As the interface contact area increases with applied pressure, more cations or anions are attracted to the contact interface, which causes a capacitance

Basic Composition and Functional Materials 55

change. Early in the development of flexible sensing technologies, all studies adopting ionic liquids in tactile sensors placed ionic liquids in closed cells, where they would be in contact with the top and bottom electrodes.[281-284] However, because the EDL forming between the ionic liquid and electrode in a closed cell has high initial capacitance, it is difficult to achieve a highly sensitive tactile sensor.

In light of these findings, Lee et al.[285] reported a flexible tactile sensor showing a super high sensitivity. Figure 33(a) illustrates the fabrication steps of the flexible tactile sensor. Here, the bottom graphene electrode was first treated with octadecyl trichlorosilane (ODTS) to obtain a low surface energy. Then, prolonged exposure with a desirable mask to UV/O$_3$ on the ODTS-treated graphene electrode successfully etched the ODTS and graphene in some parts. After that, the ionic liquid, 1-ethyl-3-methylimidazolium tetrafluoroborate ([EMIM$^+$[BF4$^-$]) was spin-coated, and the preferential formation at the contact line of the graphene grid, as depicted in Figure 34(a). Figure 34(b) aims to describe the working mechanism of the sensor which consists of the top floating graphene electrodes and ionic liquid droplets pinned on the bottom graphene grid electrodes. Under a low-pressure regime (less than a few kPa, Regime I),

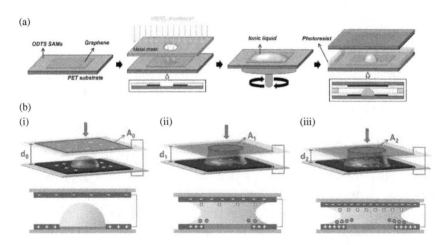

Figure 34. (a) Schematic of the preparation process of the flexible tactile sensor based on the graphene electrode layers and the ionic liquid dielectric layer. (b) Schematic of variation of the sensor geometry and the accumulation of charge carriers in the sensor when loading pressures (A: contact area, d: distance between electrodes). Reproduced with permission from Ref. 285 (Copyright 2020, Wiley-VCH).

capacitance changes under pressure are affected by the distance variation between the electrodes (the droplet size has no effect), because the ionic liquid droplet is not in contact with the top graphene electrode. With the pressure arising (Regime II and III), the spreading behavior of the ionic liquid contributes to a substantial increase in the capacitance due to the formation of the EDL. The author pointed out that this unique cell design can induce the dynamic characteristics of the droplet when the top floating graphene electrode contacts the ionic liquid droplet and generates spreading behavior under the gentle touch, thereby inducing a capacitance transition (from a few pF to over hundreds of pF).

With an in-depth exploration of ionic liquids and the improvement of processing technology, ionic gels are now formed by the combination of ionic liquids with polymer materials. Thereinto, the polymer chains form spatial network structures through cross-linking and tangling, while the liquid electrolytes fill the interspace and substantially reduce their fluidity. Ionic gels are mainly prepared by direct mixing of an ionic liquid with a thermoplastic polar polymer that exhibits high compatibility, followed by a curing step; the common thermoplastic polar polymers include PVDF-hexafluoropropylene (HFP), TPU, and PVA.[286–291]

For instance, a highly stretchable and sensitive pressure sensor was presented to provide unaltered sensing performance under stretch through the synergistic creation of the ionic capacitance sensing mechanisms of the PVDF-HFP/1-ethyl-3-methylimidazolium *bis*(trifluoromethylsulfonyl)imide ionic gel and mechanical hierarchical microstructures (98% strain insensitivity up to 50% strain), as displayed in Figure 35.[292] Through utilizing such an optimized structure, the flexible pressure sensor demonstrates a remarkable insensitivity of 98% toward 50% strain, with a low-pressure detection limit of 0.2 Pa. Equipped with the ability to deliver all the desired features for quantitative pressure sensing on a deformable surface, this sensor has been successfully employed to achieve the precise perception of physical interactions on human or soft robotic skin. Wang et al used cellulose/ionic liquid gel rich in ions as the electrolyte to further improve the sensitivity of the sensor.[293]

Additionally, inspired by the structural and functional features of biological multicellular, Kim *et al*.[294] reported a synthetic multicellular hybrid ion pump (SMHIP) (Figures 36(a) and 36(b)) and explored its potential applications in ultrasensitive ionic mechanoreceptor skin. Among them, the rationally designed SMHIP is composed of ionic liquid on the silica microstructure embedded in the TPU elastomer matrix, corresponding to

Basic Composition and Functional Materials 57

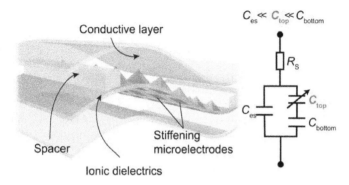

Figure 35. Schematic of the 3D structure of the stretchable and sensitive pressure sensor and its equivalent circuit (C_{es}: the electrostatic capacitance, C_{top} and C_{bottom}: EDL capacitances of the top interface and the bottom interface between the ionic dielectric and the electrode). Reproduced with permission from Ref. 292 (Copyright 2021).

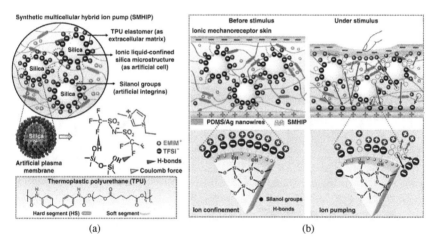

Figure 36. (a) Schematic of the designed SMHIP, in which the ionic liquid is confined on the surface of silica microstructures dispersed in TPU elastomer composed of hard and soft segments. (b) Schematic illustration of the piezocapacitive ionic skin composed of the SMHIP film sandwiched between silver nanowires/PDMS flexible electrodes. Reproduced with permission from Ref. 294 (Copyright 2021).

the physical analog of living cells and the extracellular matrix of biological multicellular structures, respectively. The corresponding SMHIP-based ion mechanoreceptor skin is ultrasensitive (48.1–5.77 kPa^{-1}) to a wide range of pressure (0–135 kPa) at ultra-low voltage (1 mV).

Interestingly, most of the materials obtained in nature contain anions and cations, such as leaves, blood, and biological tissues. The skin, being the most available natural ionic material, produces sweat that contains a large amount of ionic liquid, which provides favorable conditions for supercapacitive iontronic tactile sensors to form a natural EDL interface.

Recently, through the effective reuse of ionic gels of PVA/H_3PO_4 and electrodes of indium tin oxide/polyethylene terephthalate (ITO/PET) in a single device, Guo et al. developed an integrated hybrid device based on ionic sensing and the electrochromic display for interactive pressure perception.[295] Here, the hybrid device consists of a layered structure arranged from top to bottom, including the top ITO/PET electrode layer, WO_3 electrochromic layer, PVA/H_3PO_4 ionic sensing layer, and the bottom ITO/PET electrode layer. PVA/H_3PO_4 layer sandwiched by ITO/PET electrodes was horizontally divided into two sections, where the inner section served as the electrochromic area, while the surrounding unit functioned as the pressure-sensing area, allowing the hybrid device to detect pressure changes and change color simultaneously. First, Figure 37(a) illustrates the working principle on pressure-sensing of the hybrid device, which is attributed to the combined effect of the property and the unique

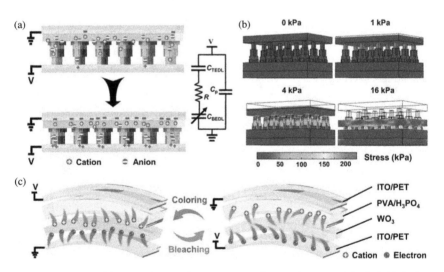

Figure 37. (a) Schematic of the ionic pressure sensing mechanism of the hybrid device and the equivalent circuit. (b) Finite element analysis on the contact stress distribution of the device under different pressures. (c) Schematic of the electrochromic mechanism of the device. Reproduced with permission from Ref. 295 (Copyright 2022, Wiley-VCH).

microstructure of the PVA/H₃PO₄ layer. When the device was connected to the measuring circuit, anions, and cations in the PVA/H₃PO₄ ionic gel layer would respectively move to the top and bottom electrodes and gather to form an EDL, resulting in an ultrahigh capacitance. It is easy to find that the capacitance value of the EDL-enabled capacitor is proportional to the contact area between the electrode and the PVA/H₃PO₄ ionic gel layer. Therefore, the microstructures designed on the PVA/H₃PO₄ ionic gel layer would contribute to the improvement of the pressure-sensing properties because of the good deformability of the microstructured PVA/H₃PO₄ ionic gel layer. Finite element analysis on the contact stress distribution of the device was also conducted to further demonstrate the changes in the contact area between the top electrode layer and the microstructured PVA/H₃PO₄ ionic gel layer with increasing the applied pressure (Figure 37(b)). Next, the author also makes the electrochromism mechanism clear in Figure 37(c). Due to the redox reactions of the WO₃ layer, the insert and release of H⁺ and electrons into the WO₃ film occurred respectively when applying the negative and positive potentials.

4.4. Metal-organic framework

As the star material, metal-organic framework (MOF) is a coordination polymer composed of metal ions (M) and organic ligands (O) that form a 3D network structure. This self-assembled structure creates crystalline porous materials with periodic skeleton arrangements, allowing for modifications and host–guest assemblies. While porous MOFs traditionally exhibit poor conductivity, recent advancements have sparked interest in their applications in sensing and electrochemistry.

Wan et al. developed a flexible resistive pressure sensor using a carbonized MOF/polyaniline nanofiber (PANIF) composite embedded in a polyurethane PU matrix, C-MOF/PANIF@PU, which was sandwiched between a breathable fabric and an interdigitated electrode-coated fabric (Figure 38).[296] The sensor exhibited remarkable performance, including a wide sensing range of up to 60 kPa, high sensitivity of 158.26 kPa⁻¹, fast response and recovery times of 22 and 20 ms respectively, reliable breathability, and excellent repeatability for up to 15,000 cycles. Notably, the MOF crystals used in this study were capable of self-assembly on various templates with complex supramolecular structures, leading to the formation of a MOF hybrid array (MHA) with a pre-designed 3D

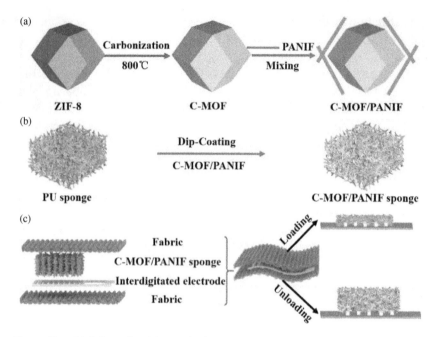

Figure 38. (a) Schematic of the synthesis C-MOF/PANIF, including the carbonization of ZIF-8 and subsequently mixing of C-MOF and PANIF. (b) Schematic of the preparation of C-MOF/PANIF@PU by dip-coating of C-MOF/PANIF on PU sponge. (c) Schematic of the assembly of the flexible sensor based on C-MOF/PANIF@PU. Reproduced with permission from Ref. 296 (Copyright 2020, Elsevier Ltd).

hierarchical orientation. This unique architecture effectively prevented nanoparticle aggregation and improved the conductivity of the sensor.

Han et al. utilized $Cu_3(HHTP)_2$ MOF as the active material for the fabrication of flexible pressure sensors, benefiting from its intrinsic electrical conductivity and easy positioning properties.[297] The researchers successfully developed a flexible pressure sensor based on a MHA by employing a well-aligned and conductive $Cu_3(HHTP)_2$ MOF design on a foil or mesh substrate (MOF MHA@mesh), as shown in Figure 39(a). The pressure sensor design incorporated a 3D hierarchical architecture of the MHA mesh, providing a large area for sensing purposes, as depicted in Figure 39(a). The MHA mesh, with its conductive properties, facilitated signal transduction through the numerous contact points between the MHA tips and the conductive thin film on the PDMS substrate under external pressure. A simplified circuit model, illustrated in Figure 39(b)

Basic Composition and Functional Materials 61

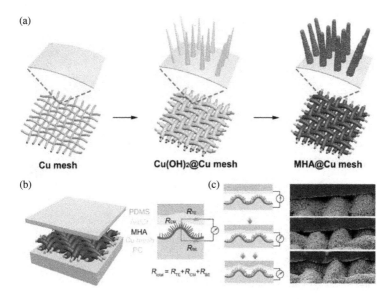

Figure 39. (a) Schematic of the synthesis of the MOF MHA on a Cu mesh substrate. (b) Schematic of the flexible sensor based on the MOF MHA@mesh and its circuit model. (c) Schematic and SEM images of the flexible sensor when loading pressures. Reproduced with permission from Ref. 297 (Copyright 2020, Wiley-VCH).

(right), can be proposed to describe the cross-section of the sensor. Furthermore, Figure 39(c) demonstrates the pressure-induced deformation of the PDMS, leading to an increase in the contact interface area (Ac), as confirmed by SEM analysis. This increase in contact area correlates with the sensor's response to pressure, which is reflected in the change in current (i.e., resistance change).

In addition to their potential in pressure sensing, MOFs have also demonstrated great promise as triboelectric materials due to their versatility in composition, size, and functionality. An exemplary illustration of this is the zeolitic imidazolate framework-8 (ZIF-8), which belongs to the subclass of MOFs with a sodalite topology and exhibits positive triboelectric behavior.[298] The unique particle geometry and size of ZIF-8 enable the adjustment of nano-roughness on triboelectric surfaces, making it a highly suitable candidate for the development of triboelectric nanogenerators (TENGs). Through the functionalization of ZIF-8 and PDMS, the typically negligible triboelectric behavior of natural wood can be significantly enhanced. This advancement paves the way for the creation of

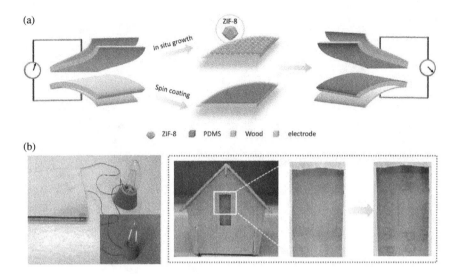

Figure 40. (a) Schematic of the fabrication method of FW-TENG-based flexible sensor and its configuration. (b) Schematic of the established intelligence system based on the FW-TENG for powering a household lamp and the smart window. Reproduced with permission from Ref. 298 (Copyright 2021, Elsevier Ltd).

functionalized wood-based TENGs (WF-TENG), which find applications in smart home technologies (Figures 40(a) and 40(b)).

In addition, MOF materials offer numerous advantageous characteristics including structural diversity, ultra-large specific surface area, controllable pore size/geometry, and host-guest interactions. These features render MOFs highly promising for applications in gas sensors, humidity sensors, and biosensors. For instance, Lee et al. devised a monolithic flexible sensor that is based on the coating of a mixed matrix membrane (MMM) composed of zeolitic imidazole framework (ZIF-7) nanoparticles (MMM(ZIF-7/PEBA) and polymers on TiO_2 sensing films.[299] Figure 41(a) illustrates the details for the preparation of the MMM(ZIF-7/PEBA) coated TiO_2 flexible sensor, and the schematic of the ZIF-7 unit cell is displayed in Figure 41(b). This innovative design effectively mitigated the interference of ethanol on TiO_2 sensing films during molecular sieving. As a result, the sensor exhibited exceptional performance with ultrahigh selectivity (response ratio >50) and response (resistance ratio >1100) to 5 ppm formaldehyde at room temperature.

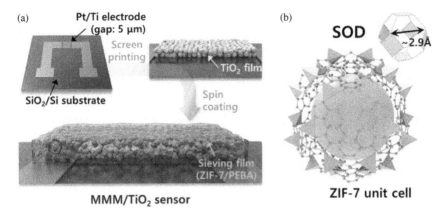

Figure 41. (a) Schematic of the fabrication steps for the MMM(ZIF-7/PEBA) coated TiO_2 flexible sensor. (b) Schematic of ZIF-7 unit cell. Reproduced with permission from Ref. 299 (Copyright 2021).

4.5. Inorganic material

Due to their excellent physical and chemical properties, inorganic materials play an important role in flexible sensors, display devices, energy storage, and conversion devices, etc.

4.5.1. One-dimensional inorganic material

In terms of flexible sensors, due to their unique morphology, 1D nanomaterials such as nanowires and nanotubes can be used to fabricate highly sensitive strain sensors and pressure sensors. These nanomaterials are highly bendable and stretchable, which can maintain stable electrical properties during deformation and are highly responsive to external strain or pressure changes. For example,[300–305] metal NWs including Au NWs, Ag NWs, and Cu NWs, have become the most common electrode materials for flexible sensors due to their excellent conductivity and mechanical flexibility. Ag NWs have been widely employed to enhance the sensing capability of strain sensors, particularly under large strains. This is attributed to the percolation networks formed by Ag NWs, which exhibit high electrical conductivity. Additionally, the unique properties of Ag NWs enable effective slippage of the network without subjecting the nanowires to significant strain. As a result, strain sensors incorporating Ag NWs demonstrate excellent sensitivity and durability, making them suitable for applications that require accurate detection and monitoring of mechanical deformations even under substantial strain conditions.

Pu *et al.* alternately and tightly coated the Ag NW/waterborne polyurethane (WPU) layers and MXene layers onto a hydrophilic polyurethane-based commercial fiber (HPUF) and constructed a stable sensing layer with strong internal interactions.[47] Here, the Ag NW/WPU guarantees the integrity of the sensing layer at large strain, while the MXene layer effectively promotes crack generation in the whole operating range. As a result, the multilayer structured fiber sensor exhibits high sensitivities (GF > 100) in a wide operating range (0–100%), with ultrahigh gauge factors (more than 1500 for strain beyond 25% and up to 1.6×10^7 for 85–100% strain), great durability for around 1000 stretching cycles, and fast response (344 ms) and relaxation (344 ms). Smart textiles fabricated by weaving such fiber strain sensors in a corset and a glove can be utilized for real-time upper body posture and hand gesture monitoring, which can help office workers make adjustments to reduce the risk of musculoskeletal disorders. The author also pointed out that such structure-designed advanced fiber strain sensors will promote the theoretical development and the innovation of high-performance strain sensors as well as the advancement of wearable bioelectronics in health monitoring.

By using a pressure-assisted imprinting technique with a polyurethane-carbon black-silver nanowire (PU-CB-Ag NW) composite mixture, regularly ordered microhenry structures are employed onto the flexible and conductive polyurethane-silver nanoparticle (PU-AgNP) fiber by Choi *et al.*, achieving a fiber-based conductive sensor with remarkable stretchability and sensitivity (Figures 42(a) and 42(b)).[306] As shown in Figure 42(c), two different types of resistance changes (single resistance and mutual resistance) are measured to distinguish external stimuli applied to the contact point of the sensor. It turns out that the fiber sensor based on these behaviors underlying the measurements of mutual resistance is capable of discerning the varying shapes of waveforms corresponding to different types of external stimuli (e.g., normal pressure, stretching, and bending), as shown in Figure 42(d). As the author discussed, with such sensing capabilities, the fiber-based conductive sensor can be used to make a commercial glove and serve as a wearable controller, enabling it to respond accurately to diverse gestures of the human

Basic Composition and Functional Materials 65

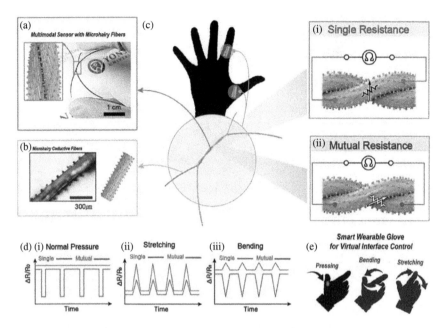

Figure 42. (a) Schematic and photographic image of the fiber- based conductive sensor. (b) Optical microscope image and schematic diagram of a single conductive fiber with micro hair arrays. (c) Schematic of a smart glove with twisted conductive fibers with hierarchical micro hairs for single resistance and mutual resistance. (d) Waveforms of resistance changes of the sensors based on twisted conductive fibers with hierarchical micro hairs (single resistance and mutual resistance) when applying pressure, stretching, and bending. (e) Demonstration of the conductive multimodal fiber-type sensor as a smart, wearable glove in the practical application of gaming interface control. Reproduced with permission from Ref. 306 (Copyright 2019, Wiley-VCH).

hand (Figure 42(e)), shedding light on advances in wearable electronics with VR interface systems, medical, and healthcare functionalities.

In current Ag NW-based flexible electrodes, achieving sufficient conductivity often requires a high dosage of Ag NWs. However, this approach increases cost and sacrifices optical transmittance, making it challenging to balance conductivity, mechanical flexibility, and transmittance.[307] Researchers have developed various strategies to address this issue. These include techniques such as cold/electroless/plasmonic welding, surface treatment, alignment/patterning of Ag NWs, and mechanical pressing, aiming to reduce the contact resistance of Ag NWs.[308–315] Despite these

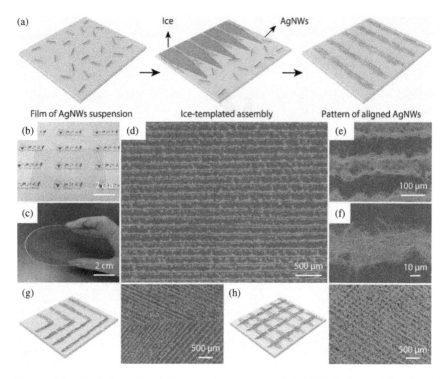

Figure 43. (a) Schematic of the assembly process of AgNWs during unidirectional freezing via an ice-templated method. (b–c) Optical images of the as-prepared electrode layer based on the ice-templated AgNW pattern, show high transmittance and excellent flexibility. (d–f) SEM images of the as-prepared electrode layer based on the ice-templated AgNW pattern. (g–h) Schematic and images of the other two complex AgNW pattern fabricated by the ice-templated method. Reproduced with permission from Ref. 316 (Copyright 2021, Wiley-VCH).

efforts, the resulting Ag NWs still fail to meet the current demands. To overcome these challenges, a facile 2D ice-templating approach has been proposed, as shown in Figure 43.[316] This method involves directionally freezing a film of Ag NW suspension at a controlled velocity. The growing ice crystals effectively align and compact the Ag NWs, leading to a high contact area and low contact resistance. Consequently, a small quantity of Ag NWs can achieve both high optical transmittance and low sheet resistance when fabricated as large-area conductive patterns on various substrates using this approach. In addition, the fabricated electrode exhibits excellent mechanical durability, even after 10,000 bending cycles and

Basic Composition and Functional Materials 67

under 50% stretching strain. As a proof of concept, they demonstrate the potential applications of the ice-templated flexible transparent electrode in touch screens and electronic skin sensors. More importantly, the authors believe that this 2D ice-templating approach is versatile to assemble various building blocks into large-area patterns with designable topology and function which are strongly demanded for advanced electronic devices. This innovative technique offers a promising solution for simultaneously achieving the desired properties of AgNW-based flexible electrodes.

Metal NWs have also received much attention for constructing active layers with excellent stretchability. There are currently two main strategies for the same. One strategy is to deposit metal NWs on elastomer surfaces to construct surface percolation networks. For instance, Cheng et al.[317] reported a standing enokitake-like Au NW film that chemically bonds to elastomeric materials. They exhibit substantially higher stretchability (up to 900%) than the conventional vacuum-evaporated bulk metal or percolating NW films, which is attributed to the standing enokitake-like NW structures and their strong adhesion to the elastomers. Typically, conventional metal films exhibit large "cliff-like" "U-shaped" cracks that cannot recover when the strain is released. However, the standing enokitake-like Au NW films exhibit tiny "V-shaped" cracks that recover when the strain is removed, thereby retaining their conductivity. The second strategy for constructing active layers with metal NWs is to embed metal NWs in an elastomer to construct internal percolation networks. For example, a highly conductive, biocompatible, and soft Ag-Au nanocomposite comprising ultralong Au-coated Ag NWs dispersed in poly(styrene-butadiene-styrene) elastomer was reported, where the high aspect ratio of Ag NWs confers high conductivity, and the inert Au shell ensures that the NWs are biocompatible and resistant to oxidation.[318] Furthermore, the phase separation in the Ag-Au nanocomposite during the solvent-drying process generates a microstructure that yields an optimized stretchability of 266% and a maximum stretchability of 840%.

In addition to the metal NWs materials, CNTs, including SWCNTs and multi-walled carbon nanotubes (MWCNTs), are regarded as nano-hollow tubular structures obtained by curling single or multi-layer graphene nanoplates. They have become ideal for flexible sensing materials due to their remarkable charge carrier mobility, robust mechanical properties, and high chemical stability. In addition, since the hollow structure allows additional energy adsorption, CNTs exhibit excellent mechanical strength and elasticity. For example, Jeong et al. combined CNTs

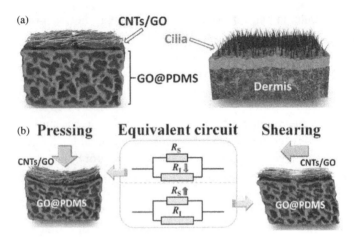

Figure 44. (a) Schematic illustrations of the structure of hairy skin and the proposed flexible sensor based on conductive architectures of the synthesized CNTs/GO@PDMS. (b) Schematic of deformation of the flexible sensor when applying the normal press and lateral shear force and the corresponding equivalent circuits of the flexible sensor. Reproduced with permission from Ref. 320 (Copyright 2018, Wiley-VCH).

composite materials with the hierarchical morphology of a micropillar-wrinkle hybrid structure to prepare a flexible and stretchable sensor that could detect pressure and strain.[319] For these two stimuli, the sensor exhibited excellent sensitivity (20.9 kPa^{-1} and gauge factor of 707), and broad sensing range (0–270 kPa and 0–50%), respectively. Besides, inspired by the hairy skin, Zhang *et al.* demonstrated a pressure–strain sensor composed of two layers of CNTs/GO hybrid 3D conductive networks, which was synthesized on a thin porous PDMS layer through a porogen-assisted self-assembling process (Figure 44).[320] Different from other works, the porous CNTs/GO@PDMS-based sensor sensitively outputted opposite resistance changes under normal and tangential forces, respectively. The sensor could not only produce responses to wrist pulse, slight feather scratching, and surfaces with different roughness, but also could monitor human breathing and music rhythm in the noncontact mode.

By a facile, stable, cost-effective, and controllable one-step extrusion method, Li *et al.* presented a MWNTs/PDMS fibers-based flexible sensor with tunable, stretchable, and thermal-sensitive properties, as depicted in Figure 45.[321] According to the author, the tunable MWNTs/PDMS fibers, with a weight ratio of 0.200 and varying parameters ranging from 150 to

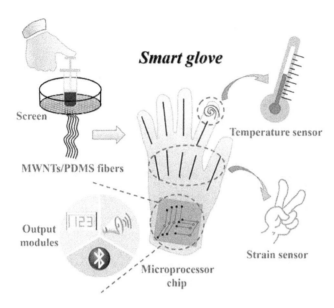

Figure 45. Schematic of the fabrication of the MWCNTs/PDMS fibers and the smart glove integrated with these fibers for gesture recognition and temperature measurement. Reproduced with permission from Ref. 321 (Copyright 2020, American Chemical Society).

400 μm, exhibited remarkable linear relationships between resistance change and strain within the range of 0% to 120%. Additionally, their resistances displayed sensitivity and linearity with temperature variations spanning from 0°C to 100°C. These MWNTs/PDMS fibers were incorporated as components of wearable sensors and woven into a fabric glove, alongside a customized microprocessor chip for decoding and transmitting finger dexterity, gesture language, and temperature information. This highly integrated smart glove enables real-time monitoring of joint movement and temperature with quantitative and visualized readings. Its multifunctional capabilities make it well-suited for applications such as motion monitoring, telemedicine (including healthcare and rehabilitation), and human-machine interfaces. By providing reliable and comprehensive data, this smart glove holds promise for enhancing various fields and improving the interaction between humans and machines.

Shen and coworkers have successfully developed a self-healable bifunctional electronic skin (Figure 46) by integrating two types of

70 F. Yin et al.

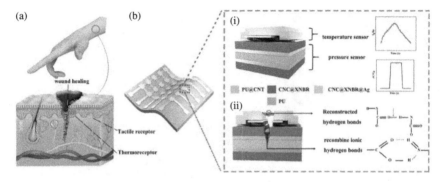

Figure 46. (a) Schematic of the self-healing human skin with tactile receptors and temperature receptors. (b) Schematics of the proposed self-healable bifunctional flexible sensor. Reproduced with permission from Ref. 322 (Copyright 2020, American Chemical Society).

sensors: a capacitive pressure sensor based on polyurethane (PU) and a resistance temperature sensor based on PU@CNT (polyurethane decorated with CNTs).[322] The device exhibits distinct responses to external pressure and temperature stimuli, with each sensor being sensitive to its respective stimulus and unaffected by the other. Furthermore, the self-recovery properties of PU and CNC@XNBR (CNT composite with cross-linked acrylonitrile butadiene rubber) enable the fabricated devices to possess stable self-healing abilities even after multiple damages. The author has thoroughly investigated the self-healing mechanism of the e-skin. Building upon this foundation, they have successfully fabricated a 5 × 5 array of multifunctional e-skin, capable of spatially distinguishing the distribution of input pressure and temperature signals. The self-healing ability of the proposed device has been demonstrated, showing the minimal impact on the sensing performance, particularly for the temperature sensors, even after cutting-off damage. This research opens up new opportunities in the field of artificial skin human–machine interfaces and biomonitoring devices, offering promising advancements in wearable technology.

Sensors with multifunctions have attracted great attention for their extensive application value, the multifunctional sensor that can simultaneously detect pressure/strain and humidity changes has also been widely developed to facilitate human health monitoring applications.[323–329] For instance, based on CNT–PDMS, Zhang *et al.* reported a dual-modal sensor with humidity and pressure-sensing capabilities by combining the

Basic Composition and Functional Materials 71

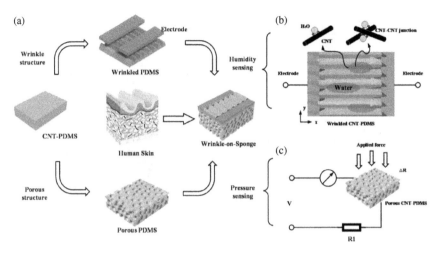

Figure 47. (a) Schematic of preparation process of human skin-inspired CNT–PDMS enabled dual-modal sensor. Schematic of the (b) humidity-sensing and (c) pressure-sensing of the dual-modal sensor. Reproduced with permission from Ref. 330 (Copyright 2019, American Chemical Society).

wrinkle structure with a porous sponge structure.[330] Figure 47(a) presents the structure schematic of the dual-modal sensor, which consists of the upper CNT–PDMS wrinkle structure and the bottom CNT–PDMS porous sponge structure. As discussed by the author, the processes of pre-stretching and the dissolution of sugar granules were performed respectively to obtain the above unique structures. Figure 47(b) illustrates the basic working mechanism of the dual-modal sensor for humidity sensing, which relies on the upper CNT–PDMS wrinkle structure. Due to the absorption of water molecules, the junction and contact area between CNTs as well as the resistance value of CNTs would change. Obviously, the bottom CNT–PDMS porous sponge structure plays an essential role in pressure sensing (Figure 47(c)), and the resistance of the CNT–PDMS is influenced by an applied force because the porous sponge structure endows good compression performance. Benefiting from that, the sensor could also be attached to the mask or the knee joint to monitor the exhaled water molecules of the human body and the movement state of the human leg, respectively, to guide people to exercise in science and health.

As shown in Figure 48, Wang *et al.* designed a dual-modal sensor with a multi-layer and internal 3D network structure by mimicking the

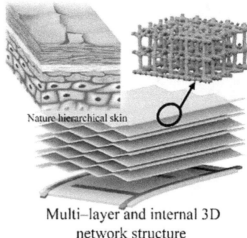

Multi-layer and internal 3D
network structure

Figure 48. (a) Schematic of the human skin and the dual-modal sensor with the multi-layer and internal 3D network structure. Reproduced with permission from Ref. 69 (Copyright 2019, Elsevier Inc).

layered structure of the skin, which could detect pressure and humidity with high sensitivity.[69] PANI-acidified MWCNTs composites (P-M) were used as the conductive material, and the basal layer and the piezoelectric layer were prepared by doping P-M to different layers of collagen aggregates (CAs). Among them, the basal layer had the function of providing cross-electrode and humidity sensing, and the piezoelectric layer served as the sensing layer of external pressure. Meanwhile, the fabricated sensor was comfortable to wear, durable, and degradable, because CA was a unique biological material.

With the development of nanotechnology processing, 1D ceramic materials, such as ceramic NWs and NFs, offers the possibility of transforming bulky and brittle ceramics into large-area flexible thin films. Additionally, due to the excellent mechanical, electrical, optical, and chemical properties, 1D ceramic materials have provided important functional and application potential for realizing high-performance, high-stability flexible electronic devices.[331,332] For example, considering the high melting point and excellent mechanical properties of TiO_2, Fu and her coworkers fabricated the large-area flexible TiO_2 nanofibrous networks by an electrospinning method.[333] And then the author adopted it as the

Basic Composition and Functional Materials 73

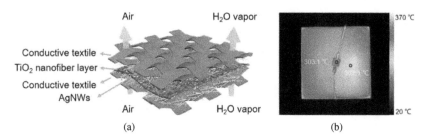

Figure 49. (a) Schematic of the structure of the high-temperature- resistant sensor. (b) Thermal image of the high-temperature-resistant sensor tested at 370°C. Reproduced with permission Ref. 333 (Copyright 2020).

dielectric layer and constructed a high-performance capacitive pressure sensor (Figure 49(a)), showing a high-temperature-resistant. Figure 49(b) presents the thermal image of the sensor tested at 370°C, and then the pressure-sensing properties of the sensor at 370°C were also investigated to illustrate the good high-temperature-resistant of the proposed flexible sensor. Furthermore, ceramic nanofibrous networks offer a highly porous structure and lightweight nature. The high degree of porosity allows for efficient airflow, enhancing breathability and comfort for the user. The lightweight characteristics of ceramic nanofibrous networks ensure that the devices remain lightweight and unobtrusive, enabling ease of wear and mobility. These properties make them well-suited for applications in wearable electronics, such as smart textiles, flexible sensors, and other wearable devices that require breathability and comfort for prolonged use.

Also, extensive researchers have introduced 1D inorganic piezoelectric materials into polymer substrates designed with microstructures to effectively enhance the deformation ability of the functional layer and achieve high-performance piezoelectric tactile sensors. For example, Ko et al. proposed an interlocked piezoelectric layer composed of hierarchical micro- and nanostructures by directly decorating PDMS micropillars with ZnO NWs arrays and then adopted such a piezoelectric layer to develop a flexible piezoelectric sensor, as illustrated in Figure 50(a).[334] Benefiting from the design of interlocked and hierarchical structures (Figure 50(b)), the developed flexible piezoelectric sensor exhibited a strong piezoelectric voltage response toward extremely small dynamic stimuli such as minute vibration and sound stimuli.

The high aspect ratio and large surface area of 1D nanoceramics facilitate efficient sensing and detection of changes in pressure or the

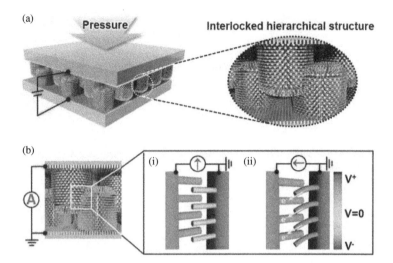

Figure 50. (a) Schematic of the flexible piezoelectric sensor based on the interlocking geometry of hierarchical ZnO NW arrays. (b) Schematic of the underlying mechanism for the piezoelectric behavior of the ZnO NWs-based flexible sensor. Reproduced with permission from Ref. 334 (Copyright 2015, Wiley-VCH).

presence of specific gases in the surrounding environment. In conclusion, the unique structures and properties of 1D nanoceramic materials make them valuable in flexible electronics, particularly in sensing and detection applications. Their flexibility allows for integration into wearable devices, enabling real-time monitoring and data collection. Overall, 1D nanoceramic materials contribute significantly to the advancement of flexible electronics and their applications in various sensing and detection fields.

4.5.2. *Two-dimensional inorganic material*

From the discovery of graphene by Andre Geim and coworkers in 2004,[335] it has attracted worldwide attention. As the most celebrated 2D material, graphene has emerged as a highly promising material for flexible sensors due to its unique properties. Its exceptional mechanical strength, high electrical conductivity, and excellent flexibility make it an ideal candidate for sensing applications. On the other hand, additional unique superiorities, such as their lightweight, good processability, as well as their good compatibility with large-area and flexible solid supports, endow these materials with great potential for the manufacturing of high-performance

Basic Composition and Functional Materials 75

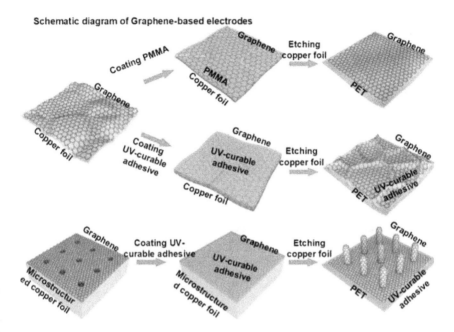

Figure 51. Schematic of the preparation of different graphene electrode layers by the traditional PMMA-mediated transfer, UVA-mediated transfer, and micro-conformal transfer methods. Reproduced with permission from Ref. 126 (Copyright 2019, American Chemical Society).

flexible sensing devices.[336–342] Graphene-based flexible sensors can detect various physical and chemical stimuli, including strain, pressure, temperature, humidity, gas molecules, and biomolecules.

For instance, Yang *et al.* presented a novel approach to fabricating flexible capacitive pressure sensors by utilizing a micro-conformal graphene electrode.[126] As illustrated in Figure 51, the authors successfully achieved controllable microstructures and conformal graphene electrode layers of smooth graphene electrodes (SGrEs), nanostructured graphene electrodes (NGrEs), and microstructured graphene electrodes (MGrEs) by exploring different methods, including the polymethylmethacrylate (PMMA)-mediated transfer method, ultraviolet-curable adhesive (UVA)-mediated transfer method, and micro-conformal transfer method. The roughness of the electrodes is known to have a significant impact on the sensitivity of capacitive pressure sensors. By improving the electrode roughness and employing controllable micro-conformal structures, the researchers were able to achieve high sensitivity, fast response speed, ultralow detection limit, and tunable sensitivity in their capacitive pressure sensor. Therefore, the proposed sensor consisted of a PDMS dielectric layer sandwiched between the top MGrE and the smooth bottom graphene electrode, resulting in a highly flexible and stable device.

The CVD method is the most widely used approach to fabricate high-quality 2D GFs, and many interesting works based on this method have been reported.[343,344] Recently, Li et al. fabricated a tactile sensor based on a CVD GF boron nitride (BN) heterostructure, wherein monolayer graphene was sandwiched between two layers of vertically stacked dielectric BN nanofilms.[345] With the protection of the BN layers, the oxidation and contamination of graphene were effectively avoided. The epidermal ridges present on the human fingertip skin, known for their ability to amplify delicate external stimuli, have served as inspiration for the design of highly sensitive fingertip skin-like pressure sensors. For example, Xia et al. demonstrated the growth of a 3D GF that mimics the morphology of fingertip skin using a CVD process,[153] as depicted in Figures 52(a)–52(c).

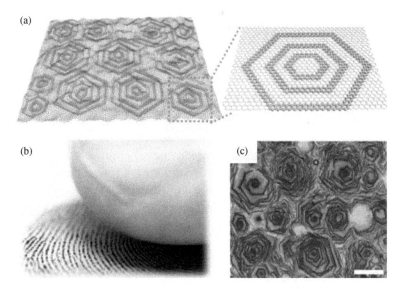

Figure 52. (a) Schematic of the structure of the 3D GF that consists of a continuous GF and closely packed concentric hexagonal graphene nanoribbon rings. (b) Optical image of the human fingertip and its fingerprint. (c) SEM image of the fabricated the 3D GF. Scale bar: 20 μm. Reproduced with permission from Ref. 153 (Copyright 2018, Springer).

Basic Composition and Functional Materials 77

The hierarchical structure of graphene, combined with PDMS films molded from a natural leaf, contributed to the exceptional performance of the pressure sensor.

Pyo et al. introduced a capacitive tactile sensor that utilized monolayer graphene electrodes separated by spacers, creating air gaps.[346] The fabrication process involved patterning and assembling the graphene electrodes on PET, with PDMS serving as the dielectric and SU-8 as the spacer between the facing graphene electrodes, as depicted in Figure 53(a). By harnessing the exceptional properties of graphene and incorporating the air gap structure, the tactile sensor exhibited remarkable mechanical flexibility

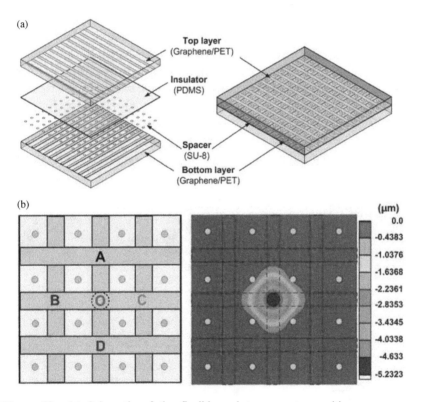

Figure 53. (a) Schematic of the flexible and transparent capacitive sensor array composed of the top and bottom PET layers with graphene-patterned electrodes, PDMS insulator, and SU-8 spacers. (b) Schematic of the flexible and transparent capacitive sensor array and the FEA result when applying an 8 kPa object on the center of cell-O. Reproduced with permission from Ref. 346 (Copyright 2017, Wiley-VCH).

and optical transparency within the visible range. The fabricated sensor demonstrated several advantageous characteristics, including high-pressure sensitivity (6.55% kPa^{-1}), the rapid response time (approximately 70 ms), and exceptional stability over 2500 cycles of loading and unloading. Additionally, the authors successfully showcased a pixelated sensor array for pressure mapping, as illustrated in Figure 53(b), without significant crosstalk between adjacent cells. According to the author, they achieved a notable advancement in the field of capacitive tactile sensors by leveraging the unique properties of graphene and employing a well-designed air gap structure. Their work showcased the sensor's mechanical flexibility, optical transparency, high-pressure sensitivity, rapid response, stability, and the potential for creating sensor arrays for pressure mapping applications.

CVD is the main method for producing high-quality graphene because of its scalability and control over layer thickness, making it a valuable technique for a variety of applications in electronics, energy storage, and optoelectronics. However, there are still challenges in this method, including high complexity, high cost, and the necessary transfer process.

Recently, Pan et al. reported a facile strategy for a high-performance piezoresistive pressure sensor that adopted a sensing layer consisting of highly conductive interfacially self-assembled graphene (ISG) and microstructured PDMS elastomer.[21] As depicted in Figure 54(a), a homogenous

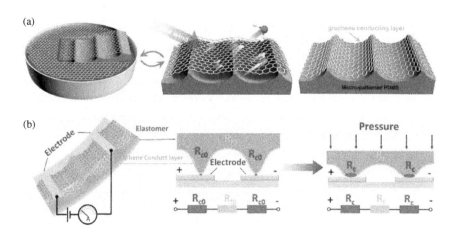

Figure 54. (a) Schematic of the fabrication process of the flexible sensor based on the ISG sensing layer and microstructured PDMS. (b) Schematic of the flexible sensor and its piezoresistive sensing mechanism. Reproduced with permission from Ref. 21 (Copyright 2019, Elsevier Inc).

Basic Composition and Functional Materials 79

graphene nanoplatelets dispersion was first sprayed onto the water surface to form uniform Langmuir monolayers at the water/air interface, and then the ultra-large conductive ISG layer were successfully assembled via the assistance of capillary force-induced compression. Next, the fabricated conductive ISG layer was transferred onto a micro structured PDMS elastomer and attached to the surface of such an elaborate 3D microstructure after the annealing process. By integrating the microstructured PDMS layer deposited with conductive ISG onto interdigitated Ni/Au electrodes (Figure 54(b)), the pressure sensor realized both high sensitivity (1875.53 kPa^{-1}) and a wide linear detection range (0–40 kPa) due to the effective contact area change under pressure loading.

In the past decade, significant advancements have been made in the development of graphene, leading to various methods for producing graphene with different qualities and costs. Among these methods, CVD has emerged as a prominent technique for growing high-quality SLG and few-layer graphene (FLG) with excellent electronic properties and optical transparency. On the other hand, liquid-phase exfoliated graphene (LEG) offers a more cost-effective solution, particularly when optical transparency is not a crucial requirement. By carefully selecting the appropriate graphene material, the potential for graphene-based flexible and wearable electronics technology can be maximized. For example, Alonso et al. proposed a roll-to-roll-compatible patterning technique and realized the integration of graphene circuits into fabrics, which can be integrated into future textile displays and devices for position-sensitive measurements by creating woven arrays from such fibers.[347] Here, the authors reported two types of electronic devices based on graphene and textile fibers, including capacitive touch sensors and light-emitting devices. For touch-sensing applications, Figures 55(a) and 55(b) illustrate the preparation method of the patterned graphene electrode on the PP textile fiber by combining a lithography process, conventional lithography, and etching processes with roll-to-roll manufacturing. Figures 55(d) and 55(f) display SEM images and the optical photos of the graphene electrode patterned on the PP textile fiber when wending for a fixed angle and fixing it on the fingers, observing the excellent mechanical property of the proposed sensor. Considering that the unique advantage of large-area flexible and foldable graphene light sources is possible, graphene was chosen as the electrode layer of the light-emitting device to realize an alternating current electroluminescent (ACEL) device configuration, as shown in Figure 55(c). Subsequently, graphene was covered by an emitter layer of commercially

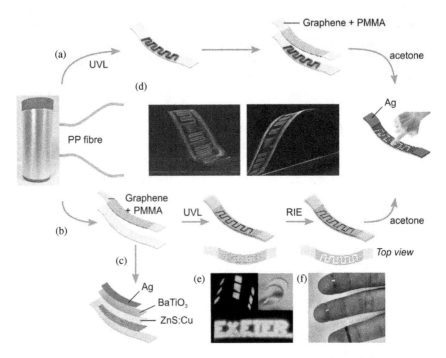

Figure 55. (a) Image of the PP textile fiber and schematic of the fabrication process of the patterned graphene electrode on the PP textile fiber with a lithography process compatible with roll-to-roll manufacturing. (b) Schematic of the fabrication process of the patterned graphene electrode on the PP textile fiber by combining the conventional lithography and etching processes with roll-to-roll manufacturing. (c) Schematic of the flexible sensor with the capacitive touch-sensing and light-emitting capabilities. (d) SEM images of the flexible sensors when wending for a fixed angle (the gap between graphene electrodes is highlighted in green). Demonstration of the flexible sensor serving as the light-emitting device (e) and the capacitive touch sensor (f). Reproduced with permission from Ref. 347 (Copyright 2018).

available Cu-doped zinc sulfide (ZnS:Cu), an insulating layer of $BaTiO_3$ and a top electrode. Upon excitation with an AC voltage, light is emitted from the ZnS:Cu layer due to impact ionization and recombination of electron-hole pairs, as depicted in Figure 55(e).

Electronic tattoos (E-tattoos) that can be intimately mounted on human skin for noninvasive and high-fidelity sensing, have attracted the attention of researchers in the field of wearable electronics. Therefore, Wang *et al.* developed a healable and multifunctional E-tattoo based on a

Basic Composition and Functional Materials 81

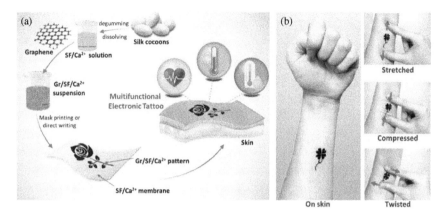

Figure 56. (a) Schematic of the fabrication process of the Gr/SF/Ca²⁺ E-tattoo. (b) Photographs showing the E-tattoo attached stably to human skin with good stretch, compress, and twist properties. Reproduced with permission from Ref. 348 (Copyright 2019, Wiley-VCH).

graphene/silk fibroin/Ca^{2+} (Gr/SF/Ca^{2+}) combination (Figure 56).[348] Such an E-tattoo can be prepared by a facile screen-printing method or a direct writing way, giving rise to the easy transfer onto human skin exactly like a temporary tattoo for further applications. The authors proved such a self-healing E-tattoo using Gr/SF/Ca^{2+} has the capability to monitor various daily life sensations, such as ECG, breathing, and temperature, with high sensitivity. In addition, the Gr/SF/Ca^{2+} composite possesses dynamic hydrogen bonds and coordination bonds, which contribute to its self-healing properties. The sensors in the E-tattoo can self-heal with a healing efficiency of approximately 100%. If mechanical damages occur, they can be repaired simply by applying a drop of water. This development opens up possibilities for the integration of flexible, wearable electronics into healthcare monitoring and other applications.

Members in the graphene family, besides graphene, include two graphene derivatives, graphene oxide (GO) and reduced graphene oxide (rGO). GO is an atomic-thick graphene segment with hydroxyl, carboxyl and epoxy functional groups. It has the advantages of high interfacial activity, simple preparation and preparation, and low cost for large-scale production. As another graphene derivative, rGO is obtained by further reduction of GO and remains some oxygen-containing functional groups. Cho *et al.* introduced a multifunctional flexible sensor matrix that was

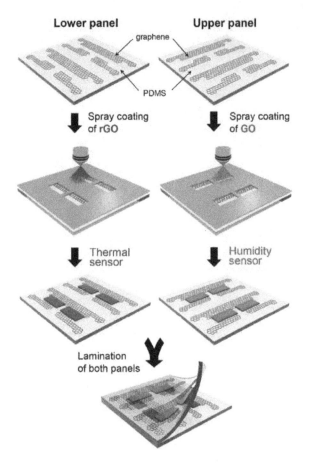

Figure 57. Schematic of the fabrication process of a multifunctional and stretchable sensor matrix based on graphene, GO, and rGO. Reproduced with permission from Ref. 349 (Copyright 2016, Wiley-VCH).

transparent and stretchable, and the electrodes as well as interconnects of such a sensor matrix were fabricated using graphene grown through CVD.[349] As illustrated in Figure 57, GO and rGO were employed as the active sensing materials for humidity and temperature sensors, respectively. By employing a simple lamination process, the humidity, temperature, and pressure sensors were integrated into a layer-by-layer geometry. Remarkably, each sensor demonstrated sensitivity solely towards its specific stimulus and was not affected by other stimuli.

Basic Composition and Functional Materials 83

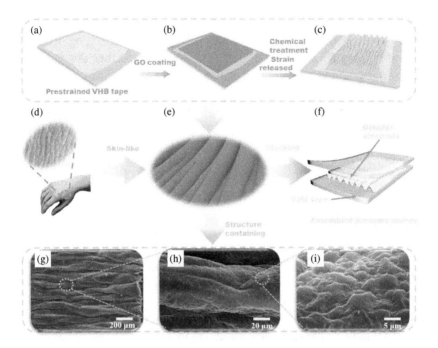

Figure 58. (a–c) Schematic of the fabrication process of the skin- like wrinkle layer of the flexible sensor. (d) Photos of human hand skin and the partial enlargement. (e) Illustration of the fabricated skin-like wrinkle pressure-sensing layer. (f) Schematic of the proposed flexible sensor based on such a skin-like wrinkle layer. (g–i) SEM images of the morphology of fabricated skin-like wrinkle pressure-sensing layer. Reproduced with permission from Ref. 154 (Copyright 2019, The Royal Society of Chemistry).

Inspired by the microstructure of the human skin surface (Figure 58(d)), Jia et al. developed a flexible pressure sensor with a hierarchical structure and gradient-rGO wrinkles.[154] The process of fabrication of the skin-like wrinkled rGO film is depicted in Figures 58(a)–58(c). First, a substrate of VHB tape was subjected to a specific strain ratio (Figure 58(a)). The VHB tape surface possesses hydrophilic groups, enabling strong bonding. In the next step, a homemade aqueous solution of graphene oxide was coated onto the VHB surface and then dried, resulting in a thin GO film (Figure 58(b)). The GO-VHB film exhibits stable interface bonding. Subsequently, the GO film undergoes a reduction process using hydrazine hydrate, which results in the formation of partially reduced graphene oxide (rGO) with an expanded interlayer structure. Upon releasing the substrate, a wrinkled structure resembling human skin is achieved, featuring dome-shaped microstructures (Figures 58(c) and 58(e)). Benefiting from the microstructures (Figures 58(a)–58(c)), the authors further designed and prepared a gradient wrinkle sensor that realizes the position and motion detection of moving objects in one dimension.

Fu et al. conducted a study where they developed a conductive glass fiber (GF) fabric by dip-coating GO onto the surfaces of GFs, followed by a reduction process using hydrogen iodide (HI).[343] The combination of GFs, known for their high mechanical strength, as a reinforcement filler, and silicone resin, renowned for its flexibility, as the matrix, resulted in the fabrication of a composite material called RGO@GFs/silicone. This composite material exhibited both high tensile strength and excellent flexibility. As we all know, graphene with the single-layer architecture of carbon atoms shows a high specific surface area (theoretically calculated value 2630 m^2g^{-1}).[350,351] As graphene gas sensing materials, the large specific surface area can provide numerous active sites to facilitate the adsorption of the target gas on their surface. Generally, the adsorption of the target gas molecules on the surface of graphene accompanies with the charge transfer, which induces the sensible variation of inside carrier concentration and the resistance value of the graphene materials. An integrated device, comprising multilayered graphene-based micro-supercapacitors (MG-MSCs) in series, MG-PANI MSCs array, and pressure/gas sensors, was developed by direct laser writing on PI/Au substrate, as demonstrated in Figures 59(a) and 59(b).[352] The device had the self-powered sensing ability that could steadily detect walking and respond quickly to NO_2 or NH_3.

Since the volatile organic compounds (VOCs) in human exhaled breath are closely related to certain diseases,[353] there is an urgent requirement for flexible dual-modal sensors that could detect VOCs markers and monitor human physiological signals. By processing the rGO film in two different ways, the interference-free detection of the above two signals was realized.[354] The sensor array made of four different porphyrin-modified rGO films was designed as the upper layer of the sensor to detect different gases in VOCs. The lower layer was a porous rGO film made with nanoparticles as a template to realize the identification of physiological signals, and a PI film was sandwiched between the two sensing layers to avoid mutual interference of signals, as shown in Figure 60. The obtained sensor was endowed with the ability to recognize eight different VOCs biomarkers, which could accurately identify the respiratory samples of patients with simulated diabetes, simulated kidney disease, and healthy

Basic Composition and Functional Materials 85

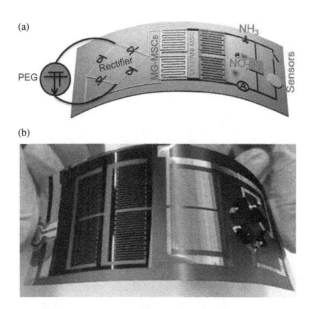

Figure 59. (a) Schematic and (b) photograph of the flexible sensor integrated on a PI substrate, enabling the energy harvesting/pressure/gas sensing capabilities. Reproduced with permission from Ref. 352 (Copyright 2019, Wiley-VCH).

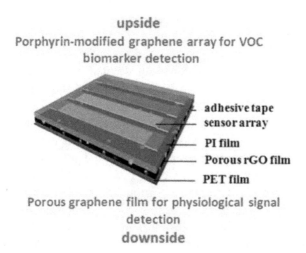

Figure 60. (a) Schematic of a multifunctional wearable sensing device made of four different porphyrin-modified rGO films. Reproduced with permission from Ref. 354 (Copyright 2018, American Chemical Society).

individuals, and could also monitor the pulse and breath signal, reflecting the great application prospect in disease prevention and diagnosis.

MXene is a type of 2D lamellar material consisting of metal carbide and metal nitride. It is obtained by selectively removing the "A" layer in the MAX phase using a wet-chemical etching method with hydrofluoric acid. The general formula for MXene is $M_{n+1}X_nT_x$ (where n = 1, 2, or 3), with M representing the transition metal site (Ti, Sc, Cr, and Mo), X representing carbon or nitrogen sites, and Tx indicating the surface terminations on the outer transition metal layers (F^-, OH^-, and O_2).

MXene has gained significant attention since its discovery in 2011 due to its hydrophilicity, high conductivity, and unique lamellar structure, making it a promising material for tactile sensors.[355,356] It can be easily dispersed in aqueous solutions and utilized in various fabrication processes such as vacuum filtration, spray coating, drop coating, 3D printing, and screen printing. For instance, Shen et al. developed a layer-by-layer self-assembly approach by inserting black phosphorus (BP) between MXene layers, creating small gaps and slightly expanding the interlayer spacing.[357] This MXene/BP layer demonstrated improved pressure sensitivity in a flexible pressure sensor, reaching a sensitivity of 77.61 kPa^{-1} with optimized elastic modulus (Figure 61). Mxene's abundance of functional groups on its surface also opens up possibilities for applications in gas sensing, biosensing, and humidity sensing.

In a recent study, Qiu et al. presented a smart soft actuator that exhibits multiple functionalities including humidity-driven actuation, humidity energy harvesting, self-powered humidity sensing, and real-time motion tracking. This innovative actuator is constructed using a composite membrane composed of MXene, cellulose, and polystyrene sulfonic acid (PSSA). It operates by utilizing the asymmetric expansion caused by a moisture gradient, effectively converting the chemical potential of humidity into mechanical power. Moreover, the gradient moisture chemistry induces directional proton diffusion, enabling the generation of electricity with high power density and open-circuit voltage. This smart soft actuator demonstrates great potential for various applications in the field of humidity-based energy conversion and self-powered sensing. Wang et al. fabricated the resistive pressure sensor by decorating MXene sheets on TPU NF. The sensitivity of the first stage is up to 6750.00 kPa^{-1} and the sensitivity of the second stage is up to 2696.25 kPa^{-1}.[358]

The preparation process of MXene materials endows them with plentiful functional groups like oxygen, hydroxyl, and fluorine groups, which

Basic Composition and Functional Materials 87

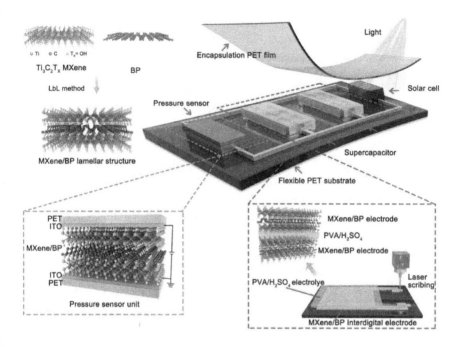

Figure 61. Schematic of the flexible and multifunctional sensor assembled via layer-by-layer self-assembly of MXene/BP multilayer film. Reproduced with permission from Ref. 357 (Copyright 2021, Wiley-VCH).

boosts the adsorption properties of MXene materials toward target VOCs gases and makes it easily tailored as a gas sensor in compliance with VOCs detection. It has been demonstrated that the $Ti_3C_2T_x$-based VOCs sensor is characterized by the capacity to respond to the VOCs analytes at room temperature.[359,360] For instance, the $Ti_3C_2T_x$ VOCs sensor proposed by Dong-Joo Kim et al. successfully measured its sensing behaviors at room temperature toward ethanol, methanol, acetone, and ammonia gas.[359] The sensing mechanism was also discussed here. It was the adsorption of the target gas caused by both defects and functional groups of the $Ti_3C_2T_x$ nanosheets that generated the rapid charge transfer between the target gas and $Ti_3C_2T_x$, and the significant resistance variation of the sensor.

For another $Ti_3C_2T_x$ sensor,[360] the interlayer swelling behavior of the $Ti_3C_2T_x$ MXene films that were induced by the target gas was used rather than the traditional charge transfer model to interpret the sensing performance toward VOCs analyte. It was worth noting that the $Ti_3C_2T_x$ MXene films showed a selective swelling behavior when exposed to ethanol vapor at room temperature, and the swelling degree of the film matched well with the sensing response (the measured resistance variation) of the device. Furthermore, $Ti_3C_2T_x$ MXene featuring highly metallic conducting channels is likely to yield a low noise for the signal, which enables the $Ti_3C_2T_x$ MXene-based gas sensor to meet the requirement of the detection of VOCs gases at the ppb level. A typical example is that Hee-Tae Jung et al. proposed the $Ti_3C_2T_x$ MXene VOCs sensor with an ultrahigh signal-to-noise ratio,[361] and such a sensor presented 2-orders of magnitude higher signal-to-noise ratio than that of other 2D materials like graphene, MoS_2, and black phosphorus.

Meanwhile, such a $Ti_3C_2T_x$ sensor exhibited a low LOD of 50–100 ppb for VOCs gases at room temperature, providing insight for developing highly sensitive sensors with a low LOD even at ppb levels. More recently, Lia Stanciu et al. fabricated the $Ti_3C_2T_x/WSe_2$ hybrid through a facile surface-treating and exfoliation-based process.[362] By performing an inkjet-printing process, the $Ti_3C_2T_x/WSe_2$ hybrid was deposited on the PI substrate with Au-interdigital electrodes and served as the sensing materials. Finally, a wirelessly-operating VOC gas sensor was obtained by incorporating a wireless monitoring system (as shown in Figure 62). The measured results demonstrated that the $Ti_3C_2T_x/WSe_2$ hybrid sensor had a low LOD, low noise level, high sensitivity, and ultrafast response/recovery times, which opened a new avenue for the development of high-performance VOC sensors for the next-generation IoT.

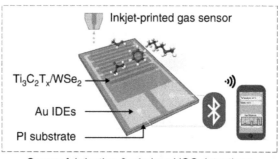

Sensor fabrication & wireless VOC detections

Figure 62. Schematic diagrams of the inkjet-printed $Ti_3C_2T_x/WSe_2$ sensor and the wireless monitoring system for the detection of VOCs gas. Reproduced with permission from Ref. 362 (Copyright 2020, published by Springer Nature).

Basic Composition and Functional Materials 89

According to a recently reported work, an emerging type of MXene material, $V_4C_3T_x$, showed a highly sensing response to acetone gas at room temperature, and Figure 63(a) demonstrated the schematic of the $V_4C_3T_x$ acetone sensor.[363] This acetone sensor was examined with a low detection limit of 1 ppm, which was lower than the threshold diabetes diagnosis (1.8 ppm). Additionally, the measured results also suggested the high selectivity of the $V_4C_3T_x$ sensor toward acetone against the water vapor. The sensing mechanism based on the experimental results and DFT calculation theory indicated that the superior selectivity for acetone was attributed to the bigger acetone molecular size than that of water molecules. As depicted in Figure 63(b), the bigger acetone molecule would

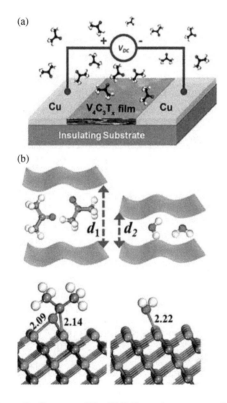

Figure 63. (a) Schematic diagram of the $V_4C_3T_x$ acetone sensor. (b) Schematic diagram of the sensing mechanism of the $V_4C_3T_x$ sensor toward acetone and H_2O and the DFT calculation results. Reproduced with permission from Ref. 363 (Copyright 2020, The Royal Society of Chemistry).

result in the strong steric effect (d1>d2) of the $V_4C_3T_x$ films, which had an influence on the contact performance of the $V_4C_3T_x$ films and yielded a bigger variation on film resistance ($\Delta R/R$). In a word, the strong points like the low detection limit and the excellent sensing selectivity made the $V_4C_3T_x$ films a promising material in the application of achieving the diagnosis of diabetes.

Apart from graphene, its derivatives, and MXene, other emerging 2D nanomaterials such as MoS,[364,365] indium selenide (In_2Se_3),[366,367] stannous sulfides (SnS),[368] tin diselenide,[369] vanadium nitride (VN),[370] and tin disulfide (SnS_2)/SnS[371] have great potential in the field of flexible sensors as well, especially in tactile sensing. Hu et al. reported a class of 2D layered materials with an identical orientation of in-plane polarization whose piezoelectric coefficients (e_{22}) increased with layer number, thereby allowing the fabrication of flexible piezotronics devices with large piezoelectric responsivity and excellent mechanical durability.[366] The piezoelectric outputs can reach up to 0.363 V for a seven-layer α-In_2Se_3 device with a current responsivity of 598.1 pA for 1% strain, which is one order of magnitude higher than the values of the reported 2D piezoelectrics.

More interestingly, a new *in situ* catalytic strategy for fabricating metallic aerogel hybrids was proposed that consists of VN nanosheets decorated with well-defined vertically aligned CNT arrays (VN/CNTs), as illustrated in Figure 64.[370] In this architecture, the 2D VN nanosheets form the main bone structure that is favorable for flexible devices due to their excellent structural compatibility during the repetitive deforming process,

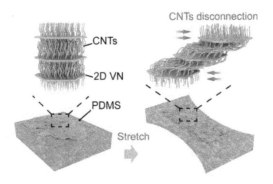

Figure 64. Schematic of the disconnection and reverse of sandwich-like VN/CNTs materials during stretch and release. Reproduced with permission from Ref. 370 (Copyright 2020, American Chemical Society).

and the sandwiched aerogel hybrids form the highly conductive 3D network that allows outstanding sensitivity for strain- responsive behaviors.

4.5.3. Three-dimensional inorganic material

Three-dimensional materials usually have good mechanical strength and flexibility and can accommodate deformations such as bending, stretching, and twisting without losing functionality. This adaptability allows for the creation of electronic devices that can conform to irregular surfaces or be worn on the body comfortably. This makes them ideal for the fabrication of flexible electronics, such as bend sensors, wearable devices, and so on. For example, graphene oxide (GO) foam exhibits both excellent elastic property and high relative dielectric permittivity, which is a novel building block for future wearable electronic devices. By using a dielectric layer of thin GO foam with a low effective elastic modulus, Wan et al. reported a flexible pressure sensor, showing excellent pressure sensitivity in a low-pressure regime (<1 kPa). Such a flexible GO foam pressure sensor demonstrates excellent reproducibility, fast response times, and significant potential for application in a wide range of fields, which is particularly well-suited for applications that require sensitive pressure detection in the low loading range, such as flexible human–computer interfaces and robotics.[372] Zhu et al. presented a highly sensitive and flexible sensor based on a 3D microstructure graphene sponge.[373] Such a graphene sponge is fabricated in a dip-coating process that stacks the graphene layers onto the PI scaffolds in a homogeneous graphene solution with graphene oxide serving as the dispersant. Such a graphene sponge has a relatively low Young's modulus of 16 kPa and is capable of enduring large strains over 60%. For the assembly of the 3 × 3 sensor array using such a graphene sponge layer, the photo etching, magnetron sputtering, and screen-printing processes are subsequently performed, as shown in Figure 65. The author pointed out that this highly sensitive and low-cost flexible sensor will have great potential in smart robotics and prostheses, wearable electronics/devices, healthcare monitors and other biomedical applications. Through the same dip-coating method, Zhu et al. modified the PI scaffolds with conductive graphene layers to fabricate a graphene sponge with a 3D microstructure and assembled a flexible and highly sensitive tactile sensor.[373]

Zhuo et al. demonstrated a highly sensitive and flexible sensor by proposing an ultralight rGO-based carbon aerogel with supercompressibility, elasticity, excellent bendability, and fatigue resistance, which

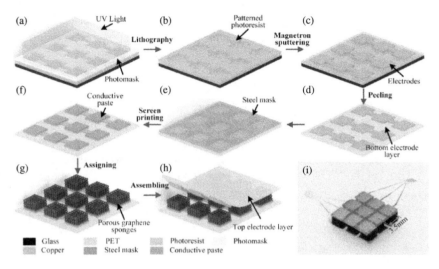

Figure 65. Schematic of the fabrication steps of the flexible sensor based on the 3D microstructure graphene sponge. (a–d) Bottom electrode array deposition. (e–f) The screen printing process of the conductive paste. (g) Assembly of the porous graphene sponges. (h) Assignment of the top electrode array. (i) Photographs of the assembled flexible sensor array. Reproduced with permission from Ref. 374 (Copyright 2018, IEEE).

achieves an ultrasensitive detection of tiny stain, pressure, and angle changes (Figure 66).[374] According to the author, there are two important factors for the fabrication of such an rGO-based carbon aerogel: (i) a low-density and wave-shape architecture that realizes high compression, and (ii) strong interaction among rGO layers via the carbonization of cellulose nanocrystals (CNCs) and low-molecular-weight carbon precursors that ensures superelasticity and fatigue resistance. Such a carbon aerogel of C-CNC/rGO was prepared by the mixing, directional freezing, freeze-drying, and carbonization processes. For obtaining the aerogels of C-CNC/rGO-glu2, C-CNC/rGO-glu5, and C-CNC/rGO-urea, glucose and urea are respectively added. The addition of CNC and glucose or urea enhances the interaction between the layers of reduced graphene oxide (rGO), leading to the formation of an ultralight, flexible, and super-stable layer structure. This unique structure contributes significantly to the mechanical strength of the carbon aerogel, and thus the carbon aerogel exhibits exceptional mechanical properties, as it can withstand extreme strains of up to 99%. This high strain capability makes it highly resilient

Basic Composition and Functional Materials 93

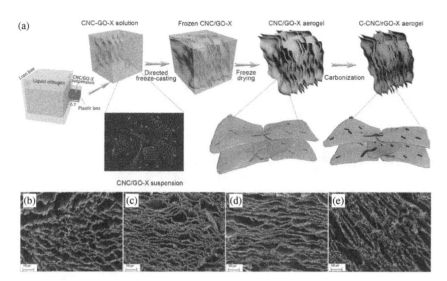

Figure 66. (a) Schematic of the fabrication process of the sensing layer of aerogel through a directional freezing method. (b–e) SEM images of the cross sections of the fabricated aerogels of (b) C-CNC/rGO, (c) C-CNC/rGO-urea, (d) C-CNC/rGO-glu2, and (e) C-CNC/rGO-glu5, respectively. Reproduced with permission from Ref. 374 (Copyright 2018, Wiley-VCH).

and resistant to deformation even under significant stress or pressure. Additionally, the aerogel demonstrates excellent fatigue resistance, meaning it can withstand repeated cycles of loading and unloading without significant degradation in its mechanical performance. Overall, the superior performance of such a carbon aerogel with ultralight, flexible, and super stable layer structure enables its application in various fields, including wearable devices, sensors, and other areas that benefit from its exceptional mechanical and sensing capabilities.

Apart from the graphene materials, MXene materials are also used as the fillers to fabricate the sponge-like functional layers for constructing high-performance flexible sensors. Wei *et al.* proposed a molybdenum microstructured electrode by helium plasma irradiation and then fabricated an MXene@PU composite sponge dielectric layer by immersing the PU sponge in the $Ti_3C_2T_x$ nanosheet dispersion (Figure 67(b)), achieving a flexible piezoresistive pressure sensor (Figure 67(g)).[375] As shown in Figure 67(a), this electrode engineering strategy enables the smooth transition between sponge deformation and MXene interlamellar

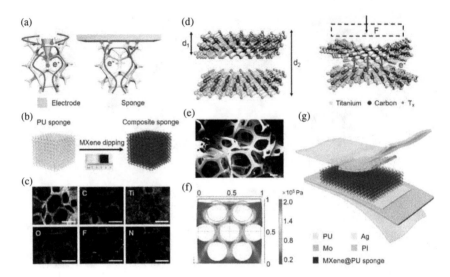

Figure 67. (a) Schematics of comparison of sensitivity enhancement principle of electrodes with conical microstructure and flat structure. (b) Schematics of the dip-coating process for fabricating the MXene-decorated PU sponge. (c) Energy dispersive spectrometer (EDS) element mapping of the MXene-coated sponge. (d) Schematic of the working mechanism of the lamellar MXene materials under pressure. (e) SEM image of the MXene-coated sponge. (f) FEA simulation on the pressure distribution of the cross-section of the MXene-coated sponge under pressure. (g) Schematics of the assembled flexible sensor by using the MXene-coated sponge layer. Reproduced with permission from Ref. 375 (Copyright 2022, Wiley-VCH).

displacement, giving rise to high sensitivity (1.52 kPa^{-1}) and good linearity (r^2 = 0.9985) in a wide sensing range (0–100 kPa) with a response time of 226 ms for pressure detection. As the authors discussed, the high performance of the pressure sensor can be attributed to the coexistence of the porous structures and the MXene materials in the MXene@PU composite sponge (Figures 67(c)–67(e)). As depicted in Figure 67(f), FEA simulation on the pressure distribution of the cross-section of the MXene-coated sponge under pressure also confirmed that the hierarchical structures modulated by pore size, plasma bias, and MXene concentration play a crucial role in improving the sensing performance.

Furthermore, considering that the calcium copper titanate (CCTO) possesses a strong dielectric property, high energy density, and thermal stability over a wide temperature range,[130] Park et al. explored an effective strategy to enhance the dielectric permittivity of the hybrid sponge

Basic Composition and Functional Materials 95

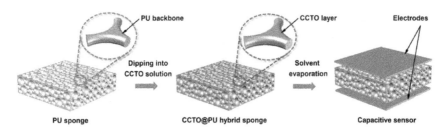

Figure 68. Fabrication for a capacitive pressure sensor based on a high permittivity dielectric layer of the hybrid sponge (CCTO@PU). Reproduced with permission from Ref. 130 (Copyright 2020, Wiley-VCH).

(CCTO@PU) by dip-coating the PU sponge with a surface-modified CCTO solution. Next, a highly sensitive capacitive pressure sensor was assembled by sandwiching such an ultrasoft CCTO@PU hybrid sponge between two electrodes (Figure 68), providing a facile and scalable route for the design of high-performance capacitive pressure sensors.

Since 3D materials can realize various properties by adjusting structure and composition, they can realize multiple functions in flexible electronics. For instance, Zhang presented a simple method to fabricate a flexible sensor based on a conductive graphene/polymer sponge, enabling to detect of multiple forms of mechanical deformations, such as tensile strain, impact, bending, vibration, and twisting (Figure 69).[376] Figure 69 illustrates the step-by-step process for preparing the conductive graphene/polymer sponge. Initially, a PU (or PVC) sponge is immersed in a solution containing hydroquinone and aqueous GO. The sponge is then repeatedly pressed until the GO aqueous adsorbed in the sponge is saturated. Next, the GO-coated PU (or PVC) sponges are transferred to an oven and heated at 100°C for 10 hours. Subsequently, the composites are soaked and cleaned in deionized water for 24 hours. Finally, after undergoing vacuum freeze-drying treatment, the composites are carefully shaken and pressed multiple times to ensure that no rGO flakes detach from the material. As the authors discussed, the shape of such a graphene composite sponge can be easily customized according to specific requirements, allowing for precise sensing capabilities in practical applications.

In addition, Park *et al.* reported a combined method to achieve the synthesis of the composite of PPy and graphene foam (GF) (PPy/GF) via CVD depositing graphene on Ni foam, electrochemical deposition of PPy, and the subsequent Ni etching process, as shown in Figure 70(a).[377]

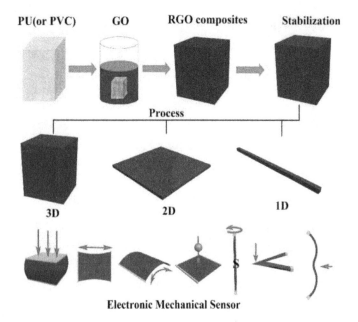

Figure 69. Schematic of the fabrication route of RGO/PU (or PVC) sponges and the multifunctional applications. Reproduced with permission from Ref. 376 (Copyright 2017, Elsevier Ltd).

According to the authors, the fabricated PPy/GF composite is easily broken because of its high modulus, thus they coated the composite with diluted PDMS. As a result, a dual-mode sensor for pressure and temperature measurement was proposed by using two indium tin oxide/polyethylene terephthalate (ITO/PET) electrodes on the top and bottom, respectively. Figure 70(b) presents the schematic of such a dual-mode sensor as well as a strain sensor assembled by the PDMS-coated PPy/GF composite. Besides, PPy/GF with a microporous structure gives high conductivity and large surface area to be highly sensitive to pressure and to have high electrochemical performance in supercapacitors. Among the various energy-storage devices, a supercapacitor has the advantages of having a wide operating temperature range, a long cyclic life, and a relatively simple structure. Therefore, the authors fabricated a supercapacitor by using the PPy/GF composite as electrode layers and gel-type electrolyte of acetonitrile–propylene carbonate–poly(methyl methacrylate)–lithium perchlorate (ACN–PC–PMMA–LiClO$_4$) as the dielectric layer, as depicted in

Basic Composition and Functional Materials 97

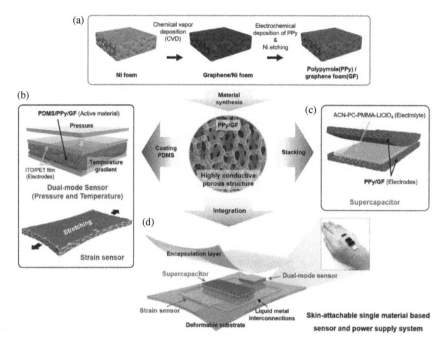

Figure 70. (a) Schematic of the fabrication route of PPy/GF composite by the CVD growth of GF and subsequent electrochemical deposition of PPy. (b) Schematics of a dual-mode sensor of pressure and temperature as well as a strain sensor by using such a PPy/GF composite as the active layers. (c) Schematic of a flexible supercapacitor by using such a PPy/GF composite coated with PDMS as the electrode layers. (d) Schematic and the optical image of the integrated dual-mode sensor, strain sensor, and supercapacitor on a flexible substrate. Reproduced with permission from Ref. 377 (Copyright 2018, Wiley-VCH).

Figure 70(c). Lastly, these individual component devices were integrated on a single flexible substrate via the liquid–metal interconnections (Figure 70(d)), in which the designed dual-mode sensor for pressure and temperature sensing, and strain sensor could be driven by using the power stored in the fabricated supercapacitor. The authors make it clear that this work demonstrates the possibility and applicability of the novel PDMS/PPy/GF with 3D microstructure in functional devices of active multifunctional sensors and supercapacitors for their wireless powering.

In addition, by exploiting the advantages of hybrid porous microstructures fabricated from Epipremnum aureum leaf and sugar templates, Zhang *et al.* proposed a highly sensitive pressure sensor based on such hybrid porous microstructures sensing layer.[159] Compared with the above complex fabrication processes, conductive porous sponge structures combining polymer elastomer with good mechanical ability and conductive materials are commonly used to fabricate the microporous piezoresistive sensing layers for developing high-performance tactile e-skin due to the facile and low-cost preparation process.[373,378–380] For example, by repeatedly dip-coating a PU sponge in a mixture solution of conductive cellulose nanofibril (CNF)/Ag NWs, Guo et al. prepared a conductive CNF/Ag NWs-coated PU (CA@PU) sponge and designed a CA@PU e-skin to demonstrate promising application prospect in detecting human body motion.[378]

Indeed, the production process of PU sponges, which are commonly used in tactile sensors, can have negative environmental impacts and may raise health concerns. To address these issues, researchers have explored alternative materials, such as natural fabrics, to achieve similar functionality without the associated environmental and health drawbacks. For instance, Liu and colleagues utilized silk as a support material to fabricate a 3D graphene structure, resulting in a graphene-silk pressure sensor.[381] This sensor demonstrated high sensitivity, good repeatability, flexibility, and comfort for use on the skin. The author said the characteristics of high sensitivity, good repeatability, flexibility, and comfort for skin provide the high possibility to fit on various wearable electronics. Yuan et al. developed a cost-effective and scalable method for fabricating high-performance strain sensors using a graphene-coated spring-like mesh network.[382] This unique 3D structure enabled the sensor to detect various deformations, including pressing, stretching, bending, and subtle vibrations. Also, Yin et al. reported a waterproof and breathable cotton fabric composite decorated by reduced graphene oxide (rGO) and CNT (Cotton/rGO/CNT) by a facile solution infiltration method.[383] Then, the authors adopt such a Cotton/rGO/CNT composite to develop a layer-by-layer structured multifunctional flexible sensor, enabling the high-sensitivity detection of pressure and temperature stimulus (Figure 71). As the authors discussed, this work provides a feasible strategy for designing cotton-based sensing layers that can effectively resist liquid water penetration and allow water vapor transmission and offers reasonable insight for constructing comfort and multifunctional wearable electronics.

Lu and coworkers utilized low-cost, commercial 3D polyester nonwoven fabrics as scaffolds to construct highly sensitive wearable piezoresistive pressure sensors.[384] Kim and collaborators fabricated strain-pressure sensors by employing a simple solution process to create rGO/SWCNT

Basic Composition and Functional Materials 99

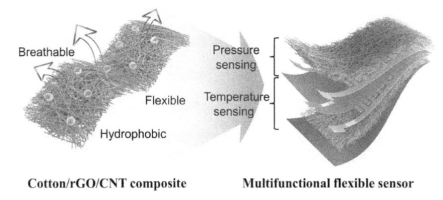

Figure 71. (a) Schematic of the multifunctional flexible sensor based on a waterproof and breathable Cotton/rGO/CNT composite. Reproduced with permission from Ref. 383 (Copyright 2022, Springer).

hybrid fabric sensors. These sensors exhibited high mechanical stability and flexibility during bending tests and maintained excellent water-resistance properties even after multiple washing tests.[385] The superior sensing performance and economical fabrication processes of these wearable tactile sensors have increased confidence in the feasibility of smart clothing applications. These sensors hold promise for practical use in household, healthcare, entertainment, and robotics fields. By exploring alternative materials and fabrication techniques, researchers are striving to develop tactile sensors that are more environmentally friendly, safer for human health, and suitable for various practical applications.

In addition to the natural texture materials, paper, as an environmentally friendly substrate, holds great potential for improving the performance of pressure sensors. Tao *et al.* presented a paradigm using multilayer tissue papers mixed with a GO solution to create a GO paper.[386] Through an annealing process and wire drawing, they constructed a graphene-paper-based pressure sensor. Also, the authors demonstrated the application of such a flexible pressure sensor in pulse detection, respiratory detection, voice recognition, and detecting various intensities of motion. In comparison to most graphene pressure sensors reported, this particular sensor achieved optimization of sensitivity and working range, making it especially suitable for wearable applications. The authors believed that the graphene-paper pressure sensor holds significant potential for use in E-skin devices, enabling health monitoring and motion detection. The use of paper as a substrate, combined with the 3D hierarchical nanostructure

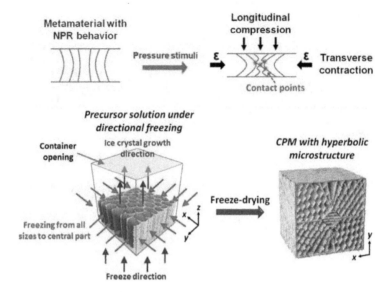

Figure 72. Illustration of the NPR effect of the materials under pressure and schematic illustration of the auxetic cellular structured piezoresistive sensor with an NPR effect. Reproduced with permission from Ref. 387 (Copyright 2022, Elsevier Inc).

and good elasticity, enhances the performance and functionality of the pressure sensor while maintaining an environmentally friendly approach.

Recently, different from the above porous piezoresistive sensing layers that presented positive Poisson's ratio (PPR) values, Liang and coworkers proposed an auxetic cellular structured piezoresistive sensing layer with negative Poisson's ratio (NPR) to improve the sensing performance of a piezoresistive sensor (Figure 72).[387] According to the authors, the NPR effect of the auxetic cellular structure significantly enhanced the mechanical elasticity and durability of the sensing layer compared with the conventional material structures with the PPR effect, giving rise to a sensor showing excellent sensing stability and reliability even after repeated compressive deformation.

4.5.4. Liquid metal

Liquid metal is a metal that flows easily at or near room temperature, deforms to some extent, and has all the useful characteristics of other solid/molten metals, such as excellent thermal conductivity, superb

Basic Composition and Functional Materials 101

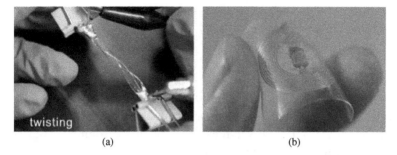

Figure 73. (a) Photograph showing the twisting test (360°) of the RTLM-based PVA-encapsulated circuit. (b) Photograph of the LM-based NFC tag. Reproduced with permission from Ref. 85 (Copyright 2019, Wiley-VCH).

conductivity, inherent high density, and low vapor pressure.[388–391] Liquid metals, such as the liquid alloy gallium indium (EGaIn), and galinstan (GaInSn), are highly conductive and deformable, making them ideal for flexible sensors. These sensors can be used to monitor and measure various physical quantities such as pressure, strain, temperature, etc. For example, Teng *et al.* introduced a straightforward method for fabricating transient and flexible circuits using room-temperature liquid metals (RTLMs), GaInSn, as the active material and water-soluble poly(vinyl alcohol) (PVA) as the packaging material (Figure 73).[85] This approach enables the creation of circuits that are capable of withstanding bending and twisting without compromising their durability or stable electric performance. The short transience times of the circuits are attributed to the excellent solubility of PVA substrates and the intrinsic flexibility of RTLM patterns. One notable advantage of the RTLM-based transient circuits is their high recycling efficiency. Up to 96% of the RTLM material used can be recovered, leading to improved economic and environmental viability of transient electronics. This feature contributes to the sustainability of the technology. To demonstrate the practicality of their concept, the researchers showcased the surface patterning of RTLMs with complex shapes. They further applied the transient circuits to create a passive near-field communication tag using a transient antenna and a transient capacitive touch sensor. Additionally, they demonstrated the use of the RTLM-based transient circuit to sequentially turn off an array of light-emitting-diode (LED) lamps. The authors proved that RTLM-based PVA-encapsulated circuits presented in this study significantly broaden the possibilities of transient electronics by enabling the development of flexible and recyclable transient systems.

Liquid-metal-based sensors have shown great promise in their ability to withstand large strains without failure. However, indeed, current liquid-metal-based strain sensors may not be suitable for applications requiring high-resolution detection of small pressure changes in the few kPa range. It makes them unsuitable for applications such as heart-rate monitoring, which requires a much lower pressure detection resolution. Gao et al. presented a microfluidic tactile diaphragm pressure sensor that utilized embedded Galinstan microchannels measuring 70 μm in width and 70 μm in height (Figures 74(a) and 74(b)).[392] This sensor demonstrated the ability to detect sub-50 Pa changes in pressure with detection limits below 100 Pa and a response time of 90 ms. As shown in Figure 74(c), the sensor is equivalent to an incorporated Wheatstone bridge circuit, which maximizes the utilization of both tangential and radial strain fields. Figure 74(d)

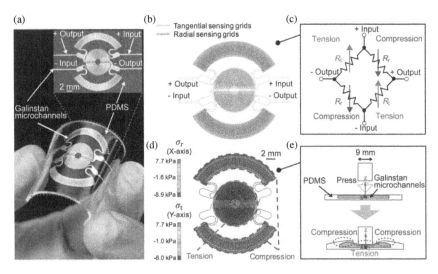

Figure 74. (a) Photograph and (b) schematic of the layout of the proposed microfluidic tactile diaphragm pressure sensor. (c) Schematic the equivalent circuit of the formed Wheatstone bridge circuit. (d) Simulation of the normal stresses for the radial sensing grids (σ_r) (along the X-axis) and the tangential sensing grids (σ_t) (along the Y-axis) of the pressure sensor under 1 kPa pressure. (e) Schematic of the pressure sensor with the indicating testing conditions. Reproduced with permission from Ref. 392 (Copyright 2017, Wiley-VCH).

presents the simulation of the pressure sensor when applying the object (1 kPa, with a diameter over 9 mm) (Figure 74(e)). This design contributed to high sensitivities, with a change in output voltage corresponding to a 0.0835 kPa^{-1} pressure change. The Wheatstone bridge configuration also enabled temperature self-compensation, allowing the sensor to operate effectively within the temperature range of 20–50°C. The researchers showcased potential applications of this sensor technology. For instance, they demonstrated a PDMS wristband featuring an embedded microfluidic diaphragm pressure sensor capable of real-time pulse monitoring. They also showcased a PDMS glove with multiple embedded sensors that provided comprehensive tactile feedback when the hand touched or held objects.

Human-interfaced electronic systems require strain-resilient circuits to withstand mechanical deformation and maintain electrical functionality. However, current integrated stretchable electronics face challenges such as electrical deterioration and difficulties in forming robust multilayered soft-rigid hybrid configurations. To address these challenges, significant efforts have been dedicated to developing solutions based on deterministic architectures, intrinsically stretchable conductors, and conductive composites. One approach involves using stretchable metal strips with wavy, serpentine, or buckling structures. Another approach is to incorporate stretchable conductive composites with materials such as silver microplates (AgMPs), AgNWs, gold-coated AgNWs, CNTs, and conductive polymers. However, there are limitations associated with these structure-based stretchable conductors.[318,393–399] They are often restricted in terms of critical strain, typically being limited to less than 200%. Additionally, the component areal density is a challenge, as achieving high-density integration of stretchable conductors is difficult. These limitations highlight the ongoing challenges in developing stretchable conductors that can maintain high conductivity, low resistance changes, and robustness under large strains.

Liquid metal can be used as conductive wire in flexible circuits. Due to their high conductivity and plasticity, liquid metal wires can be stretched, bent, and deformed without damaging their conductive properties. This makes them ideal for use in the fabrication of stretchable electronics such as wearable devices, flexible displays, and more. Chen et al. introduced a bilayer liquid-solid conductor (b-LSC) with amphiphilic properties that enable reliable interfacing with both rigid electronics and elastomeric substrates. The b-LSC consists of two layers: a top liquid metal layer and a bottom polar composite layer. The top liquid metal layer can self-solder its interface with rigid electronics, resulting in a resistance that is 30% lower compared to traditional tin-soldered rigid interfaces. This feature enhances the electrical performance and reliability of the interface. The bottom polar composite layer is composed of liquid metal particles and polymers. This layer not only facilitates the reliable interface with elastomeric substrates but also contributes to the self-healing capability of the b-LSC after breakage or damage occurs. The b-LSC can be fabricated in a scalable manner using printing and subsequent peeling strategies. It exhibits ultra-high strain-insensitive conductivity, reaching a maximum value of 22532 S cm^{-1}. The b-LSC also demonstrates extreme stretchability, with a strain capacity of up to 2260%, and negligible resistance change even under ultra-high strain conditions, showing only a 0.34-times increase in resistance under 1000% strain.

Zheng et al. introduce a coaxial wet-spinning process that enables the continuous fabrication of intrinsically stretchable, highly conductive, and conductance-stable liquid metal (LM) (EGaIn alloy) sheath-core microfibers (Figure 75).[398] The sheath-core structure and the dipole-dipole interactions between fluoroelastomer and LM make the embedded LM particles in the core conformally deform upon stretching, leading to both high stretchability and as-expected conductance stability without the concern of LM leakage. These microfibers exhibit remarkable stretchability, with the ability to be stretched up to 1170%

The microfibers produced through this process can be easily woven into everyday gloves or fabrics (Figures 75(d) and (e)), providing a range of functionalities. They serve as excellent joule heaters, capable of generating heat when an electric current is passed through them. Additionally, the microfibers exhibit electrothermochromic properties, enabling them to change color in response to applied electric current and temperature changes. Furthermore, the microfibers can act as self-powered wearable sensors, allowing for the monitoring of human activities. These findings demonstrate the potential of the coaxial wet-spinning process in creating highly stretchable and conductive microfibers with stable performance. The versatility of these microfibers makes them suitable for various applications, including wearable electronics, where they can be integrated into fabrics or garments to provide heating, display capabilities, and sensing functionalities. Zhu et al. prepared PDMS /liquid metal (LM) films with a hierarchical structure prepared via the gravity-induced deposition of LM are applied to multifunctional flexible sensors. PDMS/LM films exhibit remarkable flexibility and function as flexible electrodes for pressure

Basic Composition and Functional Materials 105

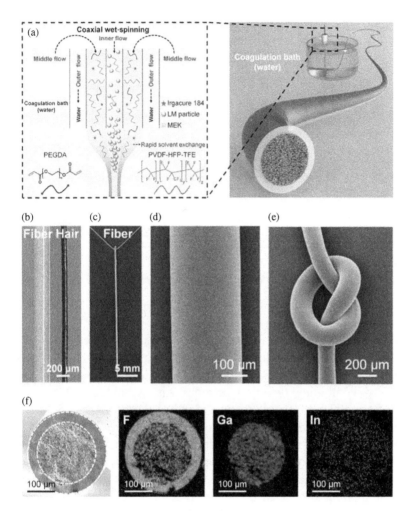

Figure 75. (a) Schematic of the coaxial wet-spinning process used to fabricate the LM sheath-core microfibers. Microscopic images of (b) the fabricated LM sheath-core microfibers and (c) human hair. SEM images of surface morphology of (d) the fabricated LM sheath-core microfibers and (e) a knotted microfiber. (f) Cross-sectional SEM image and corresponding elemental mapping images of the fabricated LM sheath-core microfibers. Reproduced with permission.[398] (Copyright 2021, The authors.)

sensors as well as a single-electrode mode triboelectric nanogenerator (TENG, Figure 76).[400]

A novel approach by integrating flexible electronics with living tissues can contribute to enhancing conventional tissue-engineered blood

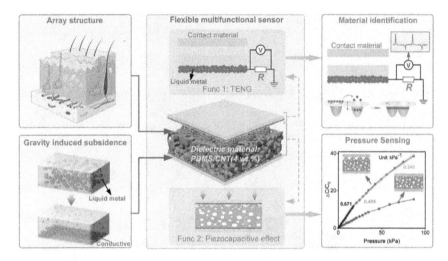

Figure 76. Schematic illustration of the multifunctional flexible sensor[400] (Copyright 2024, Wiley-VCH).

vessels (TEBVs), which offers several functionalities and capabilities that can address biomedical challenges, including precision diagnostics and treatments. To achieve this, Cheng *et al.* developed an electronic blood vessel by combining flexible electrodes with a biodegradable scaffold made from a blend of liquid metal and poly(L-lactide-co-ϵ-caprolactone) (PLC) into an multilayered porous conduit (MPC).[401] The flexible electrodes allow for *in situ* sensing of blood flow and temperature, while the biodegradable scaffold provides structural support and compatibility with living tissues. Such an electronic blood vessel is tested in vitro for electrical stimulation and electroporation. Electrical stimulation effectively promotes cell proliferation and migration in a wound-healing model, demonstrating the potential for tissue regeneration. Electroporation is used to deliver a green fluorescent protein (GFP) DNA plasmid to three types of blood vessel cells, showcasing the capability of gene delivery. To assess the efficacy and biosafety of the electronic blood vessel, a 3-month *in vivo* study is conducted using a rabbit carotid artery replacement model (Figure 77). The patency of the blood vessel is confirmed through ultrasound imaging and arteriography. These results indicate the potential of the electronic blood vessel as a platform for various treatments, including gene therapies, electrical stimulation, and electronically controlled drug release. As the authors discussed, the study paves the way for the

Basic Composition and Functional Materials 107

Figure 77. Schematic of the proposed electronic blood vessel serving as the carotid artery of a rabbit. Reproduced with permission from Ref. 401 (Copyright 2020, Elsevier Inc).

integration of flexible, degradable bioelectronics into the vascular system. This integration offers opportunities for advanced treatments and therapies, leveraging the functionalities of the electronic blood vessel to achieve improved precision diagnostics, tissue regeneration, and controlled drug delivery.

5. Summary and Outlook

Flexible sensors are a new type of sensor technology based on flexible materials, which have the advantages of being thin, soft, bendable, and stretchable, and are suitable for applications in many fields. The performance of flexible sensors is mainly determined by two key components, namely the flexible substrate and the active layer. The choice of a flexible substrate is critical to the performance of the sensor, determining the softness, reliability, and durability of the sensor. The active layer is a key part of the flexible sensor, which is responsible for sensing external stimuli and converting them into electrical signals. Flexible sensors can be divided into capacitive sensors, resistive sensors, piezoelectric/pyroelectric/triboelectric sensors, photosensitive sensors, and magnetic sensitive sensors according to the type of electrical signal detected. In this chapter, the development of flexible substrate and active layers of flexible sensors including the research progress in material selection, preparation methods, performance evaluation and application fields are concluded, aiming at a comprehensive understanding of the application potential and future development direction of flexible substrate and active layers in the field of flexible sensors. Besides, regarding the challenges of additional power supply of the e-skin device or even system and the environmental issues caused by the high volume of e-skin waste, self-powered and biodegradable e-skin devices enabled by polymer materials can provide opportunities to make e-skin devices more practical, sustainable, and reduce the environmental impact in various applications.

In summary, flexible sensors are an interdisciplinary research field integrating device physics, materials science, and micro-nano processing. And, continuous efforts focused on the above individual aspects would be further considered to improve the overall properties of flexible sensors using extensive material to the level of superiority to that of the human skin, and realizing wide commercial applications in human–machine interactions.

References

1. Lu, Y., Xu, K., Zhang, L., Deguchi, M., Shishido, H., Arie, T., Pan, R., Hayashi, A., Shen, L., Akita, S., *et al.* (2020). Multimodal plant healthcare flexible sensor system, *ACS Nano*, *14*, 10966–10975.
2. Ma, Y., Zhang, Y., Cai, S., Han, Z., Liu, X., Wang, F., Cao, Y., Wang, Z., Li, H., Chen, Y., *et al.* (2020). Flexible hybrid electronics for digital healthcare, *Advanced Materials*, *32*, 1902062.
3. Wang, Y., Zhu, W., Deng, Y., Fu, B., Zhu, P., Yu, Y., Li, J. and Guo, J. (2020). Self-powered wearable pressure sensing system for continuous healthcare monitoring enabled by flexible thin-film thermoelectric generator, *Nano Energy*, *73*, 104773.
4. Yin, F., Guo, Y., Qiu, Z., Niu, H., Wang, W., Li, Y., Kim, E. S. and Kim, N. Y. (2022). Hybrid electronic skin combining triboelectric nanogenerator and humidity sensor for contact and non-contact sensing, *Nano Energy*, *101*, 107541.
5. Wei, P., Yang, X., Cao, Z., Guo, X.-L., Jiang, H., Chen, Y., Morikado, M., Qiu, X. and Yu, D. (2019). Flexible and stretchable electronic skin with high durability and shock resistance via embedded 3D printing technology for human activity monitoring and personal healthcare, *Advanced Materials Technologies*, *4*, 1900315.

6. Xie, M., Hisano, K., Zhu, M., Toyoshi, T., Pan, M., Okada, S., Tsutsumi, O., Kawamura, S. and Bowen, C. (2019). Flexible multifunctional sensors for wearable and robotic applications, *Advanced Materials Technologies*, *4*, 1800626.
7. Pyo, S., Lee, J., Bae, K., Sim, S. and Kim, J. (2021). Recent progress in flexible tactile sensors for human-interactive systems: From sensors to advanced applications, *Advanced Materials*, *33*, 2005902.
8. Zhang, Y. and Lu, M. (2020). A review of recent advancements in soft and flexible robots for medical applications, *The International Journal of Medical Robotics and Computer Assisted Surgery*, *16*, e2096.
9. Xie, M., Zhu, M., Yang, Z., Okada, S. and Kawamura, S. (2021). Flexible self-powered multifunctional sensor for stiffness-tunable soft robotic gripper by multimaterial 3D printing, *Nano Energy*, *79*, 105438.
10. Oruganti, S. K. and Khosla, A. (2020). (Invited) 3D Printing and wireless power transfer systems for soft robotics applications, *ECS Meeting Abstracts*, *MA2020-02*, 3703.
11. Wu, Z., Wei, L., Tang, S., Xiong, Y., Qin, X., Luo, J., Fang, J. and Wang, X. (2021). Recent progress in Ti_3C_2Tx MXene-based flexible pressure sensors, *ACS Nano*, *15*, 18880–18894.
12. Yu, H., Li, N. and Zhao, N. (2021). How far are we from achieving self-powered flexible health monitoring systems: An energy perspective, *Advanced Energy Materials*, *11*, 2002646.
13. Mishra, R. B., El-Atab, N., Hussain, A. M. and Hussain, M. M. (2021). Recent progress on flexible capacitive pressure sensors: From design and materials to applications, *Advanced Materials Technologies*, *6*, 2001023.
14. Ren, Z., Yang, J., Qi, D., Sonar, P., Liu, L., Lou, Z., Shen, G. and Wei, Z. (2021). Flexible sensors based on organic–inorganic hybrid materials, *Advanced Materials Technologies*, *6*, 2000889.
15. Yin, R., Wang, D., Zhao, S., Lou, Z. and Shen, G. (2021). Wearable sensors-enabled human–machine interaction systems: From design to application, *Advanced Functional Materials*, *31*, 2008936.
16. Yang, Y. and Deng, Z. D. (2019). Stretchable sensors for environmental monitoring, *Applied Physics Reviews*, *6*, 011309.
17. Lan, L., Le, X., Dong, H., Xie, J., Ying, Y. and Ping, J. (2020). One-step and large-scale fabrication of flexible and wearable humidity sensor based on laser-induced graphene for real-time tracking of plant transpiration at biointerface, *Biosensors and Bioelectronics*, *165*, 112360.
18. Haghi, M., Neubert, S., Geissler, A., Fleischer, H., Stoll, N., Stoll, R. and Thurow, K. (2020). A flexible and pervasive IoT-based healthcare platform for physiological and environmental parameters monitoring, *IEEE Internet of Things Journal*, *7*, 5628–5647.

19. Dong, T., Simões, J. and Yang, Z. (2020). Flexible photodetector based on 2D materials: Processing, architectures, and applications, *Advanced Materials Interfaces*, *7*, 1901657.
20. Lim, H.-R., Kim, H. S., Qazi, R., Kwon, Y.-T., Jeong, J.-W. and Yeo, W.-H. (2020). Advanced soft materials, sensor integrations, and applications of wearable flexible hybrid electronics in healthcare, energy, and environment, *Advanced Materials*, *32*, 1901924.
21. He, J., Xiao, P., Lu, W., Shi, J., Zhang, L., Liang, Y., Pan, C., Kuo, S.-W. and Chen, T. (2019). A universal high accuracy wearable pulse monitoring system via high sensitivity and large linearity graphene pressure sensor, *Nano Energy*, *59*, 422–433.
22. Zhang, T., Ding, Y., Hu, C., Zhang, M., Zhu, W., Bowen, C. R., Han, Y. and Yang, Y. (2023). Self-powered stretchable sensor arrays exhibiting magnetoelasticity for real-time human–machine interaction, *Advanced Materials*, *35*, 2203786.
23. Shi, Y., Wang, F., Tian, J., Li, S., Fu, E., Nie, J., Lei, R., Ding, Y., Chen, X. and Wang, Z. L. Self-powered electro-tactile system for virtual tactile experiences, *Science Advances*, *7*, eabe2943.
24. Pang, C., Lee, G.-Y., Kim, T.-I., Kim, S. M., Kim, H. N., Ahn, S.-H. and Suh, K.-Y. (2012). A flexible and highly sensitive strain-gauge sensor using reversible interlocking of nanofibres, *Nature Materials*, *11*, 795–801.
25. Li, T., Luo, H., Qin, L., Wang, X., Xiong, Z., Ding, H., Gu, Y., Liu, Z. and Zhang, T. (2016). Flexible capacitive tactile sensor based on micropatterned dielectric layer, *Small*, *12*, 5042–5048.
26. Shi, J., Wang, L., Dai, Z., Zhao, L., Du, M., Li, H. and Fang, Y. (2018). Multiscale hierarchical design of a flexible piezoresistive pressure sensor with high sensitivity and wide linearity range, *Small*, *14*, 1800819.
27. Bae, G. Y., Han, J. T., Lee, G., Lee, S., Kim, S. W., Park, S., Kwon, J., Jung, S. and Cho, K. (2018). Pressure/Temperature sensing bimodal electronic skin with stimulus discriminability and linear sensitivity, *Advanced Materials*, *30*, 1803388.
28. Wang, Z., Zhang, L., Liu, J., Jiang, H. and Li, C. (2018). Flexible hemispheric microarrays of highly pressure-sensitive sensors based on breath figure method, *Nanoscale*, *10*, 10691–10698.
29. Zhu, B., Ling, Y., Yap, L. W., Yang, M., Lin, F., Gong, S., Wang, Y., An, T., Zhao, Y. and Cheng, W. (2019). Hierarchically structured vertical gold nanowire array-based wearable pressure sensors for wireless health monitoring, *ACS Applied Materials & Interfaces*, *11*, 29014–29021.
30. Yu, S., Li, L., Wang, J., Liu, E., Zhao, J., Xu, F., Cao, Y. and Lu, C. (2020). Light-boosting highly sensitive pressure sensors based on bioinspired multiscale surface structures, *Advanced Functional Materials*, *30*, 1907091.

31. Li, F., Wang, R., Song, C., Zhao, M., Ren, H., Wang, S., Liang, K., Li, D., Ma, X., Zhu, B., et al. (2021). A skin-inspired artificial mechanoreceptor for tactile enhancement and integration, *ACS Nano, 15*, 16422–16431.
32. Shu, Q., Pang, Y., Li, Q., Gu, Y., Liu, Z., Liu, B., Li, J. and Li, Y. (2024). Flexible resistive tactile pressure sensors, *Journal of Materials Chemistry A, 12*, 9296–9321.
33. Wang, H., Sun, B., Ge, S. S., Su, J. and Jin, M. L. (2024). On non-von Neumann flexible neuromorphic vision sensors, *npj Flexible Electronics, 8*, 28.
34. Wu, L., Xue, J., Meng, J., Shi, B., Sun, W., Wang, E., Dong, M., Zheng, X., Wu, Y., Li, Y., et al. (2024). Self-powered flexible sensor array for dynamic pressure monitoring, *Advanced Functional Materials, 34*, 2316712.
35. Xu, M., Qi, J., Li, F. and Zhang, Y. (2018). Transparent and flexible tactile sensors based on graphene films designed for smart panels, *Journal of Materials Science, 53*, 9589–9597.
36. Cheng, W., Wang, J., Ma, Z., Yan, K., Wang, Y., Wang, H., Li, S., Li, Y., Pan, L. and Shi, Y. (2018). Flexible pressure sensor with high sensitivity and low hysteresis based on a hierarchically microstructured electrode, *IEEE Electron Device Letters, 39*, 288–291.
37. Luo, S., Yang, J., Song, X., Zhou, X., Yu, L., Sun, T., Yu, C., Huang, D., Du, C. and Wei, D. (2018). Tunable-sensitivity flexible pressure sensor based on graphene transparent electrode, *Solid-State Electronics, 145*, 29–33.
38. Zhu, M., Ji, S., Luo, Y., Zhang, F., Liu, Z., Wang, C., Lv, Z., Jiang, Y., Wang, M., Cui, Z., et al. (2022). A mechanically interlocking strategy based on conductive microbridges for stretchable electronics, *Advanced Materials, 34*, 2101339.
39. Duan, L., D'Hooge, D. R. and Cardon, L. (2020). Recent progress on flexible and stretchable piezoresistive strain sensors: From design to application, *Progress in Materials Science, 114*, 100617.
40. Yao, S., Ren, P., Song, R., Liu, Y., Huang, Q., Dong, J., O'Connor, B. T. and Zhu, Y. (2020). Nanomaterial-enabled flexible and stretchable sensing systems: Processing, integration, and applications, *Advanced Materials, 32*, 1902343.
41. Cai, Y., Shen, J., Ge, G., Zhang, Y., Jin, W., Huang, W., Shao, J., Yang, J. and Dong, X. (2018). Stretchable $Ti_3C_2T_x$ MXene/Carbon nanotube composite based strain sensor with ultrahigh sensitivity and tunable sensing range, *ACS Nano, 12*, 56–62.
42. Huang, Q., Jiang, Y., Duan, Z., Wu, Y., Yuan, Z., Guo, J., Zhang, M. and Tai, H. (2024). Ion gradient induced self-powered flexible strain sensor, *Nano Energy, 126*, 109689.

43. Lee, J.-H., Cho, K. and Kim, J.-K. (2024). Age of flexible electronics: Emerging trends in soft multifunctional sensors, *Advanced Materials, 36*, 2310505.
44. Zheng, F., Jiang, H.-Y., Yang, X.-T., Guo, J.-H., Sun, L., Guo, Y.-Y., Xu, H. and Yao, M.-S. (2024). Reviews of wearable healthcare systems based on flexible gas sensors, *Chemical Engineering Journal, 490*, 151874.
45. Choong, C.-L., Shim, M.-B., Lee, B.-S., Jeon, S., Ko, D.-S., Kang, T.-H., Bae, J., Lee, S. H., Byun, K.-E., Im, J., et al. (2014). Highly stretchable resistive pressure sensors using a conductive elastomeric composite on a micropyramid array, *Advanced Materials, 26*, 3451–3458.
46. Zhang, D., Xu, S., Zhao, X., Qian, W., Bowen, C. R. and Yang, Y. (2020). Wireless monitoring of small strains in intelligent robots via a joule heating effect in stretchable graphene–polymer nanocomposites, *Advanced Functional Materials, 30*, 1910809.
47. Pu, J.-H., Zhao, X., Zha, X.-J., Bai, L., Ke, K., Bao, R.-Y., Liu, Z.-Y., Yang, M.-B. and Yang, W. (2019). Multilayer structured AgNW/WPU-MXene fiber strain sensors with ultrahigh sensitivity and a wide operating range for wearable monitoring and healthcare, *Journal of Materials Chemistry A, 7*, 15913–15923.
48. Cao, J., Liang, F., Li, H., Li, X., Fan, Y., Hu, C., Yu, J., Xu, J., Yin, Y., Li, F., et al. (2022). Ultra-robust stretchable electrode for e-skin: In situ assembly using a nanofiber scaffold and liquid metal to mimic water-to-net interaction, *InfoMat, 4*, e12302.
49. Zhao, D., Zhu, Y., Cheng, W., Chen, W., Wu, Y. and Yu, H. (2021). Cellulose-based flexible functional materials for emerging intelligent electronics, *Advanced Materials, 33*, 2000619.
50. Shi, X., Fan, X., Zhu, Y., Liu, Y., Wu, P., Jiang, R., Wu, B., Wu, H.-A., Zheng, H., Wang, J., et al. (2022). Pushing detectability and sensitivity for subtle force to new limits with shrinkable nanochannel structured aerogel, *Nature Communications, 13*, 1119.
51. Li, J., Yao, Z., Meng, X., Zhang, X., Wang, Z., Wang, J., Ma, G., Liu, L., Zhang, J., Niu, S., et al. (2024). High-fidelity, low-hysteresis bionic flexible strain sensors for soft machines, *ACS Nano, 18*, 2520–2530.
52. Liu, Y., Wang, F., Hu, Z., Li, M., Ouyang, S., Wu, Y., Wang, S., Li, Z., Qian, J., Wang, L., et al. (2024). Applications of cellulose-based flexible self-healing sensors for human health monitoring, *Nano Energy, 127*, 109790.
53. Lu, Y., Zhang, H., Zhao, Y., Liu, H., Nie, Z., Xu, F., Zhu, J. and Huang, W. (2024). Robust fiber-shaped flexible temperature sensors for safety monitoring with ultrahigh sensitivity, *Advanced Materials, 36*, 2310613.
54. Men, Y., Qin, Z., Yang, Z., Zhang, P., Li, M., Wang, Q., Zeng, D., Yin, X. and Ji, H. (2024). Antibacterial defective-ZIF-8/PPY/BC-based flexible

electronics as stress-strain and NO_2 gas sensors, *Advanced Functional Materials*, *34*, 2316633.
55. Li, Z., Zhu, M., Shen, J., Qiu, Q., Yu, J. and Ding, B. (2020). All-fiber structured electronic skin with high elasticity and breathability, *Advanced Functional Materials*, *30*, 1908411.
56. Gong, M., Wan, P., Ma, D., Zhong, M., Liao, M., Ye, J., Shi, R. and Zhang, L. (2019). Flexible breathable nanomesh electronic devices for on-demand therapy, *Advanced Functional Materials*, *29*, 1902127.
57. Zhang, L., He, J., Liao, Y., Zeng, X., Qiu, N., Liang, Y., Xiao, P. and Chen, T. (2019). A self-protective, reproducible textile sensor with high performance towards human–machine interactions, *Journal of Materials Chemistry A*, *7*, 26631–26640.
58. Xu, Y., Sun, B., Ling, Y., Fei, Q., Chen, Z., Li, X., Guo, P., Jeon, N., Goswami, S., Liao, Y., *et al.* (2020). Multiscale porous elastomer substrates for multifunctional on-skin electronics with passive-cooling capabilities, *Proceedings of the National Academy of Sciences*, *117*, 205–213.
59. Gao, G., Yang, F., Zhou, F., He, J., Lu, W., Xiao, P., Yan, H., Pan, C., Chen, T. and Wang, Z. L. (2020). Bioinspired self-healing human–machine interactive touch pad with pressure-sensitive adhesiveness on targeted substrates, *Advanced Materials*, *32*, 2004290.
60. Wu, L., Fan, M., Qu, M., Yang, S., Nie, J., Tang, P., Pan, L., Wang, H. and Bin, Y. (2021). Self-healing and anti-freezing graphene–hydrogel–graphene sandwich strain sensor with ultrahigh sensitivity, *Journal of Materials Chemistry B*, *9*, 3088–3096.
61. Lu, Q., Li, H. and Tan, Z. (2024). Physically entangled multifunctional eutectogels for flexible sensors with mechanically robust, *Journal of Materials Chemistry A*, *12*, 20307–20316.
62. Wu, P., Li, L., Shao, S., Liu, J. and Wang, J. (2024). Bioinspired PEDOT:PSS-PVDF(HFP) flexible sensor for machine-learning-assisted multimodal recognition, *Chemical Engineering Journal*, *495*, 153558.
63. Huang, X., Ma, Z., Xia, W., Hao, L., Wu, Y., Lu, S., Luo, Y., Qin, L. and Dong, G. (2024). A high-sensitivity flexible piezoelectric tactile sensor utilizing an innovative rigid-in-soft structure, *Nano Energy*, *129*, 110019.
64. Park, J.-E., Kang, H. S., Koo, M. and Park, C. (2020). Autonomous surface reconciliation of a liquid-metal conductor micropatterned on a deformable hydrogel, *Advanced Materials*, *32*, 2002178.
65. Ghosh, S. K., Park, J., Na, S., Kim, M. P. and Ko, H. (2021). A fully biodegradable ferroelectric skin sensor from edible porcine skin gelatine, *Advanced Science*, *8*, 2005010.
66. Boutry, C. M., Nguyen, A., Lawal, Q. O., Chortos, A., Rondeau-Gagné, S. and Bao, Z. (2015). A sensitive and biodegradable pressure sensor array for cardiovascular monitoring, *Advanced Materials*, *27*, 6954–6961.

67. Yang, G., Zheng, X., Li, J., Chen, C., Zhu, J., Yi, H., Dong, X., Zhao, J., Shi, L., Zhang, X., et al. (2024). Schottky effect-enabled high unit-area capacitive interface for flexible pressure sensors, *Advanced Functional Materials, 34*, 2401415.
68. Kim, H., Na, H., Noh, S., Chang, S., Kim, J., Kong, T., Shin, G., Lee, C., Lee, S., Park, Y.-L., et al. (2024). Inherently integrated microfiber-based flexible proprioceptive sensor for feedback-controlled soft actuators, *npj Flexible Electronics, 8*, 15.
69. Wang, X., Yue, O., Liu, X., Hou, M. and Zheng, M. (2020). A novel bio-inspired multi-functional collagen aggregate based flexible sensor with multi-layer and internal 3D network structure, *Chemical Engineering Journal, 392*, 123672.
70. Wang, L., Wang, K., Lou, Z., Jiang, K. and Shen, G. (2018). Plant-based modular building blocks for "Green" electronic skins, *Advanced Functional Materials, 28*, 1804510.
71. Miao, J., Liu, H., Li, Y. and Zhang, X. (2018). Biodegradable transparent substrate based on edible starch–chitosan embedded with nature-inspired three-dimensionally interconnected conductive nanocomposites for wearable green electronics, *ACS Applied Materials & Interfaces, 10*, 23037–23047.
72. Liu, H., Xiang, H., Li, Z., Meng, Q., Li, P., Ma, Y., Zhou, H. and Huang, W. (2020). Flexible and degradable multimodal sensor fabricated by transferring laser-induced porous carbon on starch film, *ACS Sustainable Chemistry & Engineering, 8*, 527–533.
73. Zhang, H., Sun, X., Hubbe, M. and Pal, L. (2019). Flexible and pressure-responsive sensors from cellulose fibers coated with multiwalled carbon nanotubes, *ACS Applied Electronic Materials, 1*, 1179–1188.
74. Kanaparthi, S. and Badhulika, S. (2016). Solvent-free fabrication of a biodegradable all-carbon paper based field effect transistor for human motion detection through strain sensing, *Green Chemistry, 18*, 3640–3646.
75. Shin, J., Noh, S., Lee, J., Jhee, S., Choi, I., Kyu Jeong, C., Heon Kim, S. and Kim, J. S. (2024). Self-powered flexible piezoelectric motion sensor with spatially aligned InN nanowires, *Chemical Engineering Journal, 486*, 150205.
76. Shi, Y., Guan, Y., Liu, M., Kang, X., Tian, Y., Deng, W., Yu, P., Ning, C., Zhou, L., Fu, R., et al. (2024). Tough, antifreezing, and piezoelectric organohydrogel as a flexible wearable sensor for human–machine interaction, *ACS Nano, 18*, 3720–3732.
77. Ma, Z., Wu, Y., Lu, S., Li, J., Liu, J., Huang, X., Zhang, X., Zhang, Y., Dong, G., Qin, L., et al. (2024). Magnetically assisted 3D printing of ultra-antiwear flexible sensor, *Advanced Functional Materials*, 2406108.

78. Liu, P., Tong, W., Hu, R., Yang, A., Tian, H., Guo, X., Liu, C., Ma, Y., Tian, H., Song, A., et al. (2024). Ultra-sensitive flexible resistive sensor based on modified PEDOT: PSS inspired by earthworm, *Chemical Engineering Journal*, 494, 152984.
79. Xu, Y., Zhao, G., Zhu, L., Fei, Q., Zhang, Z., Chen, Z., An, F., Chen, Y., Ling, Y., Guo, P., et al. (2020). Pencil–paper on-skin electronics, *Proceedings of the National Academy of Sciences*, 117, 18292–18301.
80. Jeong, H., Baek, S., Han, S., Jang, H., Kim, S. H. and Lee, H. S. (2018). Novel eco-friendly starch paper for use in flexible, transparent, and disposable organic electronics, *Advanced Functional Materials*, 28, 1704433.
81. Liao, M., Liao, H., Ye, J., Wan, P. and Zhang, L. (2019). Polyvinyl alcohol-stabilized liquid metal hydrogel for wearable transient epidermal sensors, *ACS Applied Materials & Interfaces*, 11, 47358–47364.
82. Liang, Q., Zhang, Q., Yan, X., Liao, X., Han, L., Yi, F., Ma, M. and Zhang, Y. (2017). Recyclable and green triboelectric nanogenerator, *Advanced Materials*, 29, 1604961.
83. Yoon, J., Han, J., Choi, B., Lee, Y., Kim, Y., Park, J., Lim, M., Kang, M.-H., Kim, D. H., Kim, D. M., et al. (2018). Three-dimensional printed poly(vinyl alcohol) substrate with controlled On-demand degradation for transient electronics, *ACS Nano*, 12, 6006–6012.
84. Muralidharan, N., Afolabi, J., Share, K., Li, M. and Pint, C. L. (2018). A fully transient mechanical energy harvester, *Advanced Materials Technologies*, 3, 1800083.
85. Teng, L., Ye, S., Handschuh-Wang, S., Zhou, X., Gan, T. and Zhou, X. (2019). Liquid metal-based transient circuits for flexible and recyclable electronics, *Advanced Functional Materials*, 29, 1808739.
86. Jiao, S., Yang, X., Zheng, X., Pei, Y., Liu, J. and Tang, K. (2024). Effects of charge state of nano-chitin on the properties of polyvinyl alcohol composite hydrogel, *Carbohydrate Polymers*, 330, 121776.
87. Karyappa, R., Nagaraju, N., Yamagishi, K., Koh, X. Q., Zhu, Q. and Hashimoto, M. (2024). 3D printing of polyvinyl alcohol hydrogels enabled by aqueous two-phase system, *Materials Horizons*, 11, 2701–2717.
88. Gui, T., Xiao, L.-P., Zou, S.-L., Zhang, Y., Fu, X., Liu, C.-H. and Sun, R.-C. (2024). Tough and biodegradable C-lignin cross-linked polyvinyl alcohol supramolecular composite films with closed-looping recyclability, *Chemical Engineering Journal*, 491, 151748.
89. Cui, T., Yu, J., Li, Q., Wang, C.-F., Chen, S., Li, W. and Wang, G. (2020). Large-scale fabrication of robust artificial skins from a biodegradable sealant-loaded nanofiber scaffold to skin tissue via microfluidic blow-spinning, *Advanced Materials*, 32, 2000982.
90. Kumar Kalita, N., Hazarika, D., Srivastava, R. K. and Hakkarainen, M. (2024). Faster biodegradable and chemically recyclable polycaprolactone

with embedded enzymes: Revealing new insights into degradation kinetics, *Chemical Engineering Journal*, *496*, 153982.
91. Fu, Q., He, S., Yang, J., Su, Z., Li, P., Yu, X., Jin, W., Xu, S., Yu, Z. and Zha, D. (2024). Polydopamine-modified metal-organic frameworks nanoparticles enhance the corrosion resistance and bioactivity of polycaprolactone coating on high-purity magnesium, *Journal of Magnesium and Alloys*, *12*, 2070–2089.
92. Zhang, W., Zou, C., Pan, Q., Hu, G., Shi, H., Zhang, Y., He, X., He, Y. and Zhang, X. (2024). A triple shape memory material of trans-polyisoprene/polycaprolactone with customizable response temperature controlled by crosslinking density, *Advanced Functional Materials*, 2400245.
93. Guo, Y., Zhong, M., Fang, Z., Wan, P. and Yu, G. (2019). A wearable transient pressure sensor made with MXene nanosheets for sensitive broadrange human–machine interfacing, *Nano Letters*, *19*, 1143–1150.
94. Lan, B., Chen, Y., Xiao, N., Liu, N., Juan, C., Xia, C. and Zhang, F. (2024). Efficient and selective upcycling of waste polylactic acid into acetate using nickel selenide, *Journal of Energy Chemistry*, *97*, 575–584.
95. Xiao, Y., Tao, Z., Ju, Y., Huang, X., Zhang, X., Liu, X., Volotovski, P. A., Huang, C., Chen, H., Zhang, Y., *et al.* (2024). Diamond-like carbon depositing on the surface of polylactide membrane for prevention of adhesion formation during tendon repair, *Nano-Micro Letters*, *16*, 186.
96. Zhang, Y., Liu, L., Yao, M., Feng, J., Xue, Y., Annamalai, P. K., Chevali, V., Dinh, T., Fang, Z., Liu, H., *et al.* (2024). Functionalizing lignin by *in situ* solid-phase grafting ammonium polyphosphate for enhancing thermal, flame-retardant, mechanical, and UV-resistant properties of polylactic acid, *Chemical Engineering Journal*, *495*, 153429.
97. Guan, Y., Bi, B., Qiao, D., Cao, S., Zhang, W., Wang, Z., Zeng, H. and Li, Y. (2024). Bioinspired superhydrophobic polylactic acid aerogel with a tree branch structure for the removal of viscous oil spills assisted by solar energy, *Journal of Materials Chemistry A*, *12*, 9850–9862.
98. Zheng, Q., Zou, Y., Zhang, Y., Liu, Z., Shi, B., Wang, X., Jin, Y., Ouyang, H., Li, Z. and Wang, Z. L. Biodegradable triboelectric nanogenerator as a lifetime designed implantable power source, *Science Advances*, *2*, e1501478.
99. Peng, X., Dong, K., Ye, C., Jiang, Y., Zhai, S., Cheng, R., Liu, D., Gao, X., Wang, J. and Wang, Z. L. A breathable, biodegradable, antibacterial, and self-powered electronic skin based on all-nanofiber triboelectric nanogenerators, *Science Advances*, *6*, eaba9624.
100. Khalid, M. A. U., Ali, M., Soomro, A. M., Kim, S. W., Kim, H. B., Lee, B.-G. and Choi, K. H. (2019). A highly sensitive biodegradable pressure sensor based on nanofibrous dielectric, *Sensors and Actuators A: Physical*, *294*, 140–147.

101. Chang, J.-K., Chang, H.-P., Guo, Q., Koo, J., Wu, C.-I. and Rogers, J. A. (2018). Biodegradable electronic systems in 3D, heterogeneously integrated formats, *Advanced Materials, 30*, 1704955.
102. Rodrigues, I. C. P., Orrantia Clark, L. C., Kuang, X., Sanchez Flores, R., Lopes, É. S. N., Gabriel, L. P. and Zhang, Y. S. (2024). Multimaterial coextrusion (bio)printing of composite polymer biomaterial ink and hydrogel bioink for tissue fabrication, *Composites Part B: Engineering, 275*, 111337.
103. Liu, C., Quan, X., Tian, X., Zhao, Y., Li, H.-F., Mak, J. C. W., Wang, Z., Mao, S. and Zheng, Y. (2024). Inhaled macrophage apoptotic bodies-engineered microparticle enabling construction of pro-regenerative microenvironment to fight hypoxic lung injury in mice, *ACS Nano, 18*, 13361–13376.
104. Song, M., Yu, H., Zhu, J., Ouyang, Z., Abdalkarim, S. Y. H., Tam, K. C. and Li, Y. (2020). Constructing stimuli-free self-healing, robust and ultrasensitive biocompatible hydrogel sensors with conductive cellulose nanocrystals, *Chemical Engineering Journal, 398*, 125547.
105. Wang, D., Wang, L., Lou, Z., Zheng, Y., Wang, K., Zhao, L., Han, W., Jiang, K. and Shen, G. (2020). Biomimetic, biocompatible and robust silk Fibroin-MXene film with stable 3D cross-link structure for flexible pressure sensors, *Nano Energy, 78*, 105252.
106. Guo, Y., Bae, J., Fang, Z., Li, P., Zhao, F. and Yu, G. (2020). Hydrogels and hydrogel-derived materials for energy and water sustainability, *Chemical Reviews, 120*, 7642–7707.
107. He, F., You, X., Gong, H., Yang, Y., Bai, T., Wang, W., Guo, W., Liu, X. and Ye, M. (2020). Stretchable, biocompatible, and multifunctional silk fibroin-based hydrogels toward wearable strain/pressure sensors and triboelectric nanogenerators, *ACS Applied Materials & Interfaces, 12*, 6442–6450.
108. Wang, F., Gao, Z.-Q., Feng, C.-P., Wang, D.-Y., Jin, M.-P., Zhang, F., Peng, Z.-L., Zhang, G.-M., Zhu, X.-Y. and Lan, H.-B. (2024). Flexible electronics substrate with excellent tear-resistant and high toughness using multimaterial 3D printing, *Additive Manufacturing, 81*, 103985.
109. Ding, J., Zeng, M., Tian, Y., Chen, Z., Qiao, Z., Xiao, Z., Wu, C., Wei, D., Sun, J. and Fan, H. (2024). Flexible silk-fibroin-based microelectrode arrays for high-resolution neural recording, *Materials Horizons, 11*, 4338–4347.
110. Sun, K., Zhang, Z., Tian, J., Zeng, N., Wang, B., Xing, W., Ma, L., Long, Y., Wang, C. and Fan, R. (2024). Flexible and biocompatible polyurethane/Co@C composite films with weakly negative permittivity, *Advanced Composites and Hybrid Materials, 7*, 22.
111. Zhang, H., Yin, F., Shang, S., Li, Y., Qiu, Z., Lin, Q., Wei, X., Li, S., Kim, N. Y. and Shen, G. (2022). A high-performance, biocompatible, and degradable piezoresistive-triboelectric hybrid device for cross-scale human

activities monitoring and self-powered smart home system, *Nano Energy*, *102*, 107687.
112. Le, T.-S. D., Park, S., An, J., Lee, P. S. and Kim, Y.-J. (2019). Ultrafast laser pulses enable one-step graphene patterning on woods and leaves for green electronics, *Advanced Functional Materials*, *29*, 1902771.
113. Guo, Y., Gao, S., Yue, W., Zhang, C. and Li, Y. (2019). Anodized aluminum oxide-assisted low-cost flexible capacitive pressure sensors based on double-sided nanopillars by a facile fabrication method, *ACS Applied Materials & Interfaces*, *11*, 48594–48603.
114. Niu, H., Gao, S., Yue, W., Li, Y., Zhou, W. and Liu, H. (2020). Highly morphology-controllable and highly sensitive capacitive tactile sensor based on epidermis-dermis-inspired interlocked asymmetric-nanocone arrays for detection of tiny pressure, *Small*, *16*, 1904774.
115. Niu, H., Li, H., Gao, S., Li, Y., Wei, X., Chen, Y., Yue, W., Zhou, W. and Shen, G. (2022). Perception-to-cognition tactile sensing based on artificial-intelligence-motivated human full-skin bionic electronic skin, *Advanced Materials*, *34*, 2202622.
116. Niu, H., Chen, Y., Kim, E.-S., Zhou, W., Li, Y. and Kim, N.-Y. (2022). Ultrasensitive capacitive tactile sensor with heterostructured active layers for tiny signal perception, *Chemical Engineering Journal*, *450*, 138258.
117. Yao, S. and Zhu, Y. (2014). Wearable multifunctional sensors using printed stretchable conductors made of silver nanowires, *Nanoscale*, *6*, 2345–2352.
118. Qu, X., Wu, Y., Han, Z., Li, J., Deng, L., Xie, R., Zhang, G., Wang, H. and Chen, S. (2024). Highly sensitive fiber crossbar sensors enabled by second-order synergistic effect of air capacitance and equipotential body, *Small*, *20*, 2311498.
119. Huang, Y., Bao, L., Li, Y., Zhang, X., Zhang, Z., Cui, R., Zhu, H., Wan, C. and Fu, W. (2024). Ultrasensitive quantum capacitance detector at the edge of graphene, *Materials Today*, *73*, 38–46.
120. Zhang, H.-W., Xu, X., Huang, M.-L., Wang, Y.-S., Xu, Z.-Q., Feng, Z.-S., Zhang, Y. and Wang, Y. (2024). Interlayer cross-linked MXene enables ultra-stable printed paper-based flexible sensor for real-time humidity monitoring, *Chemical Engineering Journal*, *495*, 153343.
121. Mannsfeld, S. C. B., Tee, B. C. K., Stoltenberg, R. M., Chen, C. V. H. H., Barman, S., Muir, B. V. O., Sokolov, A. N., Reese, C. and Bao, Z. (2010). Highly sensitive flexible pressure sensors with microstructured rubber dielectric layers, *Nature Materials*, *9*, 859–864.
122. Xiong, Y., Shen, Y., Tian, L., Hu, Y., Zhu, P., Sun, R. and Wong, C.-P. (2020). A flexible, ultra-highly sensitive and stable capacitive pressure sensor with convex microarrays for motion and health monitoring, *Nano Energy*, *70*, 104436.

123. Gong, M., Zhang, L. and Wan, P. (2020). Polymer nanocomposite meshes for flexible electronic devices, *Progress in Polymer Science*, *107*, 101279.
124. Schwartz, G., Tee, B. C. K., Mei, J., Appleton, A. L., Kim, D. H., Wang, H. and Bao, Z. (2013). Flexible polymer transistors with high pressure sensitivity for application in electronic skin and health monitoring, *Nature Communications*, *4*, 1859.
125. Luo, Y., Shao, J., Chen, S., Chen, X., Tian, H., Li, X., Wang, L., Wang, D. and Lu, B. (2019). Flexible capacitive pressure sensor enhanced by tilted micropillar arrays, *ACS Applied Materials & Interfaces*, *11*, 17796–17803.
126. Yang, J., Luo, S., Zhou, X., Li, J., Fu, J., Yang, W. and Wei, D. (2019). Flexible, tunable, and ultrasensitive capacitive pressure sensor with microconformal graphene electrodes, *ACS Applied Materials & Interfaces*, *11*, 14997–15006.
127. Xu, F. and Zhu, Y. (2012). Highly conductive and stretchable silver nanowire conductors, *Advanced Materials*, *24*, 5117–5122.
128. Yang, J. C., Kim, J.-O., Oh, J., Kwon, S. Y., Sim, J. Y., Kim, D. W., Choi, H. B. and Park, S. (2019). Microstructured porous pyramid-based ultrahigh sensitive pressure sensor insensitive to strain and temperature, *ACS Applied Materials & Interfaces*, *11*, 19472–19480.
129. Tay, R. Y., Li, H., Lin, J., Wang, H., Lim, J. S. K., Chen, S., Leong, W. L., Tsang, S. H. and Teo, E. H. T. (2020). Lightweight, superelastic boron nitride/polydimethylsiloxane foam as air dielectric substitute for multifunctional capacitive sensor applications, *Advanced Functional Materials*, *30*, 1909604.
130. Chhetry, A., Sharma, S., Yoon, H., Ko, S. and Park, J. Y. (2020). Enhanced sensitivity of capacitive pressure and strain sensor based on $CaCu_3Ti_4O_{12}$ wrapped hybrid sponge for wearable applications, *Advanced Functional Materials*, *30*, 1910020.
131. Nemala, S. S., Fernandes, J., Rodrigues, J., Lopes, V., Pinto, R. M. R., Vinayakumar, K. B., Placidi, E., De Bellis, G., Alpuim, P., Sampaio, R. S., et al. (2024). Sustainable graphene production for solution-processed microsupercapacitors and multipurpose flexible electronics, *Nano Energy*, *127*, 109781.
132. Dong, Z., Wang, J., Li, J., Lu, J., Tan, S., Feng, Q. and Xu, Z. (2024). Three-dimensional layered multifunctional carbon aerogel for energy storage and pressure sensors, *Chemical Engineering Journal*, *491*, 151797.
133. Yang, Y., Jung, B. K., Park, T., Ahn, J., Choi, Y. K., Oh, S., Lee, Y. M., Choi, H. J., Seo, H. and Oh, S. J. (2024). Sensory nervous system-inspired self-classifying, decoupled, multifunctional sensor with resistive-capacitive operation using silver nanomaterials, *Advanced Functional Materials*, 2405687.

134. Kang, S., Lee, J., Lee, S., Kim, S., Kim, J.-K., Algadi, H., Al-Sayari, S., Kim, D.-E., Kim, D. and Lee, T. (2016). Highly sensitive pressure sensor based on bioinspired porous structure for real-time tactile sensing, *Advanced Electronic Materials*, 2, 1600356.
135. Kim, J.-O., Kwon, S. Y., Kim, Y., Choi, H. B., Yang, J. C., Oh, J., Lee, H. S., Sim, J. Y., Ryu, S. and Park, S. (2019). Highly ordered 3D microstructure-based electronic skin capable of differentiating pressure, temperature, and proximity, *ACS Applied Materials & Interfaces*, 11, 1503–1511.
136. Nguyen, T., Dinh, T., Phan, H.-P., Pham, T. A., Dau, V. T., Nguyen, N.-T. and Dao, D. V. (2021). Advances in ultrasensitive piezoresistive sensors: From conventional to flexible and stretchable applications, *Materials Horizons*, 8, 2123–2150.
137. Chen, W. and Yan, X. (2020). Progress in achieving high-performance piezoresistive and capacitive flexible pressure sensors: A review, *Journal of Materials Science & Technology*, 43, 175–188.
138. Sang, C., Wang, S., Jin, X., Cheng, X., Xiao, H., Yue, Y. and Han, J. (2024). Nanocellulose-mediated conductive hydrogels with NIR photoresponse and fatigue resistance for multifunctional wearable sensors, *Carbohydrate Polymers*, 333, 121947.
139. Yuan, T., Yin, R., Li, C., Fan, Z. and Pan, L. (2024). Ti_3C_2Tx MXene-based all-resistive dual-mode sensor with near-zero temperature coefficient of resistance for crosstalk-free pressure and temperature detections, *Chemical Engineering Journal*, 487, 150396.
140. Liu, X., Ma, Y., Dai, X., Li, S., Li, B. and Zhang, X. (2024). Flexible pressure sensor based on Pt/PI network with high sensitivity and high thermal resistance, *Chemical Engineering Journal*, 494, 152996.
141. Fan, C., Liu, Y. and Zhang, Y. (2024). A universal, highly sensitive and seamlessly integratable textile resistive strain sensor, *Advanced Fiber Materials*, 6, 1152–1161.
142. Wei, X., Li, H., Yue, W., Gao, S., Chen, Z., Li, Y. and Shen, G. (2022). A high-accuracy, real-time, intelligent material perception system with a machine-learning-motivated pressure-sensitive electronic skin, *Matter*, 5, 1481–1501.
143. Bae, G. Y., Pak, S. W., Kim, D., Lee, G., Kim, D. H., Chung, Y. and Cho, K. (2016). Linearly and highly pressure-sensitive electronic skin based on a bioinspired hierarchical structural array, *Advanced Materials*, 28, 5300–5306.
144. Shao, Q., Niu, Z., Hirtz, M., Jiang, L., Liu, Y., Wang, Z. and Chen, X. (2014). High-performance and tailorable pressure sensor based on ultrathin conductive polymer film, *Small*, 10, 1466–1472.
145. Park, H., Jeong, Y. R., Yun, J., Hong, S. Y., Jin, S., Lee, S.-J., Zi, G. and Ha, J. S. (2015). Stretchable array of highly sensitive pressure sensors consisting of polyaniline nanofibers and Au-coated polydimethylsiloxane micropillars, *ACS Nano*, 9, 9974–9985.

146. Park, J., Lee, Y., Hong, J., Lee, Y., Ha, M., Jung, Y., Lim, H., Kim, S. Y. and Ko, H. (2014). Tactile-direction-sensitive and stretchable electronic skins based on human-skin-inspired interlocked microstructures, *ACS Nano*, *8*, 12020–12029.
147. Cao, Y., Li, T., Gu, Y., Luo, H., Wang, S. and Zhang, T. (2018). Fingerprint-inspired flexible tactile sensor for accurately discerning surface texture, *Small*, *14*, 1703902.
148. Zhang, Y., Hu, Y., Zhu, P., Han, F., Zhu, Y., Sun, R. and Wong, C.-P. (2017). Flexible and highly sensitive pressure sensor based on microdome-patterned PDMS forming with assistance of colloid self-assembly and replica technique for wearable electronics, *ACS Applied Materials & Interfaces*, *9*, 35968–35976.
149. Zhang, X., Hu, Y., Gu, H., Zhu, P., Jiang, W., Zhang, G., Sun, R. and Wong, C.-P. (2019). A highly sensitive and cost-effective flexible pressure sensor with micropillar arrays fabricated by novel metal-assisted chemical etching for wearable electronics, *Advanced Materials Technologies*, *4*, 1900367.
150. Pang, Y., Zhang, K., Yang, Z., Jiang, S., Ju, Z., Li, Y., Wang, X., Wang, D., Jian, M., Zhang, Y., *et al.* (2018). Epidermis microstructure inspired graphene pressure sensor with random distributed spinosum for high sensitivity and large linearity, *ACS Nano*, *12*, 2346–2354.
151. Qiu, Y., Tian, Y., Sun, S., Hu, J., Wang, Y., Zhang, Z., Liu, A., Cheng, H., Gao, W., Zhang, W., *et al.* (2020). Bioinspired, multifunctional dual-mode pressure sensors as electronic skin for decoding complex loading processes and human motions, *Nano Energy*, *78*, 105337.
152. Jian, M., Xia, K., Wang, Q., Yin, Z., Wang, H., Wang, C., Xie, H., Zhang, M. and Zhang, Y. (2017). Flexible and highly sensitive pressure sensors based on bionic hierarchical structures, *Advanced Functional Materials*, *27*, 1606066.
153. Xia, K., Wang, C., Jian, M., Wang, Q. and Zhang, Y. (2018). CVD growth of fingerprint-like patterned 3D graphene film for an ultrasensitive pressure sensor, *Nano Research*, *11*, 1124–1134.
154. Jia, J., Huang, G., Deng, J. and Pan, K. (2019). Skin-inspired flexible and high-sensitivity pressure sensors based on rGO films with continuous-gradient wrinkles, *Nanoscale*, *11*, 4258–4266.
155. Zhang, X., Liu, Y., Yuan, G., Wang, S., Wang, D., Zhu, T., Wu, X., Ma, M., Guo, L., Guo, H., *et al.* (2024). The synthetic NLR RGA5HMA5 requires multiple interfaces within and outside the integrated domain for effector recognition, *Nature Communications*, *15*, 1104.
156. Jiang, X., Zhou, X., Ding, K., Li, X., Huang, B. and Xu, W. (2024). Anti-swelling gel wearable sensor based on solvent exchange strategy for underwater communication, *Advanced Functional Materials*, 2400936.
157. Zhou, P., Zheng, Z., Lin, J., Gu, W., Luo, Z., Zhang, Y. and Chen, L. (2024). Wirelessly detecting of spatiotemporal mechanical stimuli by a bio-inspired

neural sensor via temperature signals, *Chemical Engineering Journal*, *494*, 152884.
158. Pan, L., Chortos, A., Yu, G., Wang, Y., Isaacson, S., Allen, R., Shi, Y., Dauskardt, R. and Bao, Z. (2014). An ultra-sensitive resistive pressure sensor based on hollow-sphere microstructure induced elasticity in conducting polymer film, *Nature Communications*, *5*, 3002.
159. Zhao, T., Li, T., Chen, L., Yuan, L., Li, X. and Zhang, J. (2019). Highly sensitive flexible piezoresistive pressure sensor developed using biomimetically textured porous materials, *ACS Applied Materials & Interfaces*, *11*, 29466–29473.
160. Yao, H.-B., Ge, J., Wang, C.-F., Wang, X., Hu, W., Zheng, Z.-J., Ni, Y. and Yu, S.-H. (2013). A flexible and highly pressure-sensitive graphene–polyurethane sponge based on fractured microstructure design, *Advanced Materials*, *25*, 6692–6698.
161. Jung, S., Kim, J. H., Kim, J., Choi, S., Lee, J., Park, I., Hyeon, T. and Kim, D.-H. (2014). Reverse-micelle-induced porous pressure-sensitive rubber for wearable human-machine interfaces, *Advanced Materials*, *26*, 4825–4830.
162. Wei, H., Li, A., Kong, D., Li, Z., Cui, D., Li, T., Dong, B. and Guo, Z. (2021). Polypyrrole/reduced graphene aerogel film for wearable piezoresisitic sensors with high sensing performances, *Advanced Composites and Hybrid Materials, 4,* 86–95.
163. Pi, Z., Zhang, J., Wen, C., Zhang, Z.-B. and Wu, D. (2014). Flexible piezoelectric nanogenerator made of poly(vinylidenefluoride-co-trifluoroethylene) (PVDF-TrFE) thin film, *Nano Energy*, *7*, 33–41.
164. Sharma, T., Je, S.-S., Gill, B. and Zhang, J. X. J. (2012). Patterning piezoelectric thin film PVDF–TrFE based pressure sensor for catheter application, *Sensors and Actuators A: Physical*, *177*, 87–92.
165. Ko, E. J., Jeon, S. J., Han, Y. W., Jeong, S. Y., Kang, C. Y., Sung, T. H., Seong, K. W. and Moon, D. K. (2019). Synthesis and characterization of nanofiber-type hydrophobic organic materials as electrodes for improved performance of PVDF-based piezoelectric nanogenerators, *Nano Energy*, *58*, 11–22.
166. Lee, J.-H., Yoon, H.-J., Kim, T. Y., Gupta, M. K., Lee, J. H., Seung, W., Ryu, H. and Kim, S.-W. (2015). Micropatterned P(VDF-TrFE) film-based piezoelectric nanogenerators for highly sensitive self-powered pressure sensors, *Advanced Functional Materials*, *25*, 3203–3209.
167. Chen, X., Shao, J., Li, X. and Tian, H. (2016). A flexible piezoelectric-pyroelectric hybrid nanogenerator based on P(VDF-TrFE) nanowire array, *IEEE Transactions on Nanotechnology*, *15*, 295–302.
168. Chen, X., Shao, J., Tian, H., Li, X., Wang, C., Luo, Y. and Li, S. (2020). Scalable imprinting of flexible multiplexed sensor arrays with distributed piezoelectricity-enhanced micropillars for dynamic tactile sensing, *Advanced Materials Technologies*, *5*, 2000046.

169. Jung, M., Kim, S., Hwang, J., Kim, C., Kim, H. J., Kim, Y.-J. and Jeon, S. (2024). Monolithic three-dimensional hafnia-based artificial nerve system, *Nano Energy*, *126*, 109643.
170. Guo, Y., Zhang, H., Fang, L., Wang, Z., He, W., Shi, S., Zhang, R., Cheng, J. and Wang, P. (2024). A self-powered flexible piezoelectric sensor patch for deep learning-assisted motion identification and rehabilitation training system, *Nano Energy*, *123*, 109427.
171. Yang, H., Bu, T., Liu, W., Liu, J., Ling, Y., Wu, M., Liu, W., Wang, C., Gao, X. and Wang, L. (2024). A novel triboelectric-optical hybrid tactile sensor for human-machine tactile interaction, *Nano Energy*, *125*, 109592.
172. Lei, H., Yin, Z.-Y., Huang, P., Gao, X., Zhao, C., Wen, Z., Sun, X. and Wang, S.-D. (2024). Intelligent tribotronic transistors toward tactile near-sensor computing, *Advanced Functional Materials*, 2401913.
173. Wang, Z. L. and Song, J. (2006). Piezoelectric nanogenerators based on zinc oxide nanowire arrays, *Science*, *312*, 242–246.
174. Dagdeviren, C., Su, Y., Joe, P., Yona, R., Liu, Y., Kim, Y.-S., Huang, Y., Damadoran, A. R., Xia, J., Martin, L. W., *et al*. (2014). Conformable amplified lead zirconate titanate sensors with enhanced piezoelectric response for cutaneous pressure monitoring, *Nature Communications*, *5*, 4496.
175. Deng, W., Jin, L., Zhang, B., Chen, Y., Mao, L., Zhang, H. and Yang, W. (2016). A flexible field-limited ordered ZnO nanorod-based self-powered tactile sensor array for electronic skin, *Nanoscale*, *8*, 16302–16306.
176. Chen, X., Li, X., Shao, J., An, N., Tian, H., Wang, C., Han, T., Wang, L. and Lu, B. (2017). High-performance piezoelectric nanogenerators with imprinted P(VDF-TrFE)/BaTiO$_3$ nanocomposite micropillars for self-powered flexible sensors, *Small*, *13*, 1604245.
177. He, J., Guo, X., Yu, J., Qian, S., Hou, X., Cui, M., Yang, Y., Mu, J., Geng, W. and Chou, X. (2020). A high-resolution flexible sensor array based on PZT nanofibers, *Nanotechnology*, *31*, 155503.
178. Wan, Y., Wang, Y. and Guo, C. F. (2017). Recent progresses on flexible tactile sensors, *Materials Today Physics*, *1*, 61–73.
179. Su, Y., Chen, C., Pan, H., Yang, Y., Chen, G., Zhao, X., Li, W., Gong, Q., Xie, G., Zhou, Y., *et al*. (2021). Muscle fibers inspired high-performance piezoelectric textiles for wearable physiological monitoring, *Advanced Functional Materials*, *31*, 2010962.
180. Wang, Z. L. (2014). Triboelectric nanogenerators as new energy technology and self-powered sensors – principles, problems and perspectives, *Faraday Discussions*, *176*, 447–458.
181. Zhang, J.-W., Zhang, Y., Li, Y.-Y. and Wang, P. (2022). Textile-based flexible pressure sensors: A review, *Polymer Reviews*, *62*, 65–94.
182. Ha, M., Park, J., Lee, Y. and Ko, H. (2015). Triboelectric generators and sensors for self-powered wearable electronics, *ACS Nano*, *9*, 3421–3427.

183. Fan, F.-R., Tian, Z.-Q. and Lin Wang, Z. (2012). Flexible triboelectric generator, *Nano Energy, 1*, 328–334.
184. Wang, Z. L. (2013). Triboelectric nanogenerators as new energy technology for self-powered systems and as active mechanical and chemical sensors, *ACS Nano, 7*, 9533–9557.
185. Fan, F.-R., Lin, L., Zhu, G., Wu, W., Zhang, R. and Wang, Z. L. (2012). Transparent triboelectric nanogenerators and self-powered pressure sensors based on micropatterned plastic films, *Nano Letters, 12*, 3109–3114.
186. Wang, S., Lin, L., Xie, Y., Jing, Q., Niu, S. and Wang, Z. L. (2013). Sliding-triboelectric nanogenerators based on in-plane charge-separation mechanism, *Nano Letters, 13*, 2226–2233.
187. Zhu, G., Chen, J., Liu, Y., Bai, P., Zhou, Y. S., Jing, Q., Pan, C. and Wang, Z. L. (2013). Linear-grating triboelectric generator based on sliding electrification, *Nano Letters, 13*, 2282–2289.
188. Lin, L., Wang, S., Xie, Y., Jing, Q., Niu, S., Hu, Y. and Wang, Z. L. (2013). Segmentally structured disk triboelectric nanogenerator for harvesting rotational mechanical energy, *Nano Letters, 13*, 2916–2923.
189. Cakir, O., Doganay, D., Cugunlular, M., Cicek, M. O., Demircioglu, O., Coskun, S. and Unalan, H. E. (2024). Post-treatment optimization for silver nanowire networks in transparent droplet-based TENG sensors, *Nano Energy, 128*, 109940.
190. Liu, S., Tong, W., Gao, C., Wang, X., Liu, Y. and Zhang, Y. (2024). Hygroscopic paper enhanced using hydroxyapatite coating for wearable TENG sensors, *Chemical Engineering Journal, 493*, 152597.
191. Fan, J., Yang, R., Du, Y., Wang, F., Wang, L., Yang, J. and Zhou, A. (2024). A triboelectric nanogenerator based on MXene/TPU composite films with excellent stretchability for self-powered flexible sensing, *Nano Energy, 129*, 109999.
192. Ning, C., Dong, K., Cheng, R., Yi, J., Ye, C., Peng, X., Sheng, F., Jiang, Y. and Wang, Z. L. (2021). Flexible and stretchable fiber-shaped triboelectric nanogenerators for biomechanical monitoring and human-interactive sensing, *Advanced Functional Materials, 31*, 2006679.
193. Wei, X., Wang, B., Wu, Z. and Wang, Z. L. (2022). An open-environment tactile sensing system: Toward simple and efficient material identification, *Advanced Materials, 34*, 2203073.
194. Chen, J., Wen, X., Liu, X., Cao, J., Ding, Z. and Du, Z. (2021). Flexible hierarchical helical yarn with broad strain range for self-powered motion signal monitoring and human-machine interactive, *Nano Energy, 80*, 105446.
195. Yang, Y., Zhou, Y. S., Zhang, H., Liu, Y., Lee, S. and Wang, Z. L. (2013). A single-electrode based triboelectric nanogenerator as self-powered tracking system, *Advanced Materials, 25*, 6594–6601.

196. Wen, Z., Yang, Y., Sun, N., Li, G., Liu, Y., Chen, C., Shi, J., Xie, L., Jiang, H., Bao, D., et al. (2018). A wrinkled PEDOT:PSS film based stretchable and transparent triboelectric nanogenerator for wearable energy harvesters and active motion sensors, *Advanced Functional Materials*, 28, 1803684.
197. Wang, X., Zhang, H., Dong, L., Han, X., Du, W., Zhai, J., Pan, C. and Wang, Z. L. (2016). Self-powered high-resolution and pressure-sensitive triboelectric sensor matrix for real-time tactile mapping, *Advanced Materials*, 28, 2896–2903.
198. Yun, J., Jayababu, N. and Kim, D. (2020). Self-powered transparent and flexible touchpad based on triboelectricity towards artificial intelligence, *Nano Energy*, 78, 105325.
199. Meng, B., Tang, W., Too, Z.-H., Zhang, X., Han, M., Liu, W. and Zhang, H. (2013). A transparent single-friction-surface triboelectric generator and self-powered touch sensor, *Energy & Environmental Science*, 6, 3235–3240.
200. Yang, Y., Zhang, H., Lin, Z.-H., Zhou, Y. S., Jing, Q., Su, Y., Yang, J., Chen, J., Hu, C. and Wang, Z. L. (2013). Human skin based triboelectric nanogenerators for harvesting biomechanical energy and as self-powered active tactile sensor system, *ACS Nano*, 7, 9213–9222.
201. Peng, W. and Wu, H. (2019). Flexible and stretchable photonic sensors based on modulation of light transmission, *Advanced Optical Materials*, 7, 1900329.
202. Xu, H., Liu, J., Zhang, J., Zhou, G., Luo, N. and Zhao, N. (2017). Flexible organic/inorganic hybrid near-infrared photoplethysmogram sensor for cardiovascular monitoring, *Advanced Materials*, 29, 1700975.
203. Khan, Y., Han, D., Pierre, A., Ting, J., Wang, X., Lochner, C. M., Bovo, G., Yaacobi-Gross, N., Newsome, C., Wilson, R., et al. (2018). A flexible organic reflectance oximeter array, *Proceedings of the National Academy of Sciences*, 115, E11015–E11024.
204. Park, S., Fukuda, K., Wang, M., Lee, C., Yokota, T., Jin, H., Jinno, H., Kimura, H., Zalar, P., Matsuhisa, N., et al. (2018). Ultraflexible near-infrared organic photodetectors for conformal photoplethysmogram sensors, *Advanced Materials*, 30, 1802359.
205. Park, J. B., Ha, J.-W., Yoon, S. C., Lee, C., Jung, I. H. and Hwang, D.-H. (2018). Visible-light-responsive high-detectivity organic photodetectors with a 1 μm thick active layer, *ACS Applied Materials & Interfaces*, 10, 38294–38301.
206. Qiu, L.-Z., Wei, S.-Y., Xu, H.-S., Zhang, Z.-X., Guo, Z.-Y., Chen, X.-G., Liu, S.-Y., Wu, D. and Luo, L.-B. (2020). Ultrathin polymer nanofibrils for solar-blind deep ultraviolet light photodetectors application, *Nano Letters*, 20, 644–651.

207. Sun, J., Liu, Y., Wei, J., Wei, P. and Chen, T. (2024). Pseudo-photoelectric cascade conversion endowing photosensitive Janus ionogel for solar energy harvesting and sensing, *Chemical Engineering Journal*, *485*, 149836.
208. Liu, J.-N., Du, K., Guo, J.-H., Wang, D., Gong, C.-B. and Tang, Q. (2024). Visual sensor with host–guest specific recognition and light–electrical co-controlled switch, *Small*, 2311823.
209. An, X., Liu, Y., Sun, Y., Zhang, X., Liu, Y., Tao, Y., Guo, L., Jiang, X. and Gao, M. (2024). Portable multifunctional sensing platform for ratiometric H_2O_2 detection and photodynamic anti-bacteria using an AIE-featured electrospinning film, *Chemical Engineering Journal*, *487*, 150675.
210. Ding, M., Jiang, T., Wang, B., Li, Y., Zhang, J., Huang, J., Ji, D. and Hu, W. (2024). Environmentally friendly and degradable organic neuromorphic vision sensors, *Matter*, *7*, 1736–1749.
211. Liu, Z., Lv, H., Li, S., Sun, Y., Chen, X. and Xu, Y. (2024). Dual regulation of hierarchical porosity and heterogeneous interfaces in Cu-BTC/Bi2MoO6 for thermally-driven and UV-light-activated selective acetone sensing, *Journal of Materials Chemistry A*, *12*, 6318–6328.
212. Zirkl, M., Haase, A., Fian, A., Schön, H., Sommer, C., Jakopic, G., Leising, G., Stadlober, B., Graz, I., Gaar, N., *et al.* (2007). Low-voltage organic thin-film transistors with high-nanocomposite gate dielectrics for flexible electronics and optothermal sensors, *Advanced Materials*, *19*, 2241–2245.
213. Zhang, Z., Geng, Y., Cao, S., Chen, Z., Gao, H., Zhu, X., Zhang, X. and Wu, Y. (2022). Ultraviolet photodetectors based on polymer microwire arrays toward wearable medical devices, *ACS Applied Materials & Interfaces*, *14*, 41257–41263.
214. Kumaar Swamy Reddy, B., Veeralingam, S., Borse, P. H. and Badhulika, S. (2022). A flexible, rapid response, hybrid inorganic–organic $SnSe_2$–PEDOT:PSS bulk heterojunction based high-performance broadband photodetector, *Materials Chemistry Frontiers*, *6*, 341–351.
215. Liu, C.-K., Tai, Q., Wang, N., Tang, G., Hu, Z. and Yan, F. (2020). Lead-free perovskite/organic semiconductor vertical heterojunction for highly sensitive photodetectors, *ACS Applied Materials & Interfaces*, *12*, 18769–18776.
216. Li, D., Du, J., Tang, Y., Liang, K., Wang, Y., Ren, H., Wang, R., Meng, L., Zhu, B. and Li, Y. (2021). Flexible and air-stable near-infrared sensors based on solution-processed inorganic–organic hybrid phototransistors, *Advanced Functional Materials*, *31*, 2105887.
217. Jun, S., Choi, K. W., Kim, K.-S., Kim, D. U., Lee, C.-J., Han, C. J., Lee, C.-R., Ju, B.-K. and Kim, J.-W. (2019). Stretchable photodetector utilizing the change in capacitance formed in a composite film containing semiconductor particles, *Composite Science and Technology*, *182*, 107773.
218. Zhao, X., Song, L., Zhao, R. and Tan, M. C. (2019). High-performance and flexible shortwave infrared photodetectors using composites of rare earth-doped nanoparticles, *ACS Applied Materials & Interfaces*, *11*, 2344–2351.

219. An, C., Nie, F., Zhang, R., Ma, X., Wu, D., Sun, Y., Hu, X., Sun, D., Pan, L. and Liu, J. (2021). Two-dimensional material-enhanced flexible and self-healable photodetector for large-area photodetection, *Advanced Functional Materials*, *31*, 2100136.
220. Hua, Q., Sun, J., Liu, H., Bao, R., Yu, R., Zhai, J., Pan, C. and Wang, Z. L. (2018). Skin-inspired highly stretchable and conformable matrix networks for multifunctional sensing, *Nature Communications*, *9*, 244.
221. Chang, H., Kim, S., Jin, S., Lee, S.-W., Yang, G.-T., Lee, K.-Y. and Yi, H. (2018). Ultrasensitive and highly stable resistive pressure sensors with biomaterial-incorporated interfacial layers for wearable health-monitoring and human–machine interfaces, *ACS Applied Materials & Interfaces*, *10*, 1067–1076.
222. Yi, P., Zou, H., Yu, Y., Li, X., Li, Z., Deng, G., Chen, C., Fang, M., He, J., Sun, X., et al. (2022). MXene-reinforced liquid metal/polymer fibers via interface engineering for wearable multifunctional textiles, *ACS Nano*, *16*, 14490–14502.
223. Liu, W., Duo, Y., Liu, J., Yuan, F., Li, L., Li, L., Wang, G., Chen, B., Wang, S., Yang, H., et al. (2022). Touchless interactive teaching of soft robots through flexible bimodal sensory interfaces, *Nature Communications*, *13*, 5030.
224. Wen, N., Zhang, L., Jiang, D., Wu, Z., Li, B., Sun, C. and Guo, Z. (2020). Emerging flexible sensors based on nanomaterials: Recent status and applications, *Journal of Materials Chemistry A*, *8*, 25499–25527.
225. Yang, M., Cheng, Y., Yue, Y., Chen, Y., Gao, H., Li, L., Cai, B., Liu, W., Wang, Z., Guo, H., et al. (2022). High-performance flexible pressure sensor with a self-healing function for tactile feedback, *Advanced Science*, *9*, 2200507.
226. Zhou, K., Xu, W., Yu, Y., Zhai, W., Yuan, Z., Dai, K., Zheng, G., Mi, L., Pan, C., Liu, C., et al. (2021). Tunable and nacre-mimetic multifunctional electronic skins for highly stretchable contact-noncontact sensing, *Small*, *17*, 2100542.
227. Li, X., Zhuang, Z., Qi, D. and Zhao, C. (2021). High sensitive and fast response humidity sensor based on polymer composite nanofibers for breath monitoring and non-contact sensing, *Sensors and Actuators B: Chemical*, *330*, 129239.
228. Zhou, L., Wang, M., Liu, Z., Guan, J., Li, T. and Zhang, D. (2021). High-performance humidity sensor based on graphitic carbon nitride/polyethylene oxide and construction of sensor array for non-contact humidity detection, *Sensors and Actuators B: Chemical*, *344*, 130219.
229. Luo, X., Wu, H., Wang, C., Jin, Q., Luo, C., Ma, G., Guo, W. and Long, Y. (2024). 3D printing of self-healing and degradable conductive ionoelastomers for customized flexible sensors, *Chemical Engineering Journal*, *483*, 149330.

230. Kim, T. H., Lee, J. H., Jang, M. H., Lee, G. M., Shim, E. S., Oh, S., Saeed, M. A., Lee, M. J., Yu, B.-S., Hwang, D. K., et al. (2024). Atto-scale noise near-infrared organic photodetectors enabled by controlling interfacial energetic offset through enhanced anchoring ability, *Advanced Materials*, 2403647.
231. Liu, H., Zhang, S., Li, Z., Lu, T. J., Lin, H., Zhu, Y., Ahadian, S., Emaminejad, S., Dokmeci, M. R., Xu, F., et al. (2021). Harnessing the wide-range strain sensitivity of bilayered PEDOT:PSS films for wearable health monitoring, *Matter*, 4, 2886–2901.
232. Lee, Y., Myoung, J., Cho, S., Park, J., Kim, J., Lee, H., Lee, Y., Lee, S., Baig, C. and Ko, H. (2021). Bioinspired gradient conductivity and stiffness for ultrasensitive electronic skins, *ACS Nano*, 15, 1795–1804.
233. Hansen, T. S., West, K., Hassager, O. and Larsen, N. B. (2007). Highly stretchable and conductive polymer material made from poly(3,4-ethylenedioxythiophene) and polyurethane elastomers, *Advanced Functional Materials*, 17, 3069–3073.
234. Teng, C., Lu, X., Zhu, Y., Wan, M. and Jiang, L. (2013). Polymer in situ embedding for highly flexible, stretchable and water stable PEDOT:PSS composite conductors, *RSC Advances*, 3, 7219–7223.
235. Oh, J. Y., Kim, S., Baik, H. K. and Jeong, U. (2016). Conducting polymer dough for deformable electronics, *Advanced Materials*, 28, 4455–4461.
236. Kee, S., Kim, N., Kim, B. S., Park, S., Jang, Y. H., Lee, S. H., Kim, J., Kim, J., Kwon, S. and Lee, K. (2016). Controlling molecular ordering in aqueous conducting polymers using ionic liquids, *Advanced Materials*, 28, 8625–8631.
237. Wang, P., Yu, W., Li, G., Meng, C. and Guo, S. (2023). Printable, flexible, breathable and sweatproof bifunctional sensors based on an all-nanofiber platform for fully decoupled pressure–temperature sensing application, *Chemical Engineering Journal*, 452, 139174.
238. Horev, Y. D., Maity, A., Zheng, Y., Milyutin, Y., Khatib, M., Yuan, M., Suckeveriene, R. Y., Tang, N., Wu, W. and Haick, H. (2021). Stretchable and highly permeable nanofibrous sensors for detecting complex human body motion, *Advanced Materials*, 33, 2102488.
239. Lu, Q., Lu, J., Sun, D. and Qiu, B. (2024). Fe3O4@PANI composite improves biotransformation of waste activated sludge into medium-chain fatty acid, *Advanced Composites and Hybrid Materials*, 7, 113.
240. Jamadi, F., Seyed-Yazdi, J., Ebrahimi-Tazangi, F. and Hosseiny, S. M. (2024). The impact of RGO and MWCNT/RGO on the microwave absorption of NiFe2O4@Fe3O4 in the presence or absence of PANI, *Journal of Materials Chemistry A*, 12, 32981–33002.
241. Zhang, F., Zang, Y., Huang, D., Di, C.-A. and Zhu, D. (2015). Flexible and self-powered temperature–pressure dual-parameter sensors using

microstructure-frame-supported organic thermoelectric materials, *Nature Communications*, *6*, 8356.
242. Li, M.-H., Ma, X., Fu, J., Wang, S., Wu, J., Long, R. and Hu, J.-S. (2024). Molecularly tailored perovskite/poly(3-hexylthiophene) interfaces for high-performance solar cells, *Energy & Environmental Science*, *17*, 5513–5520.
243. Baustert, K. N., Bombile, J. H., Rahman, M. T., Yusuf, A. O., Li, R., Huckaba, A. J., Risko, C. and Graham, K. R. (2024). Combination of counterion size and doping concentration determines the electronic and thermoelectric properties of semiconducting polymers, *Advanced Materials*, *36*, 2313863.
244. Sharma, S., Chhetry, A., Maharjan, P., Zhang, S., Shrestha, K., Sharifuzzaman, M., Bhatta, T., Shin, Y., Kim, D., Lee, S., et al. (2022). Polyaniline-nanospines engineered nanofibrous membrane based piezoresistive sensor for high-performance electronic skins, *Nano Energy*, *95*, 106970.
245. Cui, S., Zheng, Y., Zhang, T., Wang, D., Zhou, F. and Liu, W. (2018). Self-powered ammonia nanosensor based on the integration of the gas sensor and triboelectric nanogenerator, *Nano Energy*, *49*, 31–39.
246. Xiang, Y. and Chen, D. (2007). Preparation of a novel pH-responsive silver nanoparticle/poly(HEMA–PEGMA–MAA) composite hydrogel, *European Polymer Journal*, *43*, 4178–4187.
247. Devaki, S. J., Narayanan, R. K. and Sarojam, S. (2014). Electrically conducting silver nanoparticle–polyacrylic acid hydrogel by in situ reduction and polymerization approach, *Materials Letters*, *116*, 135–138.
248. Wei, Q.-B., Luo, Y.-L., Zhang, C.-H., Fan, L.-H. and Chen, Y.-S. (2008). Assembly of Cu nanoparticles in a polyacrylamide grafted poly(vinyl alcohol) copolymer matrix and vapor-induced response, *Sensors and Actuators B: Chemical*, *134*, 49–56.
249. Zhao, X., Ding, X., Deng, Z., Zheng, Z., Peng, Y. and Long, X. (2005). Thermoswitchable electronic properties of a gold nanoparticle/hydrogel composite, *Macromolecular Rapid Communications*, *26*, 1784–1787.
250. Baei, P., Jalili-Firoozinezhad, S., Rajabi-Zeleti, S., Tafazzoli-Shadpour, M., Baharvand, H. and Aghdami, N. (2016). Electrically conductive gold nanoparticle-chitosan thermosensitive hydrogels for cardiac tissue engineering, *Materials Science and Engineering: C*, *63*, 131–141.
251. Navaei, A., Saini, H., Christenson, W., Sullivan, R. T., Ros, R. and Nikkhah, M. (2016). Gold nanorod-incorporated gelatin-based conductive hydrogels for engineering cardiac tissue constructs, *Acta Biomaterialia*, *41*, 133–146.
252. Li, Q., Quan, X., Hu, R., Hu, Z., Xu, S., Liu, H., Zhou, X., Han, B. and Ji, X. (2024). A universal strategy for constructing hydrogel assemblies enabled by PAA hydrogel adhesive, *Small*, 2403844.

253. Han, Z., Lu, Y. and Qu, S. (2024). Design of fatigue-resistant hydrogels, *Advanced Functional Materials*, *34*, 2313498.
254. Hao, Z., Li, X., Zhang, R. and Zhang, L. (2024). Stimuli-responsive hydrogels for antibacterial applications, *Advanced Healthcare Materials*, 2400513.
255. Zhang, J., Wan, L., Gao, Y., Fang, X., Lu, T., Pan, L. and Xuan, F. (2019). Highly stretchable and self-healable MXene/polyvinyl alcohol hydrogel electrode for wearable capacitive electronic skin, *Advanced Electronic Materials*, *5*, 1900285.
256. Gotovtsev, P. M., Badranova, G. U., Zubavichus, Y. V., Chumakov, N. K., Antipova, C. G., Kamyshinsky, R. A., Presniakov, M. Y., Tokaev, K. V. and Grigoriev, T. E. (2019). Electroconductive PEDOT:PSS-based hydrogel prepared by freezing-thawing method, *Heliyon*, *5*, e02498.
257. Mawad, D., Artzy-Schnirman, A., Tonkin, J., Ramos, J., Inal, S., Mahat, M. M., Darwish, N., Zwi-Dantsis, L., Malliaras, G. G., Gooding, J. J., et al. (2016). Electroconductive hydrogel based on functional poly(ethylenedioxy thiophene), *Chemistry of Materials*, *28*, 6080–6088.
258. Lu, B., Yuk, H., Lin, S., Jian, N., Qu, K., Xu, J. and Zhao, X. (2019). Pure PEDOT:PSS hydrogels, *Nature Communications*, *10*, 1043.
259. Han, L., Liu, K., Wang, M., Wang, K., Fang, L., Chen, H., Zhou, J. and Lu, X. (2018). Mussel-inspired adhesive and conductive hydrogel with long-lasting moisture and extreme temperature tolerance, *Advanced Functional Materials*, *28*, 1704195.
260. Charaya, H., La, T.-G., Rieger, J. and Chung, H.-J. (2019). Thermochromic and piezocapacitive flexible sensor array by combining composite elastomer dielectrics and transparent ionic hydrogel electrodes, *Advanced Materials Technologies*, *4*, 1900327.
261. Wasserscheid, P. and Keim, W. (2000). Ionic liquids — new "solutions" for transition metal catalysis, *Angewandte Chemie International Edition*, *39*, 3772–3789.
262. Huang, Y., Fan, X., Chen, S.-C. and Zhao, N. (2019). Emerging technologies of flexible pressure sensors: Materials, modeling, devices, and manufacturing, *Advanced Functional Materials*, *29*, 1808509.
263. Matuszek, K., Piper, S. L., Brzęczek-Szafran, A., Roy, B., Saher, S., Pringle, J. M. and MacFarlane, D. R. (2024). Unexpected energy applications of ionic liquids, *Advanced Materials*, *36*, 2313023.
264. Yu, G., Dai, C., Liu, N., Xu, R., Wang, N. and Chen, B. (2024). Hydrocarbon extraction with ionic liquids, *Chemical Reviews*, *124*, 3331–3391.
265. Chang, Y., Wang, L., Li, R., Zhang, Z., Wang, Q., Yang, J., Guo, C. F. and Pan, T. (2021). First decade of interfacial iontronic sensing: From droplet sensors to artificial skins, *Advanced Materials*, *33*, 2003464.

266. Zhao, C., Wang, Y., Tang, G., Ru, J., Zhu, Z., Li, B., Guo, C. F., Li, L. and Zhu, D. (2022). Ionic flexible sensors: Mechanisms, materials, structures, and applications, *Advanced Functional Materials*, *32*, 2110417.
267. Kwon, J. H., Kim, Y. M. and Moon, H. C. (2021). Porous ion gel: A versatile ionotronic sensory platform for high-performance, wearable ionoskins with electrical and optical dual output, *ACS Nano*, *15*, 15132–15141.
268. Shen, Z., Zhu, X., Majidi, C. and Gu, G. (2021). Cutaneous ionogel mechanoreceptors for soft machines, physiological sensing, and amputee prostheses, *Advanced Materials*, *33*, 2102069.
269. Huang, Z., Chen, X., O'Neill, S. J. K., Wu, G., Whitaker, D. J., Li, J., McCune, J. A. and Scherman, O. A. (2022). Highly compressible glass-like supramolecular polymer networks, *Nature Materials*, *21*, 103–109.
270. Bai, N., Wang, L., Xue, Y., Wang, Y., Hou, X., Li, G., Zhang, Y., Cai, M., Zhao, L., Guan, F., et al. (2022). Graded interlocks for iontronic pressure sensors with high sensitivity and high linearity over a broad range, *ACS Nano*, *16*, 4338–4347.
271. Shao, B., Zhang, S., Hu, Y., Zheng, Z., Zhu, H., Wang, L., Zhao, L., Xu, F., Wang, L., Li, M., et al. (2024). Color-shifting iontronic skin for on-site, nonpixelated pressure mapping visualization, *Nano Letters*, *24*, 4741–4748.
272. Yang, J., Li, Z., Wu, Y., Shen, Y., Zhang, M., Chen, B., Yuan, G., Xiao, S., Feng, J., Zhang, X., et al. (2024). Non-equilibrium compression achieving high sensitivity and linearity for iontronic pressure sensors, *Science Bulletin*, *69*, 2221–2230.
273. Zhi, X., Ma, S., Xia, Y., Yang, B., Zhang, S., Liu, K., Li, M., Li, S., Peiyuan, W. and Wang, X. (2024). Hybrid tactile sensor array for pressure sensing and tactile pattern recognition, *Nano Energy*, *125*, 109532.
274. Zhang, P., Chen, Y., Guo, Z. H., Guo, W., Pu, X. and Wang, Z. L. (2020). Stretchable, transparent, and thermally stable triboelectric nanogenerators based on solvent-free ion-conducting elastomer electrodes, *Advanced Functional Materials*, *30*, 1909252.
275. Lin, M.-F., Xiong, J., Wang, J., Parida, K. and Lee, P. S. (2018). Core-shell nanofiber mats for tactile pressure sensor and nanogenerator applications, *Nano Energy*, *44*, 248–255.
276. Kim, Y., Lee, D., Seong, J., Bak, B., Choi, U. H. and Kim, J. (2021). Ionic liquid-based molecular design for transparent, flexible, and fire-retardant triboelectric nanogenerator (TENG) for wearable energy solutions, *Nano Energy*, *84*, 105925.
277. Nie, B., Li, R., Cao, J., Brandt, J. D. and Pan, T. (2015). Flexible transparent iontronic film for interfacial capacitive pressure sensing, *Advanced Materials*, *27*, 6055–6062.

278. Jin, M. L., Park, S., Lee, Y., Lee, J. H., Chung, J., Kim, J. S., Kim, J.-S., Kim, S. Y., Jee, E., Kim, D. W., et al. (2017). Artificial skin: An ultrasensitive, visco-poroelastic artificial mechanotransducer skin inspired by piezoprotein in mammalian merkel cells, *Advanced Materials*, *29*, 1605973.
279. Wang, J., Xiong, Z., Wu, L., Chen, J. and Zhu, Y. (2024). Highly sensitive and wide-range iontronic pressure sensors with a wheat awn-like hierarchical structure, *Journal of Colloid and Interface Science*, *669*, 190–197.
280. Han, C., Cao, Z., Hu, Y., Zhang, Z., Li, C., Wang, Z. L. and Wu, Z. (2024). Flexible tactile sensors for 3D force detection, *Nano Letters*, *24*, 5277–5283.
281. Nie, B., Xing, S., Brandt, J. D. and Pan, T. (2012). Droplet-based interfacial capacitive sensing, *Lab on a Chip*, *12*, 1110–1118.
282. Nie, B., Li, R., Brandt, J. D. and Pan, T. (2014). Iontronic microdroplet array for flexible ultrasensitive tactile sensing, *Lab on a Chip*, *14*, 1107–1116.
283. Yang, Z., Zhao, Y., Lan, Y., Xiang, M., Wu, G., Zang, J., Zhang, Z., Xue, C. and Gao, L. (2024). Screen-printable iontronic pressure sensor with thermal expansion microspheres for pulse monitoring. *ACS Applied Materials & Interfaces*, *16*, 39561–39571.
284. Ding, Z., Li, W., Wang, W., Zhao, Z., Zhu, Y., Hou, B., Zhu, L., Chen, M. and Che, L. (2024). Highly sensitive iontronic pressure sensor with side-by-side package based on alveoli and arch structure, *Advanced Science*, *11*, 2309407.
285. Kim, J. S., Lee, S. C., Hwang, J., Lee, E., Cho, K., Kim, S.-J., Kim, D. H. and Lee, W. H. (2020). Enhanced sensitivity of iontronic graphene tactile sensors facilitated by spreading of ionic liquid pinned on graphene grid, *Advanced Functional Materials*, *30*, 1908993.
286. Chen, W., Liu, L.-X., Zhang, H.-B. and Yu, Z.-Z. (2021). Kirigami-inspired highly stretchable, conductive, and hierarchical Ti3C2Tx MXene films for efficient electromagnetic interference shielding and pressure sensing, *ACS Nano*, *15*, 7668–7681.
287. Qiu, Z., Wan, Y., Zhou, W., Yang, J., Yang, J., Huang, J., Zhang, J., Liu, Q., Huang, S., Bai, N., et al. (2018). Ionic skin with biomimetic dielectric layer templated from calathea zebrine leaf, *Advanced Functional Materials*, *28*, 1802343.
288. Bai, N., Wang, L., Wang, Q., Deng, J., Wang, Y., Lu, P., Huang, J., Li, G., Zhang, Y., Yang, J., et al. (2020). Graded intrafillable architecture-based iontronic pressure sensor with ultra-broad-range high sensitivity, *Nature Communications*, *11*, 209.
289. Huang, Z., Yu, S., Xu, Y., Cao, Z., Zhang, J., Guo, Z., Wu, T., Liao, Q., Zheng, Y., Chen, Z., et al. (2024). In-sensor tactile fusion and logic for accurate intention recognition, *Advanced Materials*, 2407329.
290. Noor, A., Sun, M., Zhang, X., Li, S., Dong, F., Wang, Z., Si, J., Zou, Y. and Xu, M. (2024). Recent advances in triboelectric tactile sensors for robot hand, *Materials Today Physics*, *46*, 101496.

291. Li, B., Ge, R., Du, W., Wang, Z., Peng, T., Wang, R., Chang, Y. and Pan, T. (2024). iWood: An intelligent iontronic device for human-wood interactions, *Advanced Functional Materials*, *34*, 2314190.
292. Su, Q., Zou, Q., Li, Y., Chen, Y., Teng, S.-Y., Kelleher, J. T., Nith, R., Cheng, P., Li, N., Liu, W., et al. A stretchable and strain-unperturbed pressure sensor for motion interference–free tactile monitoring on skins, *Science Advances*, *7*, eabi4563.
293. Wang, P., Li, X., Sun, G., Wang, G., Han, Q., Meng, C., Wei, Z. and Li, Y. (2024). Natural human skin-inspired wearable and breathable nanofiber-based sensors with excellent thermal management functionality, *Advanced Fiber Materials*, *6*, 1955–1968.
294. Amoli, V., Kim, J. S., Jee, E., Chung, Y. S., Kim, S. Y., Koo, J., Choi, H., Kim, Y. and Kim, D. H. (2019). A bioinspired hydrogen bond-triggered ultrasensitive ionic mechanoreceptor skin, *Nature Communications*, *10*, 4019.
295. Guo, Y., Li, H., Li, Y., Wei, X., Gao, S., Yue, W., Zhang, C., Yin, F., Zhao, S., Kim, N.-Y., et al. (2022). Wearable hybrid device capable of interactive perception with pressure sensing and visualization, *Advanced Functional Materials*, *32*, 2203585.
296. Wang, Y., Chao, M., Wan, P. and Zhang, L. (2020). A wearable breathable pressure sensor from metal-organic framework derived nanocomposites for highly sensitive broad-range healthcare monitoring, *Nano Energy*, *70*, 104560.
297. Zhou, K., Zhang, C., Xiong, Z., Chen, H.-Y., Li, T., Ding, G., Yang, B., Liao, Q., Zhou, Y. and Han, S.-T. (2020). Template-directed growth of hierarchical MOF hybrid arrays for tactile sensor, *Advanced Functional Materials*, *30*, 2001296.
298. Sun, J., Tu, K., Büchele, S., Koch, S. M., Ding, Y., Ramakrishna, S. N., Stucki, S., Guo, H., Wu, C., Keplinger, T., et al. (2021). Functionalized wood with tunable tribopolarity for efficient triboelectric nanogenerators, *Matter*, *4*, 3049–3066.
299. Jo, Y. K., Jeong, S.-Y., Moon, Y. K., Jo, Y.-M., Yoon, J.-W. and Lee, J.-H. (2021). Exclusive and ultrasensitive detection of formaldehyde at room temperature using a flexible and monolithic chemiresistive sensor, *Nature Communications*, *12*, 4955.
300. Gong, S., Yap, L. W., Zhu, B., Zhai, Q., Liu, Y., Lyu, Q., Wang, K., Yang, M., Ling, Y., Lai, D. T. H., et al. (2019). Local crack-programmed gold nanowire electronic skin tattoos for in-plane multisensor integration, *Advanced Materials*, *31*, 1903789.
301. He, H., Liu, J., Wang, Y., Zhao, Y., Qin, Y., Zhu, Z., Yu, Z. and Wang, J. (2022). An ultralight self-powered fire alarm e-textile based on conductive aerogel fiber with repeatable temperature monitoring performance used in firefighting clothing, *ACS Nano*, *16*, 2953–2967.

302. Zhou, Q., Kim, J.-N., Han, K.-W., Oh, S.-W., Umrao, S., Chae, E. J. and Oh, I.-K. (2019). Integrated dielectric-electrode layer for triboelectric nanogenerator based on Cu nanowire-Mesh hybrid electrode, *Nano Energy, 59,* 120–128.
303. Chen, L., Xie, S., Lan, J., Chai, J., Lin, T., Hao, Q., Chen, J., Deng, X., Hu, X., Li, Y., *et al.* (2024). High-speed and high-responsivity blue light photodetector with an InGaN NR/PEDOT:PSS heterojunction decorated with Ag NWs, *ACS Applied Materials & Interfaces, 16,* 29477–29487.
304. Yan, T., Yang, W., Wu, L. and Fang, X. (2025). High-work-function transparent electrode with an enhanced air-stable conductivity based on AgNiCu core-shell nanowires for Schottky photodiode, *Journal of Materials Science & Technology, 209,* 95–102.
305. Kuo, Y.-C., Fan, J., Zong, L., Chen, F., Feng, Z., Liu, C., Wan, T., Gu, Z., Hu, L., Guan, P., *et al.* (2024). Rational design of robust Cu@Ag core-shell nanowires for wearable electronics applications, *Chemical Engineering Journal, 496,* 154001.
306. Choi, S., Yoon, K., Lee, S., Lee, H. J., Lee, J., Kim, D. W., Kim, M.-S., Lee, T. and Pang, C. (2019). Conductive hierarchical hairy fibers for highly sensitive, stretchable, and water-resistant multimodal gesture-distinguishable sensor, VR applications, *Advanced Functional Materials, 29,* 1905808.
307. Wang, Y., Liu, Q., Zhang, J., Hong, T., Sun, W., Tang, L., Arnold, E., Suo, Z., Hong, W., Ren, Z., *et al.* (2019). Giant poisson's effect for wrinkle-free stretchable transparent electrodes, *Advanced Materials, 31,* 1902955.
308. Park, J. H., Hwang, G.-T., Kim, S., Seo, J., Park, H.-J., Yu, K., Kim, T.-S. and Lee, K. J. (2017). Flash-induced self-limited plasmonic welding of silver nanowire network for transparent flexible energy harvester, *Advanced Materials, 29,* 1603473.
309. Ge, Y., Duan, X., Zhang, M., Mei, L., Hu, J., Hu, W. and Duan, X. (2018). Direct room temperature welding and chemical protection of silver nanowire thin films for high performance transparent conductors, *Journal of the American Chemical Society, 140,* 193–199.
310. Sun, Y., Chang, M., Meng, L., Wan, X., Gao, H., Zhang, Y., Zhao, K., Sun, Z., Li, C., Liu, S., *et al.* (2019). Flexible organic photovoltaics based on water-processed silver nanowire electrodes, *Nature Electronics, 2,* 513–520.
311. Meng, L., Bian, R., Guo, C., Xu, B., Liu, H. and Jiang, L. (2018). Aligning Ag nanowires by a facile bioinspired directional liquid transfer: Toward anisotropic flexible conductive electrodes, *Advanced Materials, 30,* 1706938.
312. Xiong, J., Li, S., Ye, Y., Wang, J., Qian, K., Cui, P., Gao, D., Lin, M.-F., Chen, T. and Lee, P. S. (2018). A deformable and highly robust ethyl cellulose transparent conductor with a scalable silver nanowires bundle micromesh, *Advanced Materials, 30,* 1802803.

313. Hu, L., Kim, H. S., Lee, J.-Y., Peumans, P. and Cui, Y. (2010). Scalable coating and properties of transparent, flexible, silver nanowire electrodes, *ACS Nano*, *4*, 2955–2963.
314. Yu, C., Li, H., Ding, K., Huang, L., Zhang, H., Pang, D., Xiong, Y., Yang, P.-A., Fang, L., Li, W., *et al.* (2024). Flexible and self-powered photoelectrochemical-type solar-blind photodetectors based on Ag nanowires-embedded amorphous Ga_2O_3 films, *Advanced Optical Materials*, *12*, 2400116.
315. Zhang, J., Xu, J., Gao, Y., Qin, F., Zhu, X. and Kan, C. (2024). Flexible silver nanowire and Ti_3C_2Tx MXene composite films for electromagnetic interference shielding, *ACS Applied Nano Materials*, *7*, 4960–4968.
316. Han, J., Yang, J., Gao, W. and Bai, H. (2021). Ice-templated, large-area silver nanowire pattern for flexible transparent electrode, *Advanced Functional Materials*, *31*, 2010155.
317. Wang, Y., Gong, S., Wang, S. J., Yang, X., Ling, Y., Yap, L. W., Dong, D., Simon, G. P. and Cheng, W. (2018). Standing enokitake-like nanowire films for highly stretchable elastronics, *ACS Nano*, *12*, 9742–9749.
318. Choi, S., Han, S. I., Jung, D., Hwang, H. J., Lim, C., Bae, S., Park, O. K., Tschabrunn, C. M., Lee, M., Bae, S. Y., *et al.* (2018). Highly conductive, stretchable and biocompatible Ag–Au core–sheath nanowire composite for wearable and implantable bioelectronics, *Nature Nanotechnology*, *13*, 1048–1056.
319. Sun, K., Ko, H., Park, H.-H., Seong, M., Lee, S.-H., Yi, H., Park, H. W., Kim, T.-I., Pang, C. and Jeong, H. E. (2018). Hybrid architectures of heterogeneous carbon nanotube composite microstructures enable multiaxial strain perception with high sensitivity and ultrabroad sensing range, *Small*, *14*, 1803411.
320. Mu, C., Song, Y., Huang, W., Ran, A., Sun, R., Xie, W. and Zhang, H. (2018). Flexible normal-tangential force sensor with opposite resistance responding for highly sensitive artificial skin, *Advanced Functional Materials*, *28*, 1707503.
321. Li, Y., Zheng, C., Liu, S., Huang, L., Fang, T., Li, J. X., Xu, F. and Li, F. (2020). Smart glove integrated with tunable MWNTs/PDMS fibers made of a one-step extrusion method for finger dexterity, gesture, and temperature recognition, *ACS Applied Materials & Interfaces*, *12*, 23764–23773.
322. Gao, Z., Lou, Z., Han, W. and Shen, G. (2020). A self-healable bifunctional electronic skin, *ACS Applied Materials & Interfaces*, *12*, 24339–24347.
323. Niu, H., Yue, W., Li, Y., Yin, F., Gao, S., Zhang, C., Kan, H., Yao, Z., Jiang, C. and Wang, C. (2021). Ultrafast-response/recovery capacitive humidity sensor based on arc-shaped hollow structure with nanocone arrays for human physiological signals monitoring, *Sensors and Actuators B: Chemical*, *334*, 129637.

324. Ma, L., Wu, R., Patil, A., Zhu, S., Meng, Z., Meng, H., Hou, C., Zhang, Y., Liu, Q., Yu, R., *et al.* (2019). Full-textile wireless flexible humidity sensor for human physiological monitoring, *Advanced Functional Materials*, *29*, 1904549.
325. Wu, J., Sun, Y.-M., Wu, Z., Li, X., Wang, N., Tao, K. and Wang, G. P. (2019). Carbon nanocoil-based fast-response and flexible humidity sensor for multifunctional applications, *ACS Applied Materials & Interfaces*, *11*, 4242–4251.
326. Yu, X., Shi, Z., Xiong, C. and Yang, Q. (2024). Nanopaper-based sensors with ultrahigh and stable conductance for wearable sensors and heaters, *Chemical Engineering Journal*, *493*, 152797.
327. Ahmadi, N., Lee, J., Godiya, C. B., Kim, J.-M. and Park, B. J. (2024). A single-particle mechanofluorescent sensor, *Nature Communications*, *15*, 6094.
328. Qiao, Y., Elhady, A., Arabi, M., Abdel-Rahman, E. and Zhang, W. (2024). Thermal noise-driven resonant sensors, *Microsystems & Nanoengineering*, *10*, 90.
329. Hu, Y., Chatzilakou, E., Pan, Z., Traverso, G. and Yetisen, A. K. (2024). Microneedle sensors for point-of-care diagnostics, *Advanced Science*, *11*, 2306560.
330. Miao, L., Wan, J., Song, Y., Guo, H., Chen, H., Cheng, X. and Zhang, H. (2019). Skin-inspired humidity and pressure sensor with a wrinkle-on-sponge structure, *ACS Applied Materials & Interfaces*, *11*, 39219–39227.
331. Qi, K., He, J., Wang, H., Zhou, Y., You, X., Nan, N., Shao, W., Wang, L., Ding, B. and Cui, S. (2017). A highly stretchable nanofiber-based electronic skin with pressure-, strain-, and flexion-sensitive properties for health and motion monitoring, *ACS Applied Materials & Interfaces*, *9*, 42951–42960.
332. Kim, S. J., Mondal, S., Min, B. K. and Choi, C.-G. (2018). Highly sensitive and flexible strain–pressure sensors with cracked paddy-shaped MoS2/graphene foam/ecoflex hybrid nanostructures, *ACS Applied Materials & Interfaces*, *10*, 36377–36384.
333. Fu, M., Zhang, J., Jin, Y., Zhao, Y., Huang, S. and Guo, C. F. (2020). A highly sensitive, reliable, and high-temperature-resistant flexible pressure sensor based on ceramic nanofibers, *Advanced Science*, *7*, 2000258.
334. Ha, M., Lim, S., Park, J., Um, D.-S., Lee, Y. and Ko, H. (2015). Bioinspired interlocked and hierarchical design of ZnO nanowire arrays for static and dynamic pressure-sensitive electronic skins, *Advanced Functional Materials*, *25*, 2841–2849.
335. Novoselov, K. S., Geim, A. K., Morozov, S. V., Jiang, D., Zhang, Y., Dubonos, S. V., Grigorieva, I. V. and Firsov, A. A. (2004). Electric field effect in atomically thin carbon films, *Science*, *306*, 666–669.

336. Fu, W., Jiang, L., van Geest, E. P., Lima, L. M. C. and Schneider, G. F. (2017). Sensing at the surface of graphene field-effect transistors, *Advanced Materials*, *29*, 1603610.
337. Singh, E., Meyyappan, M. and Nalwa, H. S. (2017). Flexible graphene-based wearable gas and chemical sensors, *ACS Applied Materials & Interfaces*, *9*, 34544–34586.
338. Suvarnaphaet, P. and Pechprasarn, S. (2017). Graphene-based materials for biosensors: A review. In *Sensors*, 2017; Vol. 17, p. 2161.
339. Tan, R. K. L., Reeves, S. P., Hashemi, N., Thomas, D. G., Kavak, E., Montazami, R. and Hashemi, N. N. (2017). Graphene as a flexible electrode: Review of fabrication approaches, *Journal of Materials Chemistry A*, *5*, 17777–17803.
340. Ahn, J., Jeong, Y., Kang, M., Ahn, J., Padmajan Sasikala, S., Yang, I., Ha, J.-H., Hwang, S. H., Jeon, S., Gu, J., et al. (2024). Nanoribbon yarn with versatile inorganic materials, *Small*, 2311736.
341. Qiu, P., Deng, T., Chen, L. and Shi, X. (2024). Plastic inorganic thermoelectric materials, *Joule*, *8*, 622–634.
342. Shi, J., Sun, X., Song, L., Hong, M., Yuan, Q. and Zhang, Y. (2024). Inorganic persistent luminescence materials: Emerging optical theranostic agents, *Progress in Materials Science*, *142*, 101246.
343. Fu, Y.-F., Li, Y.-Q., Liu, Y.-F., Huang, P., Hu, N. and Fu, S.-Y. (2018). High-performance structural flexible strain sensors based on graphene-coated glass fabric/silicone composite, *ACS Applied Materials & Interfaces*, *10*, 35503–35509.
344. Yin, F., Li, X., Peng, H., Li, F., Yang, K. and Yuan, W. (2019). A highly sensitive, multifunctional, and wearable mechanical sensor based on RGO/synergetic fiber bundles for monitoring human actions and physiological signals, *Sensors and Actuators B: Chemical*, *285*, 179–185.
345. Li, M., Wu, C., Zhao, S., Deng, T., Wang, J., Liu, Z., Wang, L. and Wang, G. (2018). Pressure sensing element based on the BN–graphene–BN heterostructure, *Applied Physics Letters*, *112*, 143502.
346. Pyo, S., Choi, J. and Kim, J. (2018). Flexible, transparent, sensitive, and crosstalk-free capacitive tactile sensor array based on graphene electrodes and air dielectric, *Advanced Electronic Materials*, *4*, 1700427.
347. Torres Alonso, E., Rodrigues, D. P., Khetani, M., Shin, D.-W., De Sanctis, A., Joulie, H., de Schrijver, I., Baldycheva, A., Alves, H., Neves, A. I. S., et al. (2018). Graphene electronic fibres with touch-sensing and light-emitting functionalities for smart textiles, *npj Flexible Electronics*, *2*, 25.
348. Wang, Q., Ling, S., Liang, X., Wang, H., Lu, H. and Zhang, Y. (2019). Self-healable multifunctional electronic tattoos based on silk and graphene, *Advanced Functional Materials*, *29*, 1808695.

349. Ho, D. H., Sun, Q., Kim, S. Y., Han, J. T., Kim, D. H. and Cho, J. H. (2016). Stretchable and multimodal all graphene electronic skin, *Advanced Materials*, *28*, 2601–2608.
350. Fei, H., Dong, J., Chen, D., Hu, T., Duan, X., Shakir, I., Huang, Y. and Duan, X. (2019). Single atom electrocatalysts supported on graphene or graphene-like carbons, *Chemical Society Reviews*, *48*, 5207–5241.
351. Cui, G., Bi, Z., Zhang, R., Liu, J., Yu, X. and Li, Z. (2019). A comprehensive review on graphene-based anti-corrosive coatings, *Chemical Engineering Journal*, *373*, 104–121.
352. Ye, J., Tan, H., Wu, S., Ni, K., Pan, F., Liu, J., Tao, Z., Qu, Y., Ji, H., Simon, P., et al. (2018). Direct laser writing of graphene made from chemical vapor deposition for flexible, integratable micro-supercapacitors with ultrahigh power output, *Advanced Materials*, *30*, 1801384.
353. Yuan, W., Yang, K., Peng, H., Li, F. and Yin, F. (2018). A flexible VOCs sensor based on a 3D Mxene framework with a high sensing performance, *Journal of Materials Chemistry A*, *6*, 18116–18124.
354. Xu, H., Xiang, J. X., Lu, Y. F., Zhang, M. K., Li, J. J., Gao, B. B., Zhao, Y. J. and Gu, Z. Z. (2018). Multifunctional wearable sensing devices based on functionalized graphene films for simultaneous monitoring of physiological signals and volatile organic compound biomarkers, *ACS Applied Materials & Interfaces*, *10*, 11785–11793.
355. Ho, D. H., Choi, Y. Y., Jo, S. B., Myoung, J.-M. and Cho, J. H. (2021). Sensing with MXenes: Progress and prospects, *Advanced Materials*, *33*, 2005846.
356. Naguib, M., Kurtoglu, M., Presser, V., Lu, J., Niu, J., Heon, M., Hultman, L., Gogotsi, Y. and Barsoum, M. W. (2011). Two-dimensional nanocrystals produced by exfoliation of Ti_3AlC_2, *Advanced Materials*, *23*, 4248–4253.
357. Zhang, Y., Wang, L., Zhao, L., Wang, K., Zheng, Y., Yuan, Z., Wang, D., Fu, X., Shen, G. and Han, W. (2021). Flexible self-powered integrated sensing system with 3D periodic ordered black phosphorus@MXene thin-films, *Advanced Materials*, *33*, 2007890.
358. Wang, P., Liu, G., Sun, G., Meng, C., Shen, G. and Li, Y. (2024). An integrated bifunctional pressure–temperature sensing system fabricated on a breathable nanofiber and powered by rechargeable zinc–air battery for long-term comfortable health care monitoring, *Advanced Fiber Materials*, *6*, 1037–1052.
359. Lee, E., VahidMohammadi, A., Prorok, B. C., Yoon, Y. S., Beidaghi, M. and Kim, D.-J. (2017). Room temperature gas sensing of two-dimensional titanium carbide (MXene), *ACS Applied Materials & Interfaces*, *9*, 37184–37190.
360. Koh, H.-J., Kim, S. J., Maleski, K., Cho, S.-Y., Kim, Y.-J., Ahn, C. W., Gogotsi, Y. and Jung, H.-T. (2019). Enhanced selectivity of MXene gas sensors through metal ion intercalation: *In situ* x-ray diffraction study, *ACS Sensors*, *4*, 1365–1372.

361. Kim, S. J., Koh, H.-J., Ren, C. E., Kwon, O., Maleski, K., Cho, S.-Y., Anasori, B., Kim, C.-K., Choi, Y.-K., Kim, J., et al. (2018). Metallic Ti$_3$C$_2$Tx MXene gas sensors with ultrahigh signal-to-noise ratio, *ACS Nano*, *12*, 986–993.
362. Chen, W. Y., Jiang, X., Lai, S.-N., Peroulis, D. and Stanciu, L. (2020). Nanohybrids of a MXene and transition metal dichalcogenide for selective detection of volatile organic compounds, *Nature Communications*, *11*, 1302.
363. Zhao, W.-N., Yun, N., Dai, Z.-H. and Li, Y.-F. (2020). A high-performance trace level acetone sensor using an indispensable V$_4$C$_3$Tx MXene, *RSC Advances*, *10*, 1261–1270.
364. Dai, M., Zheng, W., Zhang, X., Wang, S., Lin, J., Li, K., Hu, Y., Sun, E., Zhang, J., Qiu, Y., et al. (2020). Enhanced piezoelectric effect derived from grain boundary in MoS2 monolayers, *Nano Letters*, *20*, 201–207.
365. Naqi, M., Kim, B., Kim, S.-W. and Kim, S. (2021). Pulsed gate switching of MoS2 field-effect transistor based on flexible polyimide substrate for ultrasonic detectors, *Advanced Functional Materials*, *31*, 2007389.
366. Dai, M., Wang, Z., Wang, F., Qiu, Y., Zhang, J., Xu, C.-Y., Zhai, T., Cao, W., Fu, Y., Jia, D., et al. (2019). Two-dimensional van der Waals materials with aligned in-plane polarization and large piezoelectric effect for self-powered piezoelectric sensors, *Nano Letters*, *19*, 5410–5416.
367. Cai, W., Wang, J., He, Y., Liu, S., Xiong, Q., Liu, Z. and Zhang, Q. (2021). Strain-modulated photoelectric responses from a flexible α-In2Se3/3R MoS2 heterojunction, *Nano-Micro Letters*, *13*, 74.
368. Li, L., Lou, Z. and Shen, G. (2018). Flexible broadband image sensors with SnS quantum dots/Zn$_2$SnO$_4$ nanowires hybrid nanostructures, *Advanced Functional Materials*, *28*, 1705389.
369. Wang, Y.-Y., Chen, D.-R., Wu, J.-K., Wang, T.-H., Chuang, C., Huang, S.-Y., Hsieh, W.-P., Hofmann, M., Chang, Y.-H. and Hsieh, Y.-P. (2021). Two-dimensional mechano-thermoelectric heterojunctions for self-powered strain sensors, *Nano Letters*, *21*, 6990–6997.
370. Zhang, H., Han, W., Xu, K., Zhang, Y., Lu, Y., Nie, Z., Du, Y., Zhu, J. and Huang, W. (2020). Metallic sandwiched-aerogel hybrids enabling flexible and stretchable intelligent sensor, *Nano Letters*, *20*, 3449–3458.
371. Cao, V. A., Kim, M., Hu, W., Lee, S., Youn, S., Chang, J., Chang, H. S. and Nah, J. (2021). Enhanced piezoelectric output performance of the SnS2/SnS heterostructure thin-film piezoelectric nanogenerator realized by atomic layer deposition, *ACS Nano*, *15*, 10428–10436.
372. Wan, S., Bi, H., Zhou, Y., Xie, X., Su, S., Yin, K. and Sun, L. (2017). Graphene oxide as high-performance dielectric materials for capacitive pressure sensors, *Carbon*, *114*, 209–216.
373. Zhu, L., Wang, Y., Mei, D. and Wu, X. (2019). Highly sensitive and flexible tactile sensor based on porous graphene sponges for distributed tactile

sensing in monitoring human motions, *Journal of Microelectromechanical Systems*, 28, 154–163.
374. Zhuo, H., Hu, Y., Tong, X., Chen, Z., Zhong, L., Lai, H., Liu, L., Jing, S., Liu, Q., Liu, C., et al. (2018). A supercompressible, elastic, and bendable carbon aerogel with ultrasensitive detection limits for compression strain, pressure, and bending angle, *Advanced Materials*, 30, 1706705.
375. Wei, Q., Chen, G., Pan, H., Ye, Z., Au, C., Chen, C., Zhao, X., Zhou, Y., Xiao, X., Tai, H., et al. (2022). MXene-sponge based high-performance piezoresistive sensor for wearable biomonitoring and real-time tactile sensing, *Small Methods*, 6, 2101051.
376. Zhang, B.-X., Hou, Z.-L., Yan, W., Zhao, Q.-L. and Zhan, K.-T. (2017). Multi-dimensional flexible reduced graphene oxide/polymer sponges for multiple forms of strain sensors, *Carbon*, 125, 199–206.
377. Park, H., Kim, J. W., Hong, S. Y., Lee, G., Kim, D. S., Oh, J. H., Jin, S. W., Jeong, Y. R., Oh, S. Y., Yun, J. Y., et al. (2018). Microporous polypyrrole-coated graphene foam for high-performance multifunctional sensors and flexible supercapacitors, *Advanced Functional Materials*, 28, 1707013.
378. Zhang, S., Liu, H., Yang, S., Shi, X., Zhang, D., Shan, C., Mi, L., Liu, C., Shen, C. and Guo, Z. (2019). Ultrasensitive and highly compressible piezoresistive sensor based on polyurethane sponge coated with a cracked cellulose nanofibril/silver nanowire layer, *ACS Applied Materials & Interfaces*, 11, 10922–10932.
379. Pang, Y., Tian, H., Tao, L., Li, Y., Wang, X., Deng, N., Yang, Y. and Ren, T.-L. (2016). Flexible, highly sensitive, and wearable pressure and strain sensors with graphene porous network structure, *ACS Applied Materials & Interfaces*, 8, 26458–26462.
380. Samad, Y. A., Li, Y., Alhassan, S. M. and Liao, K. (2015). Novel graphene foam composite with adjustable sensitivity for sensor applications, *ACS Applied Materials & Interfaces*, 7, 9195–9202.
381. Liu, Y., Tao, L.-Q., Wang, D.-Y., Zhang, T.-Y., Yang, Y. and Ren, T.-L. (2017). Flexible, highly sensitive pressure sensor with a wide range based on graphene-silk network structure, *Applied Physics Letters*, 110, 123508.
382. Rinaldi, A., Tamburrano, A., Fortunato, M. and Sarto, M. S. A flexible and highly sensitive pressure sensor based on a PDMS foam coated with graphene nanoplatelets. In *Sensors*, 2016; Vol. 16, 2148.
383. Yin, F., Guo, Y., Li, H., Yue, W., Zhang, C., Chen, D., Geng, W., Li, Y., Gao, S. and Shen, G. (2022). A waterproof and breathable Cotton/rGO/CNT composite for constructing a layer-by-layer structured multifunctional flexible sensor, *Nano Research*, 15, 9341–9351.
384. Lu, Y., Tian, M., Sun, X., Pan, N., Chen, F., Zhu, S., Zhang, X. and Chen, S. (2019). Highly sensitive wearable 3D piezoresistive pressure sensors based on graphene coated isotropic non-woven substrate, *Composites Part A: Applied Science and Manufacturing*, 117, 202–210.

385. Kim, S. J., Song, W., Yi, Y., Min, B. K., Mondal, S., An, K.-S. and Choi, C.-G. (2018). High durability and waterproofing rGO/SWCNT-fabric-based multifunctional sensors for human-motion detection, *ACS Applied Materials & Interfaces, 10*, 3921–3928.
386. Chun, S., Hong, A., Choi, Y., Ha, C. and Park, W. (2016). A tactile sensor using a conductive graphene-sponge composite, *Nanoscale, 8*, 9185–9192.
387. Shi, X., Zhu, Y., Fan, X., Wu, H.-A., Wu, P., Ji, X., Chen, Y. and Liang, J. (2022). An auxetic cellular structure as a universal design for enhanced piezoresistive sensitivity, *Matter, 5*, 1547–1562.
388. Zhao, S., Zhang, J. and Fu, L. (2021). Liquid metals: A novel possibility of fabricating 2D metal oxides, *Advanced Materials, 33*, 2005544.
389. Wang, Y., Zhao, C., Chen, L., Wu, Q., Zhao, Z., Lv, J.-J., Wang, S., Pan, S., Xu, M., Chen, Y., *et al.* (2024). Flexible, multifunctional, ultra-light and high-efficiency liquid metal/polydimethylsiloxane sponge-based triboelectric nanogenerator for wearable power source and self-powered sensor, *Nano Energy, 127*, 109808.
390. Lin, Y., Yin, Q., Jia, H., Ji, Q. and Wang, J. (2024). Ultrasensitive and highly stretchable bilayer strain sensor based on bandage-assisted woven fabric with reduced graphene oxide and liquid metal, *Chemical Engineering Journal, 487*, 150777.
391. Li, N., Yuan, X., Li, Y., Zhang, G., Yang, Q., Zhou, Y., Guo, M. and Liu, J. (2024). Bioinspired liquid metal based soft humanoid robots, *Advanced Materials*, 2404330.
392. Gao, Y., Ota, H., Schaler, E. W., Chen, K., Zhao, A., Gao, W., Fahad, H. M., Leng, Y., Zheng, A., Xiong, F., *et al.* (2017). Wearable microfluidic diaphragm pressure sensor for health and tactile touch monitoring, *Advanced Materials, 29*, 1701985.
393. Kim, A., Ahn, J., Hwang, H., Lee, E. and Moon, J. (2017). A pre-strain strategy for developing a highly stretchable and foldable one-dimensional conductive cord based on a Ag nanowire network, *Nanoscale, 9*, 5773–5778.
394. Xu, S., Zhang, Y., Cho, J., Lee, J., Huang, X., Jia, L., Fan, J. A., Su, Y., Su, J., Zhang, H., *et al.* (2013). Stretchable batteries with self-similar serpentine interconnects and integrated wireless recharging systems, *Nature Communications, 4*, 1543.
395. Liu, Y., Wang, X., Xu, Y., Xue, Z., Zhang, Y., Ning, X., Cheng, X., Xue, Y., Lu, D., Zhang, Q., *et al.* (2019). Harnessing the interface mechanics of hard films and soft substrates for 3D assembly by controlled buckling, *Proceedings of the National Academy of Sciences, 116*, 15368–15377.
396. Ohm, Y., Pan, C., Ford, M. J., Huang, X., Liao, J. and Majidi, C. (2021). An electrically conductive silver–polyacrylamide–alginate hydrogel composite for soft electronics, *Nature Electronics, 4*, 185–192.

397. Song, P., Qin, H., Gao, H.-L., Cong, H.-P. and Yu, S.-H. (2018). Self-healing and superstretchable conductors from hierarchical nanowire assemblies, *Nature Communications*, *9*, 2786.
398. Liu, Z. F., Fang, S., Moura, F. A., Ding, J. N., Jiang, N., Di, J., Zhang, M., Lepró, X., Galvão, D. S., Haines, C. S., *et al.* (2015). Hierarchically buckled sheath-core fibers for superelastic electronics, sensors, and muscles, *Science*, *349*, 400–404.
399. Liu, Y., Liu, J., Chen, S., Lei, T., Kim, Y., Niu, S., Wang, H., Wang, X., Foudeh, A. M., Tok, J. B. H., *et al.* (2019). Soft and elastic hydrogel-based microelectronics for localized low-voltage neuromodulation, *Nature Biomedical Engineering*, *3*, 58–68.
400. Zhu, L., Xu, P., Chang, B., Ning, J., Yan, T., Yang, Z. and Lu, H. (2024). Hierarchical structure by self-sedimentation of liquid metal for flexible sensor integrating pressure detection and triboelectric nanogenerator, *Advanced Functional Materials*, 2400363.
401. Cheng, S., Hang, C., Ding, L., Jia, L., Tang, L., Mou, L., Qi, J., Dong, R., Zheng, W., Zhang, Y., *et al.* (2020). Electronic blood vessel, *Matter*, *3*, 1664–1684.

© 2025 World Scientific Publishing Company
https://doi.org/10.1142/9789811266867_0003

Chapter 3

Flexible Sensors

Yunjian Guo[*], Peng Wang[†], Li Yang[‡,¶], and Shen Guozhen[§,**]

[*]Department of Electronic Convergence Engineering, Kwangwoon University, Seoul, 01897, South Korea
[†]School of Mechanical Engineering, University of Jinan, Jinan 250022, China
[‡]School of Integrated Circuits, Shandong University, Jinan 250101, China
[§]School of Integrated Circuits and Electronics, Beijing Institute of Technology, Beijing 100081, China

[¶]yang.li@sdu.edu.cn

[**]gzshen@bit.edu.cn

1. Flexible Tactile Sensor

In human skin, tactile information is recorded by an array of mechanoreceptors and transmitted to the somatosensory cortex for processing and understanding.[1–7] Inspired by these transduction mechanisms, a variety of flexible tactile sensors play a key role in enhancing and/or replacing motor, sensory, or cognitive modalities. Slowly adapting (SA) receptors (SA-I and SA-II) are characterized by a regular static phase firing rate that facilitates the measurement of static forces.[8,9] SA-I receptors have high spatial resolution and sensitivity to normal forces, while SA-II receptors are more responsive to skin stretch. However, rapid adaptation (RA) receptors fire action potentials only at the initial and final contact of a stimulus, providing dynamic force and vibration sensitivity from low to high frequencies.

1.1. Static force recognition

1.1.1. Flexible pressure sensor

Pressure is one of the most common mechanical stimuli that needs to be monitored in nature and humans. Since a variety of applications require the detection of pressures ranging from a few pascals to hundreds of kilopascals, a large population of flexible pressure sensors has been developed. Common measurable pressures fall into four ranges including subtle pressure (0–1 kPa), low pressure (1–10 kPa), medium pressure (10–100 kPa) and high pressure regimes (>100 kPa).[10,11] The subtle pressure regime (0–1 kPa) covers the pressure generated by weak interactions and the weight of many small objects. For example, placing a pencil on a flat surface induces a pressure of about 300 Pa, a layer of pencil shavings induces a pressure of about 40 Pa, and a water drop produces a pressure of about 13 Pa.[12] Sensitive responses in subtle pressure regions are critical for the development of pressure sensors assembled in highly sensitive touchscreen devices. Many of the pressures caused by gentle manipulation of objects and pressures within the human body (such as intraocular and intracranial pressures) typically fall in the low-pressure region (1–10 kPa). Pressure sensors that have shown excellent performance in this system have received significant attention for use in electronic skin and health monitoring/diagnostic systems.[13] The medium pressure system (10–100 kPa) includes atmospheric pressure at high altitudes and average plantar pressure at rest. During human movement, higher plantar pressure distribution values can easily exceed this range, so flexible pressure sensors are preferred for monitoring plantar pressures for movement and gait analysis that still work well when high-pressure regimes (>100 kPa) are reached.[14] Typical design strategies for flexible pressure sensors include the use of porous structures and surface microstructures to sense pressure stimuli with high sensitivity. Not only do these sensors show good agreement with the way SA receptors transmit static forces, but adjusting the degrees of freedom of the microstructure provides customizable pressure response profiles. For example, by varying the size, spacing, and arrangement of the microstructures, the pressure sensors exhibit different sensitivities.

Using the strategy of engineering microstructures, Park et al. demonstrated highly sensitive capacitive tactile sensors based on porous pyramidal dielectric elastomer, which can induce large deformations in response to subtle external stimuli.[15] As shown in Figure 1, porous materials have

Flexible Sensors 145

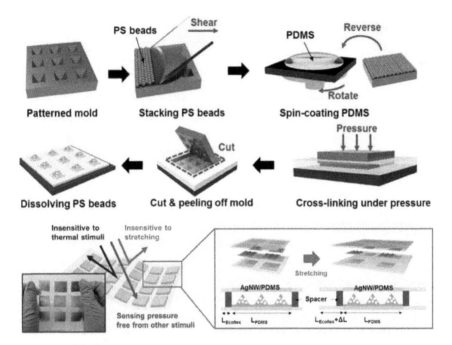

Figure 1. Schematic depiction of porous pyramid dielectric layer fabrication process, and working principle of porous pyramid dielectric layer-based capacitive sensor on hard islands, embedded in soft Ecoflex[15] (Copyright 2019, American Chemical Society).

low compressive modulus; therefore, they are susceptible to dielectric deformation, which leads to capacitance changes. In addition, due to the pyramidal structure, the applied stress is concentrated at the apex, further reducing the effective compressive modulus and increasing the sensitivity of the sensor. The sensitivity of the sensor reaches 44.5 kPa^{-1} at a pressure of 100 Pa with a detection limit of 0.14 Pa. By placing the pressure sensor on a hard elastomer embedded in a soft elastomer substrate, the sensor exhibits insensitivity to strain. The pressure sensor also does not respond to temperature. As shown in Figure 2, by constructing fingerprint ring array microchannels, MXene can be well confined in microchannels and form 3D stacked structures, resulting in accordion microstructured MXene with larger deformation space and more sensitive micromotor capability.[16] Due to the microchannel confinement effect, the sensor has excellent sensitivity and low detection limits to sense micro-motions such as acoustic waves.

Figure 2. General schematic diagram and detailed structure of pressure sensor[16] (Copyright 2020, Wiley-VCH).

In flexible pressure sensors, the electrodes must have excellent bending stability, maintaining their original conductivity even after thousands of bending cycles. Figure 3 illustrates a high-performance MXene/polyacrylonitrile (PAN) composite flexible pressure sensor with uniform $Ti_3C_2T_x$ electrodes. When pressure is applied, the contact between MXene nanosheets is tighter, the number of current paths increases, and the contact area of the composite fiber network increases significantly.[17] When the pressure is removed, the fiber network quickly returns to its original state due to the good flexibility of the nanofibers. The sensors with MXene electrodes improve the sensing performance and mechanical properties of wearable electronic devices compared to other electrodes. The use of 3D printing technology allows for rapid fabrication of TENG structures with simple and efficient conversion. Pour's team developed a wearable sensing system that integrates a self-powered TENG with a pressure response using 3D printing technology (Figure 4).[18] The self-powered haptic sensor has excellent sensitivity for continuous, real-time and on-demand radial pulse waveform monitoring without external power.

1.1.2. *Flexible strain sensor*

Flexible strain sensors are used to measure the deformation of objects. Flexible strain sensors play an important role in monitoring human body

Flexible Sensors 147

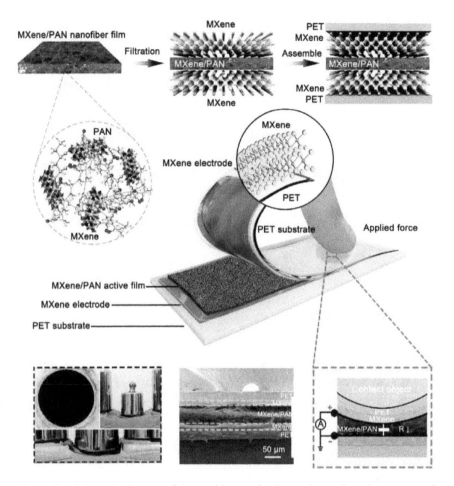

Figure 3. Schematic diagram of the working mechanism and manufacturing process of the MXene/PAN-based flexible pressure sensor[17] (Copyright 2021, Wiley-VCH).

motions in different positions, which can be divided into two categories: (1) motions with large skin deformation, including bending movements of fingers, wrists, arms, legs, and spinal; (2) motions with small skin deformation, including subtle movements of the face, chest, and neck which are directly related to emotional expression, breathing, speaking/swallowing activities, respectively.[19–21] Flexible strain sensors are usually attached directly to human skin or on clothes and they enable wide applications, such as recording hand gestures, capturing body movements, analyzing facial expressions, diagnosing throat diseases, and monitoring skin sclerosis.[22–25]

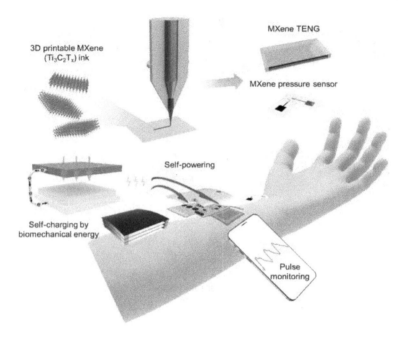

Figure 4. Schematic of the self-powered physiological sensing system, which comprises an MXene-based TENG, an MXene-based pressure sensor and a power-management, data collecting, and wireless data/power transmitting modules[18] (Copyright 2022, Elsevier Inc).

For example, a novel skin-inspired hydrogel–elastomer hybrid with a sandwich structure and strong interfacial bonding for mechanical–thermal multimode sensing applications is developed.[26–28] An inner-layered ionic hydrogel with a semi-interpenetrating network is prepared using sodium carboxymethyl cellulose (CMC) as a nanofiller, lithium chloride (LiCl) as an ionic transport conductor, and polyacrylamide (PAM) as a polymer matrix. The outer-layered polydimethylsiloxane (PDMS) elastomers fully encapsulating the hydrogel endow the hybrids with improved mechanical properties, intrinsic waterproofness, and long-term water retention (>98%). The silane modification of the hydrogels and elastomers imparts the hybrids with enhanced interfacial bonding strength and integrity. The assembled sensors can detect multi-scale human movements in various environments and recognize superimposed temperature and strain signals.

The integration of conductive materials and textiles is an effective approach for the development of flexible strain sensors and related applications. Luo *et al.* have demonstrated a waterproof and breathable smart textile with MXene as the sensing material, where the sensor was

Flexible Sensors 149

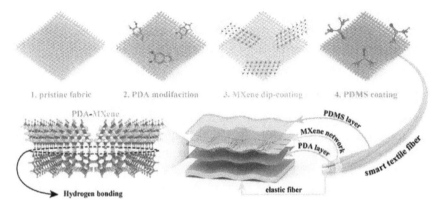

Figure 5. Schematic illustrations of the preparation process and structure of the PDA/MXene/PDMS textile[29] (Copyright 2021, Elsevier Inc.).

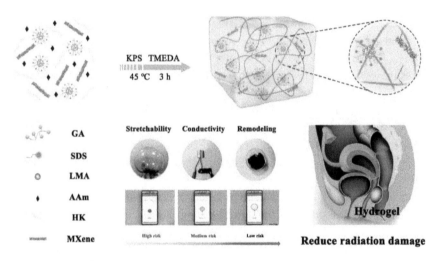

Figure 6. Schematic illustration of the preparation process and application diagram of hydrogels[30] (Copyright 2022, Wiley-VCH).

fabricated by a simple dip-coating method.[29] As shown in Figure 5, MXene has abundant functional groups that can be firmly adsorbed onto the PDA-modified textile through hydrogen bonding, forming a complete conductive pathway. This wearable smart textile electronic device has excellent waterproof and breathable performance and can monitor body temperature, human movement and personal heat management. Inspired by balloon inflation, Guan et al. reported a strain sensor based on the hydrophobically bound hydrogel, as illustrated in Figure 6. The sensor has great potential in tumor therapy because of the good self-healing properties and remodeling ability of hydrogels, which can mimic the decay process of radiation signals by monitoring the change of resistance.[30]

1.2. Dynamic force sensing

Flexible sensors to perceive shear stress play an important role in monitoring fluidic dynamics, robotics and the biomedical field.[31–33] Real-time shear-stress information is critical in estimating the airflow situation on the surface of aircraft to adjust the flight control correspondingly. For example, a 1D array of flexible shear sensors has been developed to detect the leading-edge flow separation point of unmanned aerial vehicles to guide the independent flight control of pitching, rolling, and yawing via force imbalance.[34] Similarly, shear-stress monitoring in blood vessels promotes the understanding of the relationship between blood flow and vascular disease.[35] The increasing demand for measuring shear stress in the medical community can also be found in the significance of analyzing interfacial forces between the human body and external objects, such as measuring the friction by flexible sensors between a prosthesis and a stump to check its fitness.[36] Moreover, tactile sensors with the capability of sensing shear stress can provide robotics with direct information on textures or slip detection.[37–39] Typical design strategies of flexible shear sensors include using a bump on the sensor's surface with four distributed underlying sensing elements and deformable surface/internal microstructures to perceive shear stimuli.

Shear sensing is usually accompanied by shear and normal segments. For example, Pang *et al.* used interlocking nanofibers to measure shear loading generated by amplified van der Waals forces, as shown in Figure 7.[40] The highly flexible, multiplex, real-time sensor can detect pressure, shear and torsion. When different sensing stimuli are applied, the degree of interconnection and resistance of the sensor changes in a reversible, directional manner, with specific, identifiable strain coefficients. The small deformation of the hair can be transferred to the detector by a corresponding change in resistance, thus the detectable shear force can be as small as 0.001 N. Another study by Choi *et al.* proposed a pyramidal ion gel activated by neuromechanical receptors and sculpted electrodes, as shown in Figure 8.[41] They found that the shear force sensitivity was 22.1 times higher than the normal force, and when subjected to shear force, it was pressed against the pyramidal-shaped ionic gel. The gel

Flexible Sensors 151

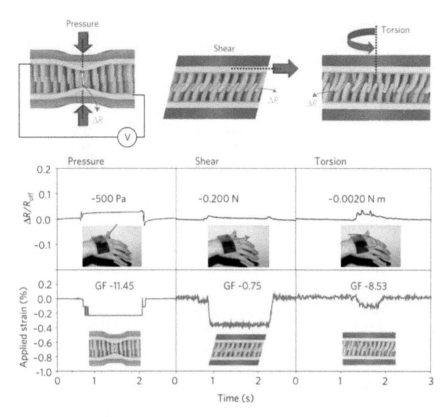

Figure 7. Schematic illustrations of the pressure, shear and torsion loads and their possible geometric distortions of the paired hairs, and the detection and decoupling of pressure, shear and torsion loads with each time-dependent signal pattern of resistance ratio and applied compressive strain as a function of time[40] (Copyright 2012, The Nature Publishing Group).

would move to the edge, deform, and then contact the edge region of the bottom electrode. This is due to the initial contact area of the deformed gel and the reduction of free ions, caused by the lower normal stress, which greatly provides the opportunity for gel-free ions when shear is applied. Besides, Gao *et al.* focused on the detection of shear angle by introducing four planar capacitive sensing units, and their detection method is explained in Figure 9.[42] In general, capacitive tactile sensors are capable of detecting shear angles when more than one capacitor is introduced symmetrically or asymmetrically. During shear operation, the upper

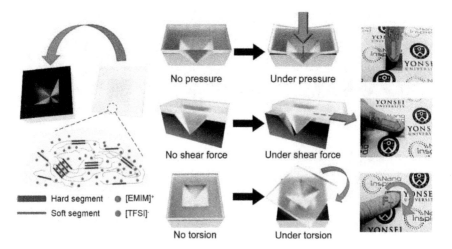

Figure 8. Tactile sensor based on geometric pyramid-plug structure of the ionic gel and iron electrode under three different mechanical stimuli for pressure, shear force, and torsion (Copyright 2018, Wiley-VCH).

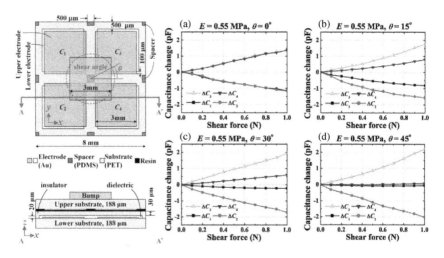

Figure 9. Schematic plot of the sensing unit from the top and side, and its detection setup for shear angle sensing mode at (a) 0°, (b) 15°, (c) 30°, and (d) 45°[42] (Copyright 2019, Elsevier Inc).

electrode is repositioned at a certain angle. The four capacitance changes (ΔCs) in the sensing unit vary depending on the contribution of the intentional offset of these four capacitors. In addition, when Young's modulus

Flexible Sensors 153

of the material decreases, the sensitivity to shear force increases regardless of the shear angle.

1.3. Multimodal sensing

Flexible multimodal sensors capable of sensing and decoupling different types of stimuli are one of the current research priorities for a wide range of applications, including robotics, health monitoring, and human-machine interfaces.[43–46] Flexible multimodal mechanical sensing platforms offer significant advantages over single-stimulus sensors in capturing the combined information of pressure, strain, vibration, shear, and other mechanical stimuli to achieve complex tasks where cross-sensitivity must be attenuated to achieve accurate measurements. For example, haptic sensors capable of sensing both normal pressure and shear stress are an important component of electronic skins that provide robots with information for complex object recognition and dexterous object manipulation.[37] Constructing array layouts of identical sensing units, integrating different sensing units, and developing novel materials with multiple sensing mechanisms are effective ways to decouple different mechanical signals.[47]

There are various types of mechanoreceptors embedded in the human skin that are responsible for transducing external stimuli into action potentials. To mimic the role of neural systems in interpreting the data, a further effort for artificial mechanoreceptor networks to understand the objects was achieved by recording and analyzing the tactile information transduced from the sensor. Sundaram *et al.* designed a tactile glove with 548 pressure sensors located evenly on the palm and all the fingers operating at seven frames per second, as shown in Figure 10.[48] By manipulating the objects and recording the pressure maps during different frames of grasping, the temporal and spatial relationships of the interactive maps were analyzed through deep convolutional neural networks. The biomimetic glove demonstrated the ability to distinguish various objects and estimate their weights.

To simulate SA and RA adapters in a single sensory device, Chun *et al.* developed a PVDF-layered polyaniline (PANI) electrolyte separated by a nanoporous membrane to detect static and dynamic pressures.[49] As shown in Figure 11, the applied pressure triggers the piezoelectric effect of the PVDF layer, which is sensitive to high-frequency signals, while the accumulated piezoelectric potential induces the ions in the PANI to flow through the nanomembrane, causing a piezoresistive signal. Thus, static and dynamic signals can be measured simultaneously. In another

154 Y. Guo et al.

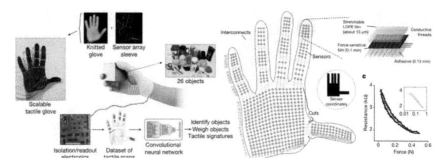

Figure 10. Scalable tactile glove as a platform to learn from the human grasp[48] (Copyright 2019, The Nature Publishing Group).

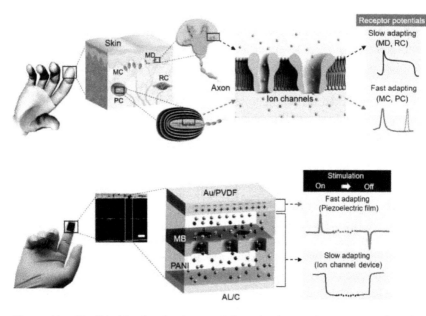

Figure 11. Flexible biomimetic slow- and fast-adaptive mechanoreceptors based on piezoelectric and ion channel layers, which are composed of a bottom panel with an SA-mimicking sensor, a top panel with an FA-mimicking sensor, and an artificial fingerprint structure with microlines[49] (Copyright 2018, Wiley-VCH).

approach, inspired by human fingertips, Chun *et al.* developed a bio-inspired layered e-skin with triboelectric and piezoresistive layers to mimic RA and SA receptors, and an artificial fingerprint to directly dock the texture, as shown in Figure 12.[50] Signals from both the triboelectric

Flexible Sensors 155

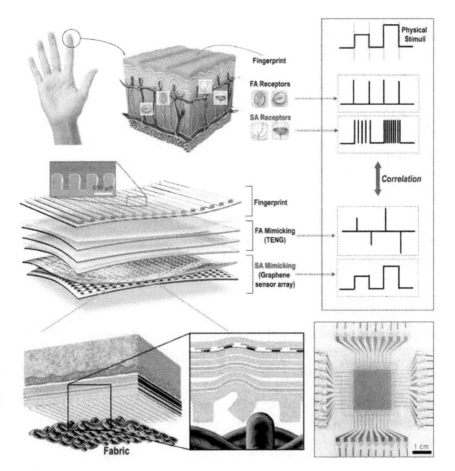

Figure 12. Schematics of a biological somatosensory system and an artificial mechanoreceptor cutaneous sensor for texture detection with triboelectric and piezoresistive layers[50] (Copyright 2019, American Chemical Society).

and piezoresistive layers were recorded. The combined analysis of RA and SA signals, aided by machine learning algorithms, showed the ability to distinguish between different fabric textures with an accuracy of 99.1%, an improvement over the single component analysis from RA (95.1%) and SA (92.3%), respectively.

Previous work on multidirectional force sensing has been mainly based on basic interlocking structures, where the shape of the pressure-time curve changes with the applied normal, shear, and torsional

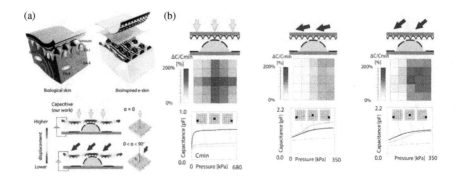

Figure 13. (a) Graphic illustration of biological skin structure and the bioinspired hierarchical structure, and concepts for deflections at different locations upon the microtome under external pressure. (b) sensor fields resolved by a 5 × 5 map and the characteristic pixel-level capacitance curve versus pressure from different directions[57] (Copyright 2024, Wiley-VCH).

forces.[51–56] However, the deformation caused by pressure in different directions cannot be distinguished from the electrical signal readout, limiting quantitative analysis at the microscopic level. To address this limitation, as shown in Figure 13(a), an electronic skin with layered patterns for detecting multidirectional forces was recently reported by Bao's group.[57] In contrast to previous designs, this bionic skin uses 3D top and bottom electrodes to mimic the epidermal-dermal interface, leaving the dielectric membrane unstructured. This novel capacitance-based sensory network has periodic arrays of microspheres on the bottom electrode. Since a microtome corresponds to 25 sensing capacitors (micro-pyramids) from different locations of the microtome (top, slope, corners, and perimeter), it provides pixel-level analysis of the pressure response for better quantification as these sensing capacitors experience different strains under multifunctional forces. This unique sensor configuration results in a non-uniform sensor field where each sensing pixel exhibits a characteristic pressure response profile, providing a more accurate and quantitative assessment to determine the direction of force (Figure 13(b)).

1.4. Multifunctional sensing

1.4.1. Stretchable sensors

To fully reconstruct the skin's tactile sensation, additional functions such as the ability to be stretchable, self-healing, biocompatible and biodegrade need to be considered.[58] Skin not only accommodates tactile receptors but also plays a key role in social communication with people and objects. This is because it readily forms conformal interfaces over the entire body. The flexibility and stretchability of the skin enable body movements such as stretching, bending, and twisting. To create an artificial interface for prosthetics and robotics inspired by skin, mechano-electronic systems that can stretch and bend beyond the conventional rigid silicon devices must be developed. Recently, tremendous efforts have been made to develop stretchable electronics compatible with human skin, robotic end effectors, and human–machine interfaces.[59,60]

To date, buckling methods have enabled many flexible but stiff materials stretchable, such as gold, silver, and graphene-based materials, which demonstrate potential applications such as electrodes and stretchable sensors.[61,62] Recently, buckling has been used to realize more complex 3D conducting networks comparable to the human biological system (cytoskeletal webs, vasculature networks) from 2D precursors. The helix is a familiar structure in nature, for example, in plant tendrils. The 3D helical structure takes advantage of the flexibility of wires with high elastic modulus, like copper and gold, to wrap into stretchable devices such as smart springs, conductors, and robotic systems.[63–65] As stretching of the helices does not change the length of the wires significantly, the interconnect remains conductive with invariant conductivity. Systems using the metal helix show higher conductivity than intrinsically stretchable materials. For example, Tee *et al.* developed a helical copper encapsulated elastomer interconnect, as shown in Figure 14, which possessed conductivity as high as 5.9 × 105 S cm⁻¹, without showing conductivity variation when

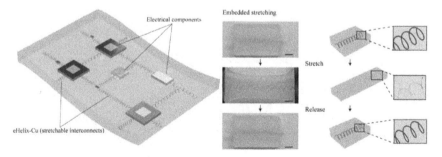

Figure 14. Schematic of a stretchable electronic device with Helical copper interconnects connecting multiple electronic components encapsulated by the elastomer[66] (Copyright 2019, the American Institute of Physics).

stretched up to 170%.[66] Such systems can be useful as another option where conductivity and reliability of electrical signals are important, in microelectronic systems that require embedding within an elastic matrix with tolerance for thicker substrates. Inspired by the traditional art of kirigami and origami, Xu et al. developed stretchable 3D micro/nanostructures from more than 40 2D geometries through compressive buckling.[67] It is noted that with 2D serpentine networks, buckling can be implemented to self-assemble 3D structured helical circuits, which better distribute the stress (Figure 15).[68] Under 50% strain, the 3D helices showed much smaller peak Mises stress (≈100 MPa) than that of 2D serpentine (>200 MPa). In addition, the elastic stretchability of 3D structures increased linearly up to 400% with prestrain, while that of the 2D structure was only limited to 50%. In addition, their group further exploited and modified buckling procedures (e.g., buckling and twisting, step-dependent buckling) to achieve more versatile stretchable 3D mesostructured such as near-field

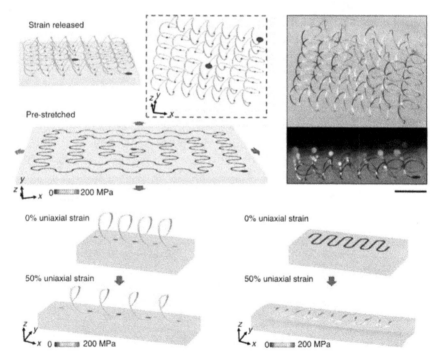

Figure 15. Assembly of conductive 3D helical coils, and buckled serpentine structure for higher circuit-level stretchability[68] (Copyright 2017, The Nature Publishing Group).

Flexible Sensors 159

communication (NFC) coils, and a sophisticated piezoelectric device, providing stretchability not only for interconnects but also the device itself.[69,70] Adopting this concept, a scalable sensor array was fabricated from their 2D precursor through pretrain buckling and origami design (Figure 16), indicating the promise of self-assembled 3D electronics.[71] The sensor uses a single-crystal silicon nanomembrane as a piezoresistive element in a configuration that can individually and simultaneously measure multiple mechanical stimuli, such as normal force, shear and bending, and temperature. These solutions naturally extend to an array of devices for mapping. In all cases, the operation is compatible with standard Bluetooth electronics, allowing data to be collected, transmitted and analyzed in a completely wireless manner, compatible with envisioned applications in health monitors, biomedical devices, and man–machine devices.

The designed stretchable structures can provide strain-independent conductivity, making them more reliable for signal input and output. Various approaches serve as an alternative to achieve circuit-level stretchable electronics, by forming stretchable interconnects between the flexible devices.[72]

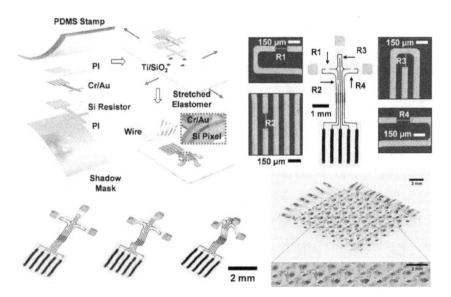

Figure 16. Materials and construction of a 3D piezoresistive sensor network on an elastomer substrate[72] (Copyright 2016, American Association for the Advancement of Science).

1.4.2. Self-healing sensors

One problem with flexible sensors is physical damage, such as cracks caused by external forces and fractures caused by aging over long periods of use. Even stretchable materials can be damaged and fractured when cut by sharp objects or subjected to large deformations. Therefore, there is an expectation to design and develop functional materials to solve this problem.[73-76] The emergence of self-healing materials has provided a solution to this thorny problem. During the long evolutionary process in nature, plants and animals have acquired the ability to heal themselves. For example, the leaves of plants and the skin of animals can repair themselves after injury, which allows them to survive in harsh environments. Inspired by nature, self-healing materials that can restore their physical properties after damage can be used to improve the lifetime of flexible sensors. The self-healing process is achieved through dynamic bond exchange or supramolecular interactions in polymer chain segments or by releasing encapsulated healing reagents.[77-80]

For example, based on ion–dipole interactions, Cao et al. combine an amorphous polymer with a chemically compatible ionic species to create gel-like, aquatic, stretchable and self-healing electronic skin, as shown in Figure 17.[81] After damage, the ion-dipole interactions between ionic liquids and polymer matrix offered fast self-healing ability to recover the mechanical properties under wet, acidic, and basic conditions. In addition, the electrical conductivity is also maintained after healing. Self-healing chemistry has driven important advances in deformable and reconfigurable electronics, particularly with self-healing electrodes as a key enabler. Unlike polymeric substrates with self-healing dynamic properties, a disrupted conductive network cannot recover its extensibility after damage. The discovery of a reconstructed carbon nanotubes (CNTs) network in a self-healing polymer, as shown in Figure 18, gives hope for electromechanical self-healing nanomaterial/self-healing polymer systems.[82] Based on this discovery, the researchers developed an integrated multifunctional electronic system including a strain sensor, an electrocardiogram (ECG) sensor, and a light-emitting capacitor in which each component, including interconnections, is fully self-healing. This work shows great promise for integrated self-healing systems, but in the future, we may expect to develop higher levels of multilayer device density for more complex devices and functions.

Compared to stretchable skin, which acts as a protective layer and sensory organ, muscle has a higher mechanical strength and pulls bones by contraction to produce joint movements. Figure 19 shows the structure

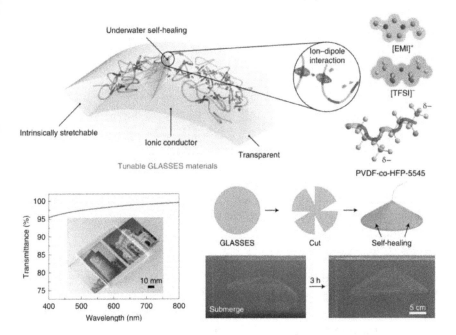

Figure 17. Design of a gel-like, aquatic, stretchable and self-healing electronic skin.

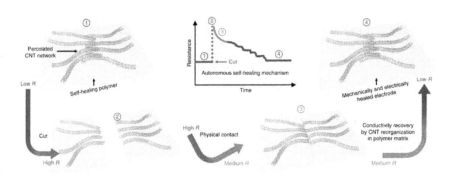

Figure 18. Recovery mechanism for CNTs embedded in self-healing polymer matrix.

of human muscle, where the densely connected epidermis muscle contains many muscle fiber bundles (fascicles), giving it strong mechanical properties. Inspired by the muscle structure, Ge et al. doped polyaniline nanofibers (PANI NFs) into polyacrylic acid (PAA) hydrogels to fabricate multifunctional self-healing hydrogels through hydrogen bonding and

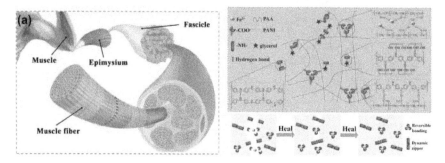

Figure 19. Schematic of the human muscle structure, in which the epimysium is tightly connected and contains many bundles (fascicles) of muscle fibers, and structure of the multifunctional hydrogel and its reversible interactions[83] (Copyright 2020, American Chemical Society).

metal–ligand interactions.[83] After the addition of glycerol, the PAA-PANI binary network hydrogel (PPBN-hydrogel) exhibits excellent freeze resistance and maintains electrical conductivity at −26°C. PPBN-hydrogels can withstand fracture stresses over 35 kPa and achieve nearly 1000% of the fracture strain. After 6 hours of healing at room temperature, the healed strain can reach 90% of the original strain. Compared to the PAA unitary network hydrogel, the resistance of the PPBN-hydrogel increases exponentially with strain due to the embedding of PANI NFs to form a porous network. When the luminal pores are squeezed under large strains, the cross-linked percolation joints rupture, blocking the conductivity pathway and causing significant changes in resistance. In addition, the resistance of the PPBN-hydrogel decreases with increasing temperature, exhibiting a linear response with high sensitivity and a wide sensing range. This was achieved by using 1D-conducting PANI NF as the channel material, which can be attached to the human forehead as a "fever indicator" temperature sensor. However, the mechanical strength of PPBN-hydrogels is not ideal due to the inherent properties of hydrogels, which are inevitably reduced by the presence of solvents such as water and glycerol. Polymeric elastomers offer a solution to this problem with stretchability, toughness and self-healing properties for durable electronics. Ying *et al.* reported a self-healing PU that self-assembles from a donor–acceptor (D-A), similar to the skeletal muscle protein titin (Figure 20).[84] DA-PUs have superb mechanical properties and can withstand a fracture stress of ≈26 MPa and reach a fracture strain of 1900%. After cutting, the mechanical properties

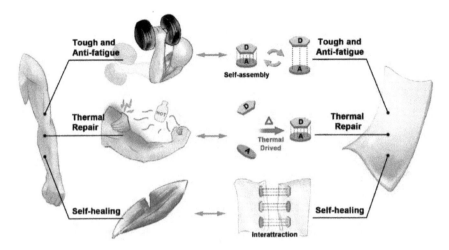

Figure 20. Diagram of the muscle-inspired self-healing elastomer. D-A self-assembly endows the elastomer with muscle-like toughness, thermal repair, and self-healing capabilities[84] (Copyright 2021, Wiley-VCH).

could be recovered to 400% after 97 min. The researchers also performed cyclic tensile tests on the self-healing DA-PU, revealing its excellent fatigue resistance. Finally, capacitive sensors were prepared using DA-PU as the dielectric and filler layers and Cu powder containing GaInSn liquid metal as the conductive layer. When stretched, the distance between the two conductive layers decreases, thus increasing the capacitance. A linear relationship exists between the capacitance and the tensile strain of the original and repaired sensor devices.

Chameleons can adjust their body color in response to changes in their environment to hide and capture prey because their unique structural coloration is produced by microstructures and nanostructures. When stimulated, chameleons stretch their skin, causing the distance between guanine nanocrystals under the skin to widen, further driving changes in reflectance and thus changing the color they display. As shown in Figure 21, inspired by the structural color of chameleons and mussel adhesion proteins, Wang *et al.* prepared self-healing membranes by doping conductive CNT-PDA fillers into a PU inverse opal matrix.[85] First, silica nanoparticles were self-assembled into ordered hexagonal arrays, and then PU solution was introduced into the arrays and the hybrid films were cured after solvent evaporation.

164 Y. Guo et al.

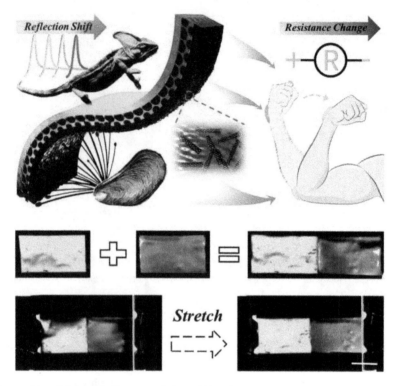

Figure 21. Schematic diagram of the self-healing structural color film inspired by chameleon[85] (Copyright 2020, Wiley-VCH).

Finally, the silica nanoparticles were etched to obtain the PU inverse opal matrix. In this system, the PU film provides stretching and self-healing ability, while the CNT provides electrical conductivity. When the hybrid film is stretched, the resistance of the film increases, while the reflection spectrum is blue-shifted due to the decrease in the distance between diffraction planes. As a demonstration, the hydrogel was adhered to human fingers, wrists and elbows, respectively, for human motion detection. Taking the finger as an example, the structural color of the film changes from orange-red to green with increasing bending angle of the finger with good repeatability. At the same time, the increase in resistance corresponds to the change in wavelength. This dual-signal sensing extends the signal expression compared to conventional soft electronic sensors. In addition, thanks to the self-healing capability, films with different structural colors can be stitched together to obtain complex patterned films.

1.4.3. Visualization sensors

The human–machine interaction of pressure visualization devices is reflected in the fact that pressure signals released by humans (e.g., finger press or regular touch) can be accepted by pressure sensors and the sensing results of the machine can be uploaded to the display device for further interaction with the human brain.[86–89] Through continuous experimentation, humans can eventually correct their pressing behavior and machines can better analyze and process human-initiated pressing behavior. Easily interpretable visual feedback allows the user to receive information about the state of the device and its measurements more effectively. With advances in flexible pressure sensing modules and the development of various visualization techniques based on the principles of electroluminescence (EL) and electrochromic, pressure visualization devices with unique patterns and self-powered characteristics are emerging.[90,91]

Flexible visualization devices based on the piezoelectric photoelectric effect have been extensively studied. Wang *et al.* proposed a pressure visualization device based on the piezoelectric photoelectric effect.[92–95] The grounded end of a piezoelectric semiconductor nanowire composed of ZnO and GaN generates a piezoelectric potential under pressure or strain. This internal piezoelectric potential is used as a "grid" voltage to easily regulate/control the charge separation and compounding process of each NW-LED in the optoelectronic process. Under forward bias, the luminescence intensity of the nanowires under compressive strain increases significantly. One of the advantages of this method is its ultra-high resolution. The resolution of the device is highly dependent on the diameter of the nanowires and offers a spatial resolution of up to 2.7 mm, corresponding to a pixel density of 6,350 DPI. In addition, the technique is easily integrated with photonics technologies for fast data transfer, processing and recording. A triboelectrification-induced EL is proposed that can convert dynamic motion into luminescence under mild mechanical interactions (Figure 22).[96] The strong electric field generated by the transient charge from the upper triboelectrically charged layer uniformly stimulates the lower phosphorescence layer to emit light along the motion trajectory, thus directly visualizing the pressure on the device. Compared to conventional pressure visualization techniques, this method does not require complex circuitry and can be simply fabricated to be highly integrated, thus providing excellent stability, repeatability and durability. In addition, it can be excited by very weak stimuli and produces far stronger pressure at low

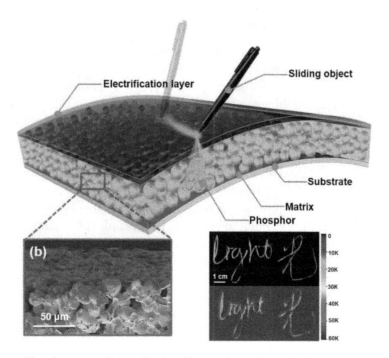

Figure 22. Structure of the triboelectrification-induced EL composite material and induced luminescence along the sliding trajectory[96] (Copyright 2016, Wiley-VCH).

stresses (less than 20 kPa) than elastico-triboluminescence (ETL). Therefore, it enables high-resolution localization and motion tracking of the contact. However, the disadvantage is that the triboelectric-generated electrostatic field tends to disperse laterally because the luminescence phenomenon is not controlled by the driving circuit. The individual pixels of a passive driver light-emitting device need to be able to achieve relatively high luminance to meet the needs of the display. Crosstalk between pixels is typically prevented by increasing the rejection ratio (ratio of on-current to leakage current). Wearable flexible pressure visualization arrays designed to facilitate human-computer interaction require sufficient brightness and temperature control to significantly reduce potential user discomfort while reducing mechanical and electrical failures of the device. Kim's group prepared a quantum dot-based 8 × 8 pixel integrated pressure visualization device with an ultra-thin thickness (~5.5 μm), as depicted in Figure 23.[97] The device was able to wrinkle, roll, and repeatedly bend for up to 1000 cycles and functioned properly under extreme deformations

Flexible Sensors 167

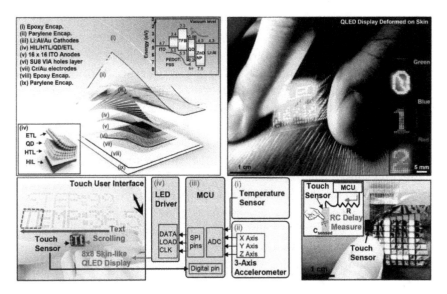

Figure 23. View of an ultra-thin quantum dot light-emitting diode display showing layer information, and its photo on the morphing skin[97] (Copyright 2017, Wiley-VCH).

with a local bending radius of 68 μm. At the same time, the device shows a controlled operating temperature (below 28.5°C).

Electrochromic-based pressure visualization devices can be constructed by integrating flexible pressure sensors with electrochromic devices (ECDs).[98–101] They still have potential in human–machine interfaces, electronic skins, and wearable devices due to their low-power characteristics and tight connection to supercapacitors.[102–104] For example, by covering the surface of a human or robot with pressure-visualized ECDs, it is possible to mimic epidermal stress responses similar to those of chameleons and brachiopods, which is a major achievement. Bao et al. demonstrated an electrochromic-based electronic skin.[105] The upper part has an electrochromic module and the lower part has a resistance-based pressure-sensing module integrated in a single circuit. The device can control the changes in the electronic skin by applying pressure of varying intensity and duration, simulating novel haptic sensing. At pressures of 0, 50, and 200 kPa, the device appears dark red, blue and blue-gray, respectively. When the pressure is removed, the device reverts to its initial dark red state. Thus, this unique color-changing technology can also be used to distinguish the different intensities of applied pressure. High transparency

168 Y. Guo et al.

and fit to human skin are trends in ECDs that will simultaneously add to the technological advances of recent years while looking to a bright future. Park *et al.* designed a transparent and stretchable electrochromic display with pressure sensing.[106] PANI and V were used to form ECDs V_2O_5 and transparent resistive strain sensors based on nanocomposites that were integrated into the skin and driven by an Arduino circuit. They display a variety of color variations from light yellow to blue-black. By optimizing the PDMS in the device, its ECDs can withstand elastic deformation of up to 50% of the original surface area without causing delamination and subsequent degradation.

Besides, a compressible ECD using a 3D compressible sponge electrolyte layer was designed, as shown in Figure 24.[107] The 3D porous sponge was prepared using a simple and environmentally friendly method of leaching sugar (a pore-creating agent) from a UV-cured skeletal structure. The simple structure of the compressible ECD facilitates its

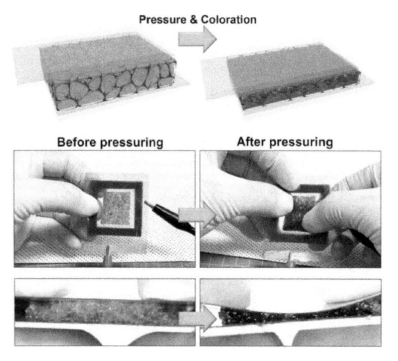

Figure 24. Schematic diagram of ECD with a porous sponge and compressive stress–strain curve of the mercaptan sponge[109] (Copyright 2017, Wiley-VCH).

Flexible Sensors 169

suitability as an optical haptic sensor device. Optical measurements demonstrated that low-pressure operation ensures long-term sensor performance. In addition, the color change of the ECD that occurs when pressure is applied can be used as an effective pressure-sensing mechanism for the system in question. Array displays are also important for electrochromic pressure visualization devices. Han et al. developed a pressure visualization and recording (PVR) system with a resolution of 500 μm (50.8 DPI).[108] As illustrated in Figure 25, a nanowire pressure sensing matrix at the bottom of the device transmits pressure information to a WO_3 thin film ECDs array in the upper layer through a piezoelectric effect. Each piezoelectric nanowire and each electrochromic pixel are connected one by one through gold electrodes to produce a pressure sensor with real-time pressure mapping. In addition, due to the color memory effect, the distribution of the external pressure field can be recorded independently with a color contrast of >300% after 85 cycles.

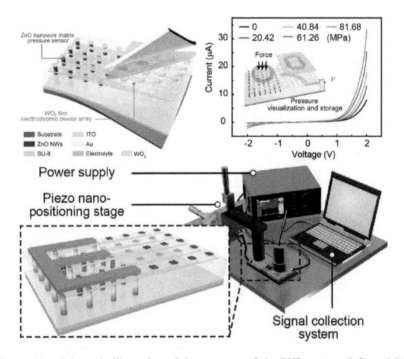

Figure 25. Schematic illustration of the structure of the PVR system (left) and I–V characteristics of a single ZnO NW pixel under various applied pressures (right)[108] (Copyright 2022, Elsevier Inc).

1.4.4. Biocompatible and biodegradability

Living organisms have specific requirements for the materials used in biosensors to ensure no allergy, toxicity, decomposition, and physical damage during direct or indirect contact.[110–112] Stable and biocompatible flexible materials, including polymers, metals, carbon and biomass, exhibit great friendliness to body tissues.[113] Photolithography and printing (3D printing and inkjet printing) technologies can improve device functionality while reducing the size of the device. This size reduction will also promote portability and reduce the body's rejection of biodegradable devices. A substrate-less nanomesh artificial mechanoreceptor was reported, which is coupled with an unsupervised meta-learning framework and can provide user-independent, data-efficient recognition of different hand tasks (Figure 26).[114] The nanomesh, which is made from biocompatible materials and can be directly printed on a person's hand, mimics human cutaneous receptors by translating electrical resistance changes from fine skin stretches into proprioception. A single nanomesh can simultaneously measure finger movements from multiple joints, providing a simple user implementation and low computational cost. The developed time-dependent contrastive learning algorithm that can differentiate between different unlabeled motion signals. This meta-learned information is demonstrated

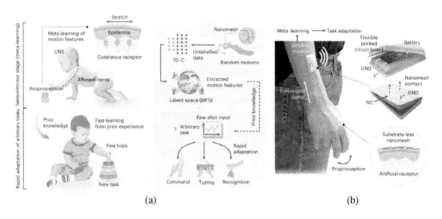

Figure 26. (a) Illustration of human sensorimotor stage that includes the meta-learning of motions through cutaneous receptors and its RA to unknown tasks. (b) An artificial sensory intelligence system that consists of printed, biocompatible nanomesh cutaneous receptors directly connected with a wireless Bluetooth module through a nanomesh connector, and is further trained through few-shot meta-learning[114] (Copyright 2023, The Nature Publishing Group).

to rapidly adapt to various users and tasks, including command recognition, keyboard typing and object recognition.

As an emerging class of electronic devices, transient electronics reduce the environmental impact, which is desirable for an eco-friendly and sustainable future. The degradability of devices is of great importance for bioelectronics. Initially, to expand the application of inorganic materials in transient bioelectronics, a platform including components such as transistors, inductors, and capacitors was demonstrated by Hwang *et al.* as shown in Figure 27.[115] Electronic components based on inorganic materials (Si, SiO_2, Mg, and MgO) underwent hydrolysis and dissolved in water. However, biodegradable polymers based on naturally derived polymers and synthetic polymers have become an area of intensive research due to the non-compliant mechanical properties and possible toxicity of inorganic materials.[116] For example, substrates (e.g., poly(lactic acid-glycolic acid) (PLGA)) and dielectric layers (e.g., polyglycerol sebacate (PGS)) and semiconductor films are widely used for biodegradable applications.[117,118] Research in transient wearable and skin-like electronics is expected to proliferate with a wide selection of materials to achieve degradability.

1.4.5. *Implantable sensors*

Implantable sensors have promising applications in post-surgical vascular repair and blood flow monitoring, and they offer irreplaceable advantages over non-implantable devices in detecting deep tissues.[119–122] Implantable sensors that can be used for continuous monitoring of biomechanical strain need to address three key challenges before they can be used in

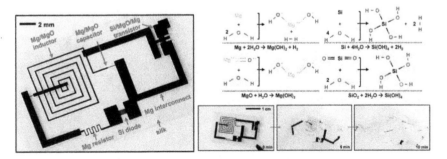

Figure 27. Components of the inorganic materials-based degradable electronics, and its degradation mechanisms and photographs of the dissolution in deionized water[115] (Copyright 2012, American Association for the Advancement of Science).

clinical practice: structural mismatches between sensors and tissues or organs should be eliminated; practical suture attachment processes should be developed; and sensors should be equipped with wireless readout capabilities.[123] A bioresorbable triboelectric sensor (BTS), which can be implanted *in vivo* and can identify vascular occlusion events successfully in large animals based on the triboelectric initiation effect, was reported to convert the biomechanical signal of vascular changes with the diastole and systole phases into an electrical signal (Figure 28).[124] The main part

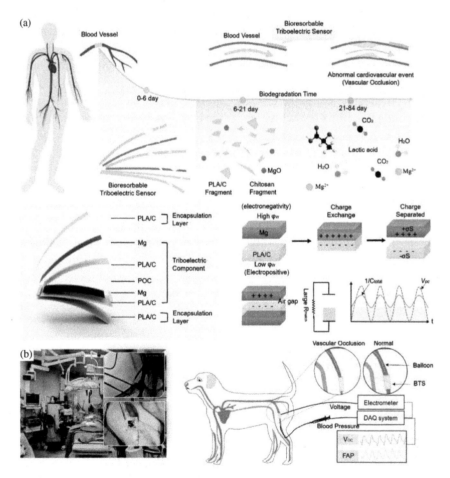

Figure 28. (a) Overview and principle of the bioresorbable triboelectric sensor. (b) Image of the human-scale animal cardiac monitoring experiment, and schematic of *in vivo* experimental electrical characterization and physiological signal monitoring[124] (Copyright 2021, Wiley-VCH).

of the sensor is based on a triboelectric electrode layer constructed of polylactic acid-(4% chitosan) (PLA/C) and magnesium, with stable output performance under mechanical stimulation, good sensitivity, linearity, and good durability to remain stable at 450,000 cycles, as well as antimicrobial properties. The service life after implantation is 5 days, and *in vitro*, tests have verified that complete degradation can be achieved in an average of 84 days. The limitations are the short service life and the fact that it still requires magnesium wires to be threaded out of the body for data transmission and does not achieve full wireless implantability.

To overcome this limitation, Bao *et al.* reported a biodegradable pressure sensor for real-time monitoring of arterial blood flow that can be wrapped outside arteries of different diameters and does not need to be removed after use (Figure 29).[125] Based on fringe-field capacitor technology, arterial blood flow was measured by contact or non-contact mode. Due to the changes in arterial diameter with blood flow pulsation, capacitance changes and is measured by the capacitive sensor, resulting in a change in the resonant frequency of the inductance–capacitance–resistance (LCR) circuit, which is coupled inductively to an external reader coil for wireless monitoring through the skin without a battery. The sensor can be easily wrapped around arteries less than 1 mm in diameter. The sensor was subsequently implanted in rats for up to 12 weeks and verified to have good biocompatibility. The application of bioabsorbable devices can avoid secondary surgery and is getting more attention in the research of implantable sensors. Similarly, as shown in Figure 29, a wireless and saturable fiber

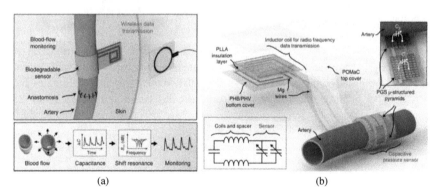

Figure 29. (a) Post-operative monitoring of arterial pulsations after surgery. (b) Illustration of the sensor with an exposed view of the bilayer coil structure for wireless data transmission and the cuff-type pulse sensor wrapped around the artery[125] (Copyright 2023, Wiley-VCH).

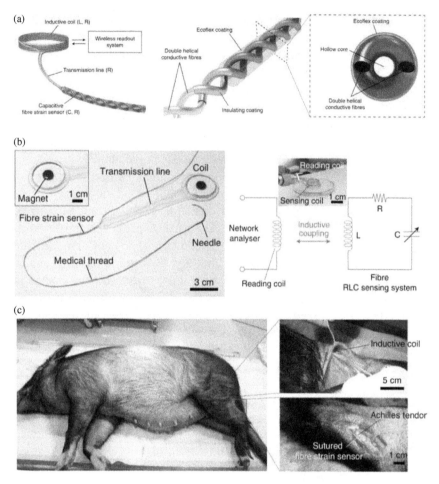

Figure 30. (a) Schematic illustration of a passive wireless strain-sensing system based on a fiber strain sensor, and design for the fiber strain sensor based on double helical stretchable conductive fibers. (b) Photograph of the suturing system consisting of the wireless fiber strain-sensing system and medical suturing thread. (c) Photograph of the minipig used in the *in vivo* experiment. The fiber strain sensor was sutured onto the Achilles tendon of a posterior leg and the inductive coil was placed under the skin of the thigh near the implantation site of the sensor[126] (Copyright 2021, The Nature Publishing Group).

strain-sensing system was created by combining a capacitive fiber strain sensor with an inductive coil for wireless readout.[126] The sensor is composed of two stretchable conductive fibers organized in a double helical structure with an empty core and has a sensitivity of around 12.

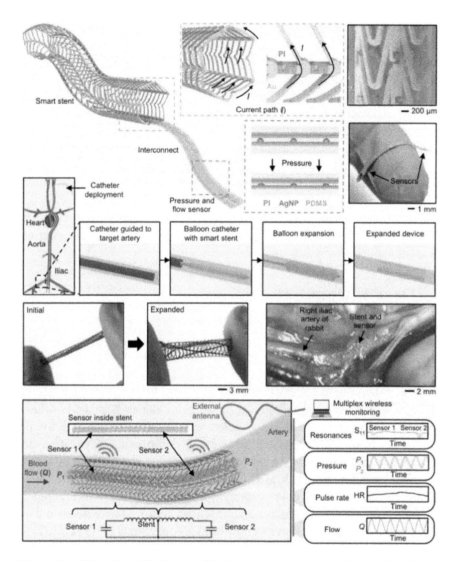

Figure 31. Illustration of the implantable electronic components and the wireless design and sensing scheme to simultaneously monitor pressure, heart rate, and flow[127] (Copyright 2022, American Association for the Advancement of Science).

Mathematical analysis and simulation of the sensor can effectively predict its capacitive response and can be used to modulate performance according to the intended application. The system is further used to continuously monitor the biomechanical strain of an Achilles tendon during movements using an *ex vivo* and *in vivo* porcine leg model. In addition, a fully implantable, wireless vascular electronic system, consisting of a multi-material inductive stent and printed soft sensors capable of real-time monitoring of arterial pressure, pulse rate, and flow without batteries or circuits.[127] The design, materials, and fabrication strategies of the inductive stent are developed to enhance wireless capabilities while maintaining key aspects of conventional stents. The fully printed capacitive pressure sensors demonstrate fast response times, high durability, and sensing at small bending radii. The device is monitored via inductive coupling at communication distances notably larger than prior vascular sensors. The wireless electronic system is validated in artery models, while minimally invasive catheter implantation is demonstrated in an *in vivo* rabbit study. The vascular system offers an adaptable framework for comprehensive monitoring of hemodynamics.

2. Flexible Optoelectronic Sensor

The basic idea of optical sensing technology is to convert light modulation into electrical signals to reconstruct static or dynamic images, motion, biosignals, etc.[128–132] Benefiting from the non-invasive, versatile and high-speed nature of light, optoelectronic sensing has been explored in areas such as military or space exploration, while already being widely adopted in commercial products (e.g., food production, cameras, smart watches, etc.).[133] In recent years, flexible optoelectronic sensors that can conform to curved surfaces and withstand mechanical deformation have attracted even more interest, as they hold great promise for applications in smart and wearable devices, such as flexible displays, artificial skin, etc.[134–136] In this section, the basic index parameters and representative technologies of flexible photoelectric sensors are presented, followed by a detailed discussion of their representative applications by functional classification.

2.1. Basic structures and characteristics

To achieve a high-performance flexible photodetector, there are generally several aspects to weigh up, as shown in Figure 32, from the device to the material, to the design, to the system, in line with the requirements of the application environment.[137] Of these, three main aspects need to be considered in a practical study, the photodetector, the flexible substrate, and the design layout.

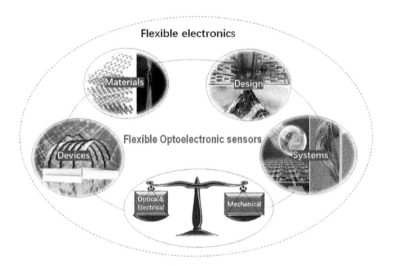

Figure 32. Development of flexible optoelectronic sensor.[137] (Copyright 2023, The Royal Society of Chemistry).

Firstly, the photodetector is the device that converts light energy into electrical energy and is the core of the photoelectric sensor. Table 1 shows the overall performance indicators for the photodetectors.[138–140] There are two main types of photodetectors: photodiodes and photoconductors. The former is usually PN junctions, which can also be extended to PIN or Schottky structures, and consist of a built-in electric field in which excess carriers are shielded and dark currents are suppressed, while the presence of a potential barrier prevents carrier multiplication (e.g., avalanche photodiodes) from occurring. As a result, photodiodes generally have a low response but a large bandwidth, which facilitates high-speed communication. Photoconductors, on the other hand, are photosensitive materials with ohmic contacts that operate on the basis of the photoconductivity effect, and if one type of charge carrier (electron or hole) can cycle many times before recombining, photoconductivity gain can be achieved.[141] This may be due to differences in mobility between electrons and holes (i.e., one type moves faster) or/and more commonly due to the tendency of traps to trap one type of carrier.

Substrates provide mechanical support and usually form the bulk of the weight/thickness of the equipment, in principle, any sufficiently thin material can be flexible.[142–144] Mechanical flexibility or tensile properties

Table 1. Figures of merit of photodetectors.

Figure of merit	Definition	Explanation
Responsivity (R)	$R = \dfrac{I_{ph}}{P_{in}}$ Iph: produced photocurrent Pin: incoming light power	R indicates the sensitivity of photodetector.
External quantum efficiency (EQE)	$EQE = \dfrac{I_{ph}}{P_{in}}\dfrac{h\nu}{e} = R\dfrac{h\nu}{e}$	EQE directly correlates to R
Internal quantum efficiency (IQE)	IQE = EQE/Aa Aa: Light absorption ratio	IQE indicates the photon-to-electron conversion efficiency of the absorbed photons
Response time (τ_r)	$\tau_r = \sqrt{\tau_{diff}^2 + \tau_{drift}^2 + \tau_{RC}^2}$	τ_r shows the operational speed of the device
Diffuse time (τ_{diff})	$\tau_{diff} = \dfrac{L_{diff}^2}{D} = \dfrac{eL_{diff}^2}{k_B T \mu}$ D: diffusion coefficient; L_{diff}: diffusion length μ: carrier mobility	τ_{diff} is the diffusion time of carriers at non-equilibrium
Drift time (τ_{drift})	$\tau_{drift} = \dfrac{W}{\mu E}$ W: width of region with electrical field;	τ_{drift} is the time to travel under electrical field
3 dB bandwidth (f_c)	$f_c = \dfrac{1}{2\pi\tau_r}$	f_c shows the frequency characteristics (maximal operating frequency) of the photodetector
Noise equivalent power (NEP)	$NEP = \dfrac{I_{noise}}{R}$	NEP equals to the input light power at 1 Hz bandwidth, signal-to-noise ratio (SNR) of 1, which describes the minimal power can be detected

Table 1. (Continued)

Figure of merit	Definition	Explanation
Linear dynamic range (LDR)	$LDR = 10 \times \log_{10}\left(\dfrac{P_{sat}}{NEP}\right)$ [dB] P_{sat}: the saturation light power under which the photocurrent starts to deviate its linearity	LDR is used to characterize the light intensity range where the detector has a constant responsivity
Detectivity (D^*)	$D^* = \dfrac{\sqrt{AB}}{NEP} = \dfrac{R\sqrt{AB}}{I_{noise}} = \dfrac{R\sqrt{A}}{S_n}$ S_n: noise spectral density	D^* is more universal parameter that allows comparison among different types of photodetectors

depend to a large extent on the substrate, which needs to have low oxygen and moisture permeability, high optical transparency, low surface roughness and excellent bending ability.[145]

Most inorganic substrate materials such as metals and glass have a high modulus of elasticity and a low strain at break and are therefore unsuitable for applications where a high degree of deformability is required. Similarly, metallic foils with good mechanical flexibility are usually opaque and are not suitable for bottom emission and double-sided emission in lighting or display applications. In addition, thin glass tends to break during repeated bending and the application of polymers is limited by the coefficient of thermal expansion. In contrast, organics/polymers such as polyethylene naphthalate (PEN), polyimide (PI), polyethylene terephthalate (PET), PDMS and hydrogels have a lower modulus of elasticity and a higher elongation and are often used as flexible substrates.[146–149]

Amongst these, paper substrates offer many unique advantages, including renewable, biodegradable characteristics. Compared to conventional paper substrates, which have a rough surface that leads to degraded performance, the new paper substrates are smooth and highly transparent, making them more reliable for use in flexible organic optoelectronic devices.[150–152] A simple template-transfer method was used to combine plastic and paper to form ultra-flat substrates with optical coupling capabilities that are lightweight, cost-effective and highly transparent.[153,154] Also, PDMS is widely used as a substrate due to its high flexibility and stretchability. Oh et al. investigated a PDMS-based transistor by introducing 2,6-pyridinedicarboxamide to the part of the conjugated polymer backbone of a stretchable semiconductor polymer developed to produce hydrogen bonding units.[155] The organic transistor made from the stretchable semiconductor polymer exhibited a mobility of up to 1.3 cm 2 V^{-1} s^{-1} and a high on/off ratio of over 10.[6] The transistor can be strained up to 100% with a slow linear decrease in mobility. After releasing the strain, the mobility almost returns to the pre-strain value.

In addition to the inherent flexibility of the material, the design of the electronic device/circuit layout is crucial to the ultimate flexibility of the overall system.[156,157] Depending on the degree of flexibility (or type of deformation), flexible devices can often be characterized by bendability and/or stretchability.[158,159] Flexibility can be easily achieved by thinning a single material. However, electrode devices are often composed of multiple layers and the complexity increases exponentially; therefore the design of the architecture becomes important to design an overall flexible and stretchable system.[160,161]

2.2. Epidermal sensing

The most common application of flexible wearable devices is epidermal sensing attached to the skin surface. Optoelectronic devices use optical signals to interact with biological systems with unique features such as high temporal/spatial/spectral resolution, deep tissue penetration and wireless energy/signal transmission for the detection of various physiological signals.[162–165] Application scenarios in this category require a focus on the degree of fit to the skin and the effect of skin flexion on sensing accuracy, so when integrated using flexible and stretchable substrates with mechanical properties similar to human skin, they can be conformally attached to the epidermis to perform various sensing functions.[166,167]

Pulse, blood oxygen and muscle contractions are often detected by photoelectric sensors.[168] A battery-free and stretchable photonic system was reported in which multiple photodetectors and LEDs of different colors are transferred onto a flexible tape and interconnected with external circuitry, as shown in Figure 33(a).[169] The infrared LEDs transmit light

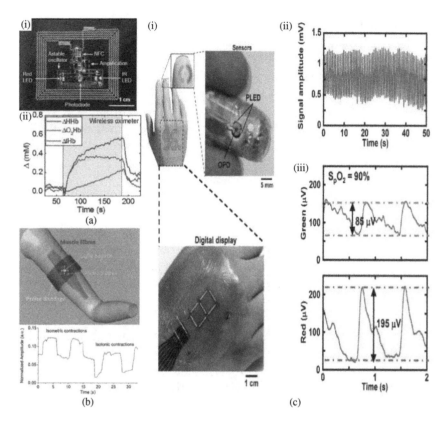

Figure 33. Epidermal sensing applications. (a) Tissue oxygenation levels detection[169] (Copyright 2016, American Association for the Advancement of Science). (b) Muscle contraction sensor[170] (Copyright 2015, Wiley-VCH). (c) Ultra-flexible multi-sensing wearable device[171] (Copyright 2016, American Association for the Advancement of Science).

signals to penetrate the skin and the transmission response is measured by adjacent photodiodes to determine tissue oxygenation levels in real-time based on the difference in spectral response between oxygenated and deoxygenated haemoglobin and read by a coupling chip in a smartphone. The results of the assay are quite consistent with commercial oximeters. Samuel et al. have developed a muscle contraction sensor consisting of four photodiodes around a centrally located light source as shown in Figure 33(b).[170] The design makes the OLED and OPD more flexible to improve the accuracy of the results and better track muscle movement and tissue oxygenation. The result is able to measure and differentiate between isometric and isotonic muscle contractions in order to control the movement of the robotic arm. Someya *et al.* have developed an ultra-flexible multi-sensing wearable device with tricolor OLEDs and OPD for tracking and displaying pulse, blood oxygen levels and other information on human skin.[171] The device includes a smart optoelectronic skin system with a pulse oximeter and display that provides long-term stable detection of pulse waves as shown in Figure 33(c).

2.3. *Implantable sensing*

In general, the penetration depth of light signals in the visible and near-infrared wavelengths into biological tissues ranges from a few tens of micrometers to a few millimeters.[172] Penetration is limited due to complex light-biomass interactions such as scattering and absorption.[173,174] Therefore, advanced implantable optoelectronic devices and systems are required in order to achieve deep tissue light transmission (>1 cm).[175] Implantable optical solutions offer the advantage of using light to stimulate and detect specific neuronal activity via genetic coding compared to conventional electrophysiological (EP) techniques.[176] In recent years, thin-film, miniature optoelectronic devices consisting of detectors, sensors, LEDs and lasers have enabled advanced optical neural interfaces.[177,178] These packaged and ultra-miniaturized devices can be injected into tissues for direct biointegration, with multi-point recording/stimulation, wireless operation and multimodal sensing/modulation capabilities.

Micro-optical systems have made considerable progress in implantable sensing.[179] Bruchas *et al.* designed a cellular-level implantable photopole as shown in Figure 34(a).[180] The InGaN LED, microelectrode, silicon photodiode and temperature sensor are all integrated on an ultra-thin flexible needle (<10 μm) and powered by an RF driver circuit board, enabling optogenetic stimulation of channel retinoid 2-expressing neurons and control of neural activity *in vivo*. Further and more complete integration, such as laminating a nerve-type electrode, photodetector, temperature sensor and LED onto a single injection needle, enables extremely small dimensions and opens the way for real-time neural activity monitoring and biosensing as shown in Figure 34(b).[181] Meanwhile, with the advancement of microfluidics, microfluidics, and miniature LEDs are integrated to allow further intelligent sensing through closed-loop feedback for simultaneous

Flexible Sensors 183

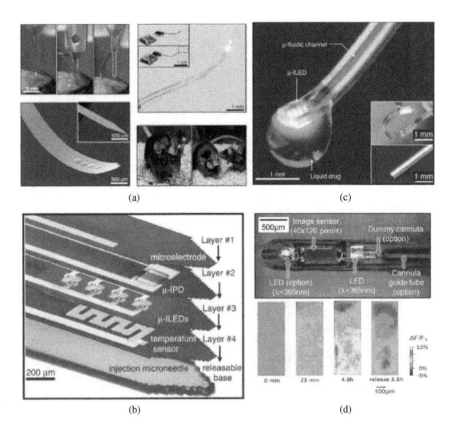

Figure 34. Implantable sensing applications. (a) Implantable photopole[180] (Copyright 2013, American Association for the Advancement of Science). (b) Single injection needle combined nerve-type electrode, photodetector, temperature sensor and LED[181] (Copyright 2015, Elsevier Inc). (c) Ultra-flexible sensors consist of microfluidics and micro-LEDs[182] (Copyright 2015, Elsevier Inc.). (d) Custom Si CMOS imaging sensor with a micro-LED[183] (Copyright 2017, IEEE).

neural stimulation and drugs, as shown in Figure 34(c).[182] A custom Si CMOS imaging sensor and a micro-LED are combined on a flexible sheet to be implanted in tissue for high-resolution deep brain fluorescence imaging.[183] The micro-LED adjacent to the imaging sensor acts as a light source to excite green fluorescent protein-expressing cells and the imaging sensor captures and processes the fluorescence signal. This implantable imaging platform offers great opportunities for advanced neural activity imaging and other biosignal sensing in deep tissues in Figure 34(d).

2.4. Bionic system

The successful formation of devices with high-performance photon detection, modulation and emission capabilities on flexible and stretchable substrates has made it possible to develop advanced optoelectronic platforms that mimic the various structures and functions of biological systems.[184–187] In particular, as a key application for flexible image sensors, artificial vision consisting of a photosensitive unit and an optical lens can, with the right design, simulate the remarkable functions of the natural eye.[188,189]

A thin-film monocrystalline silicon photodiode array is transferred to a compressible elastic substrate to form a hemispherical camera similar to a human eye (Figure 35(a)).[157] This electronic eye camera does not have the complexity of a multi-lens system consisting of a planar imaging

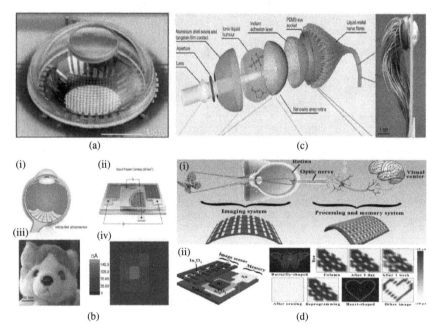

Figure 35. Bionic System. (a) Monocrystalline silicon-based photodiodes human[157] (Copyright 2008, The Nature Publishing Group). (b) Retina-like organic photosensor structure[190] (Copyright 2017, Wiley-VCH). (c) Hemispherical artificial retina[191] (Copyright 2020, The Nature Publishing Group). (d) Biological vision system[192] (Copyright 2018, Wiley-VCH).

sensor like a conventional digital camera. With a simple imaging lens, it achieves a wide viewing angle with low aberration. With the further implementation of dynamically deformable lenses and substrates, the e-eye camera system will have adjustable zoom capabilities. Similarly, analysis of the principles of image construction in the human retina focuses on the use of photoreceptor cells to distinguish between different wavelengths of photons and convert them into electronic signals. To mimic such a system and further extend the spectral range, Wang et al. demonstrated a retina-like organic photosensor structure with color differentiation and memory functions, as shown in Figure 35(b).[190] A voltage driver is used for light sensing, while a floating grid organic field effect transistor (FET) is used for data readout. An ultra-thin (800 nm) pixelated sensor array was fabricated to fit the eye of a husky plushie, which senses spatially resolved optical excitation, mimicking the imaging process in mammals. To simulate an artificial retina, Fan et al. fabricated a hemispherical artificial retina made from an array of calcium titanite nanowires, as shown in Figure 35(c).[191] This electrochemical device has a high degree of structural similarity to the human eye and has the potential to achieve high resolution imaging. In addition, the spherical design of this artificial retina ensures a more consistent distance between the pixels and the lens, resulting in a wider Field of View (FoV) and better focus on each pixel. In addition to the human eye, other chambered eye structures in the compound eyes of fish, birds and insects have been successfully mimicked for specific applications requiring dark vision/telescopic vision, wide FoV and deep Depth of Field (DoF). Shen et al. integrated In_2O_3 photodetectors with Al_2O_3 thin-film memristors, where light modulates the transition between high/low resistance modes to simulate a biological vision system (Figure 35(d)).[192] In addition to amnesic resistors, devices such as thin-film transistors, phase-change memories and ferroelectric devices can integrate photosensitive elements in a single pixel for simultaneous sensing-storage-processing.

3. Flexible Acoustic Sensor

Flexible acoustic sensors offer advantages in terms of wearability and portability for personalization, but can also clearly detect human voices in noisy environments or when a mask covers the mouth.[193–195] They exhibit an excellent frequency response and a wide range of sound pressure, improving the sound-sensing quality, which is crucial in

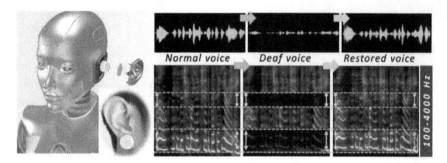

Figure 36. Triboelectric auditory sensor-based sound receiver used for a robot and worn as ear stud. Sound waves and corresponding acoustic spectrograms of normal, weakened, and restored voices[199] (Copyright 2018, American Association for the Advancement of Science).

user-friendly sound recognition systems for medical monitoring, intelligent human–computer interaction and the Internet of Things (IoT).[196,197] Flexible acoustic sensors for collecting and recognizing human voices need to work well within the basic speech frequency range (80–255 Hz) and standard telephone bandwidth (300–3400 Hz).[198] In particular, their vibration characteristics including frequency, amplitude and acceleration are of greater interest.

For instance, a soft triboelectric acoustic sensor was utilized as a light, user-friendly hearing aid (Figure 36).[199] To demonstrate the feasibility of the sensor as a hearing aid, hearing impairments in 207−837 and 1594−2090 Hz frequency regions were artificially introduced. The original voice file was treated to weaken the sound amplitudes of these frequency regions to −30 dB. The treated voice file was then played by a loudspeaker and the sensor was used to record the sound information. Thanks to its resonant property, the triboelectric acoustic sensor could naturally enhance the signals of the impaired frequency regions, successfully functioning as a hearing aid. Moreover, the triboelectric acoustic sensor is self-powered, single-channel, and cost-efficient, which could reduce power consumption and prolong the robot's operation. As shown in Figure 37, a soft microparticle-based acoustic sensor is waterproof and exhibits a significant enhancement in the working frequency range of 0.1–20 kHz due to the sound responsivity of its internal microparticles, almost covering the whole human audible range.[200] Additionally, the closed structure design and waterproof membrane contribute to superior waterproofness, enabling

Flexible Sensors 187

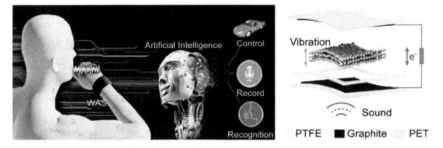

Figure 37. Schematics of the waterproof acoustic sensor as a sound-sensitive wristband for wearable human–machine interaction applications[200] (Copyright 2022, Wiley-VCH).

stability against human perspiration. The sensor could recognize five vocal commands ("left", "go", "right", "back" and "stop") with the use of an AI classification algorithm. When the speaker inputted vocal commands through the sensor and the mel-frequency cepstral coefficient features were extracted for each vocal command. These features were labeled to the specific vocal commands and were then used to train the learning algorithm. After ten rounds of training, unlabeled vocal commands could be classified into one of the aforementioned word commands with 98% recognition accuracy, successfully controlling a motor car.

Similarly, an ultra-thin, comfortable, and vibration-responsive device was further proposed which detects the acceleration of the skin, which is highly linearly correlated with voice pressure.[201] The device consists of a cross-linked ultra-thin polymer film and a perforated diaphragm structure that can quantitatively sense sound. It has an excellent sensitivity of 5.5 V Pa^{-1} in the speech frequency range. In addition, this ultra-thin device exhibits excellent skin consistency, which allows it to accurately recognize sounds. Accurate speech recognition as it eliminates vibrational distortion from rough and curved skin surfaces. The device is suitable for several promising speech recognition applications such as security authentication, remote control systems and acoustic healthcare. When attached to a throat, the sensor was connected to a speech-recognition module and the phrase "Siyoung login" was set as a vocal password for an automatic door lock. Interestingly, the sensor could distinguish different speech waveforms and frequency spectrums of the voices of different individuals. When different individuals uttered the vocal password "Siyoung login", only an authorized user whose voiceprints matched the authorized database could open the automatic door lock. More importantly, the sensor could recognize the voice of the authorized user even when a mask was worn because its sensing mechanism was based on skin vibration, greatly reducing the risk of exposing the vocal password to others. The sensor was further used to remotely control a lamp. Because the sensor could recognize human voices by exploiting neck skin vibrations instead of sound waves transmitted through air, it could control the lamp effectively in both quiet and noisy environments. The sensor acquired identical voice sound waveforms of the phrase "light on" even under the presence of ~62 dB SPL noise whereas a commercial microphone could not filter out the noise and obtained dissimilar voice sound waveforms in quiet and noisy environments. Moreover, an ultrathin eardrum-like triboelectric acoustic sensor, consisting of silver-coated nanofibers with a thickness of only 40 μm, demonstrated its potential as a hearing aid (Figure 38).[202] When used synergistically with deep learning, the sensor

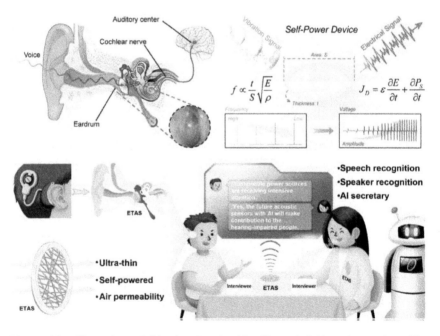

Figure 38. Illustration of biomimetic ultrathin film mimicking the eardrum-like membrane of the human cochlea for converting analog sound waves into electrical signals, and application of the sensor with hearing aids for voice-text conversion to allow a hearing-impaired person to interview[202] (Copyright 2022, Wiley-VCH).

performed highly accurate voice-to-text conversions, potentializing an easier way to conduct interviews for individuals with hearing impairments. The algorithm extracted features from the raw output of the sensor and took them as inputs to train the learning algorithm. After 70 rounds of training, eight different spoken words were recognized and converted into texts on a computer screen with 92.64% accuracy. This work presents a strategy for self-powered auditory systems that could significantly accelerate the miniaturization of self-powered systems for wearable electronics, augmented reality, virtual reality and automated control centers. In addition, a highly sensitive and miniaturized piezoelectric mobile acoustic sensor (PMAS) is demonstrated by exploiting an ultrathin membrane for biomimetic frequency band control (Figure 39).[203] Simulation results prove that the resonant bandwidth of a piezoelectric film can be

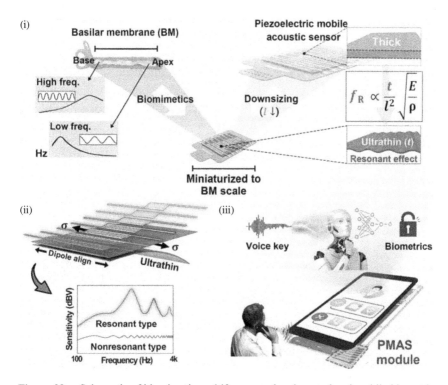

Figure 39. Schematic of biomimetic multifrequency band control and mobile biometric authentication of miniaturized PMAS[203] (Copyright 2021, American Association for the Advancement of Science).

broadened by adopting a lead-zirconate-titanate membrane on the ultrathin polymer to cover the entire voice spectrum. Machine learning–based biometric authentication was demonstrated by the integrated acoustic sensor module with an algorithm processor and customized Android App. Exceptional error rate reduction in speaker identification is achieved by a PMAS module with a small amount of training data, compared to a conventional microelectromechanical system microphone.

Skin-mounted soft electronics, including high-bandwidth triaxial accelerometers, capture a wide range of physiologically relevant information, including the mechano-acoustic signature of underlying body processes (such as those measured by stethoscopes) and the precise kinematics of core body movements. Rogers *et al.* proposed a soft patch-based wireless device for continuous measurement of mechano-acoustic sensing ranging from fine skin vibrations with accelerations of approximately 10^{-3} m s^{-2} to large movements of the entire body at frequencies up to approximately 800 Hz at approximately 10 m s^{-2} while compliantly fixing the device to the suprasternal notch (Figure 40(a)).[204] A schematic layout of the entire device is shown in Figure 40(b), illustrating the design of the device consisting of a serpentine interconnect, electronics, an integrated accelerometer sensor and a silicone elastomer-based patch. For wireless access to physiological information, a low-energy Bluetooth module is assembled with a processing unit integrated circuit (IC) for visualizing biomedical data on a smartphone, together with a wireless charging circuit, the whole process of which is depicted in the block diagram in Figure 40(c). By placing the device around the neck, mechanical acoustic sensing (X, Y, and Z axes) was monitored in real-time, demonstrating sensing changes under different conditions (sitting, holding breath, talking, swallowing, walking, and jumping) (Figure 40(d)). Different types of mechanical acoustic signal information were measured during the experiment, advancing the development of wearable device technology in the biomedical field. In addition, an electret-powered, perforated polymer diaphragm is utilized in a skin-attachable auditory transducer.[205] The optimized charged electret diaphragm induces a voltage bias of >400V to the counter electrode, which reduces the need for bulky power supplies and allows the capacitive sensor to display high sensitivity (2.2V Pa^{-1}) with the addition of an elastic nanodrop seismic mass. The precision capacitive construction with low mechanical damping gives a flat frequency response and good linearity. A pore-type electret diaphragm helps this skin-attachable transducer to detect only the vibrations of the skin of the neck, not the

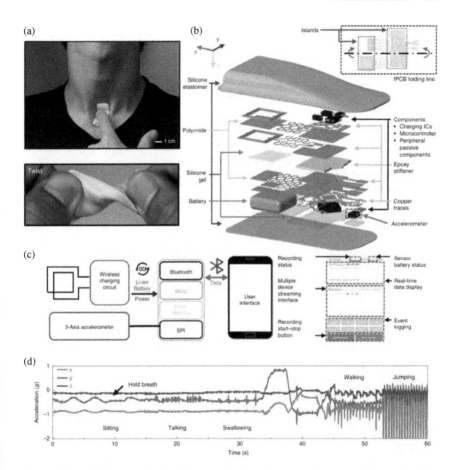

Figure 40. (a) Images of soft-device mechanics during movements of the neck while interfaced to the suprasternal notch and its twisted (right) configuration. (b) Exploded illustration of the active components, interconnect schemes and enclosure architectures. (c) Block diagram of the system operation. (d) Three-axis time series data simultaneously recorded over a 60 s interval as a subject engages in various activities that include sitting at rest, talking, drinking water, changing body orientation, walking and jumping[204] (Copyright 2020, The Nature Publishing Group).

dynamic air pressure, allowing the human voice to be detected in harsh acoustic environments. The sensor operates reliably even in the presence of surrounding noise and when the user is wearing a gas mask. As illustrated in Figure 41, the sensor shows strong potential as a communication tool for disaster response and quarantine activities, as well as a diagnostic

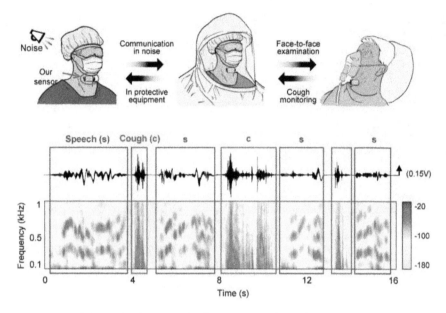

Figure 41. Schematic of communication system and cough monitoring in infectious-disease outbreak response, and vocal waveform and frequency spectrum of a female participant measured by the device when the participant coughed while speaking[205] (Copyright 2022, Wiley-VCH).

tool for vocal health applications such as cough monitoring and voice dosimetry. Wen et al. designed a sensor that can detect and recognize weak muscle movements and skin vibrations, such as word pronunciation and carotid pulsations.[206] By combining the proposed deep learning model based on a number recognition convolutional neural network (NR-CNN), speech recognition of different number pronunciations that occur frequently in everyday conversations can be achieved. By training and testing the proposed NR-CNN and the large amount of data recorded by the sensors, a high recognition accuracy (91%) can be achieved.[141] Flexible acoustic sensors should have sensing performance comparable to that of human hearing. Considering human voice (100–8,000 Hz, 40–70 dB_{SPL}) and ambient quiet sound (30 dB_{SPL}) and noise (120 dB_{SPL}, e.g., rock concerts) environments, the sensor must have sufficient bandwidth (20–10,000 Hz) and a sound pressure range (30–120 dB_{SPL}). Furthermore, acoustic sensors should be able to compensate for the inherent deficiencies in human hearing ability and the degradation of hearing due to aging

Flexible Sensors 193

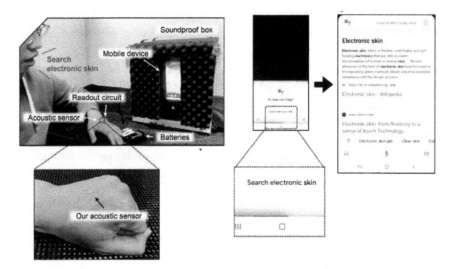

Figure 42. Photograph of the voice-recognition application setup for demonstration of the skin-attachable acoustic sensor with a readout circuit, which was directly connected to a mobile electronic device[207] (Copyright 2022, Wiley-VCH).

or declining health conditions. A skin-applicable acoustic sensor with higher sensing accuracy in a wider hearing field than the human ear, a flat frequency response (15–10,000 Hz) and a good linearity range (29–134 dB_{SPL}), as well as high conformality to flexible surfaces and human skin, is presented (Figure 42).[207] This high acoustic sensing quality is achieved by exploiting the low residual stress and high processability of the polymer material in a diaphragm structure designed using acoustic-mechanical-electrical modeling. As a result, this acoustic sensor shows high acoustic fidelity by sensing human audible sounds, even loud and low frequency sounds that cannot be detected by the human ear without distortion.

4. Flexible Biosensor

Biosensors have been widely used for disease detection, diagnosis, treatment, patient health monitoring and human health management, and have attracted significant attention in the biomedical and healthcare fields.[208,209] In particular, flexible and wearable forms of biosensing are able to function at various biological interfaces that are soft, curved, and irregular, enabling real-time and effective signal detection for versatile and intelligent real-time sensing.[210–212]

In this section, the basic characteristics, sensing principles, and main technologies of flexible biosensors are discussed; in addition, the development and applications of flexible biosensors in biomarkers, physiological signals, and pharmaceutical therapeutics are mainly discussed according to functional classification. This chapter is expected to provide insights into building flexible biosensor architectures and systems and to inspire and improve flexible biosensors in the future for continuous long-term health monitoring in clinical and daily healthcare.

4.1. Basic features and representative technology

4.1.1. Fundamental structure

Biosensors are defined as an analytical device that detects biological signals and converts them into recognizable electrical signals. Therefore, like other sensors, a biosensor consists of several fundamental structures, such as the analyte, biorecognition marker, transducer, and reading device, etc., as shown in Figure 43(a).[213]

- **Analytes:** The biological signal is being recognized and measured. As shown in Figure 43(b), this includes antibodies, aptamers, enzymes, MIPS, membrane proteins, nucleic acids, etc.
- **Bio-receptor:** The bio-receptor is a substance that recognizes an analyte and binds or reacts specifically with it only.
- **Transducer:** The transducer is a device that converts the analyte directly or indirectly through the binding of a bioreceptor to another signal, can be considered as a signal converter. The transducer produces proportional signals originating from the amount of analyte or interaction between the bioreceptor and analyte. Several techniques have been developed including electrochemical, electrical, optical, Perfect Electric Conductor (PEC), Quartz Crystal Microbalance (QCM), cantilever, etc., as shown in Figure 43(b).
- **Reading device:** A reading device is a device that reads the signal generated by the transducer and visualizes it, sometimes integrated with the transducer or with a signal processing module.

Flexible Sensors 195

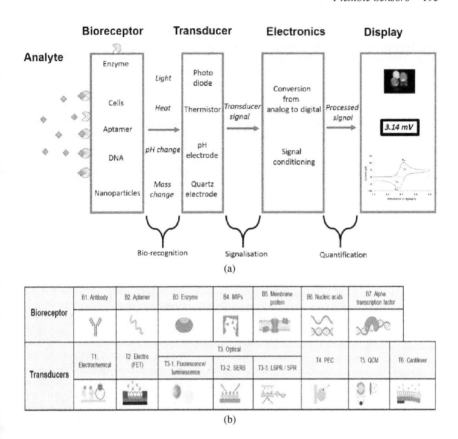

Figure 43. Development of flexible biosensors[213] (Copyright 2016, Springer Inc).

4.1.2. Fundamental characteristics

When developing biosensors, there are a variety of important parameters that need to be taken into account and focused on, and these characteristics directly determine the strengths and weaknesses of the biosensor.

4.1.2.1. Selectivity

Selectivity represents the ability and sensitivity of the biosensor to specifically identify the analyte. Selectivity is an important characteristic to consider while selecting a bio receptor for a biosensor. Here, the bio receptor must detect a specific analyte element even in a sample mixture consisting of both analyte and unwanted contaminants. It can generate positive results only when the interaction occurs between the target molecule and the bio receptor and are not affected by other contaminants. In clinical applications, it is still hard for biosensors to present high selectivity because of the complex biological sample components. Sometimes, biosensors generate false-positive result rates. To prevent it, selectivity is critical for the creation of reliable point-of-care biosensors and requires more effort.

4.1.2.2. Sensitivity

Sensitivity represents the relationship between the concentration of the substance to be measured and the intensity of the electrical signal converted by the transducer, and the quality of this performance directly determines the detection capability of the biosensor. It also determines the upper and lower limits of detection and therefore affects whether certain biomarkers can be detected, as some of them have a very low concentration range.

4.1.2.3. Reproducibility

Reproducibility is the consistency of the results of multiple detections of the same analyte concentration and is one of the fundamental properties that characterize a biosensor's performance. This performance is particularly affected by biological activity, such as enzymes, antibodies, etc., which are highly susceptible to changes in parameters such as temperature, pH, etc. Once inactivated, the sensing process and results can be inaccurate. Therefore, especially when developing biosensors for the market, this property needs to be taken into account.

4.1.2.4. Stability

Stability represents the continuous monitoring capability of a biosensor. Biosensors represented by enzymes can be limited by the loss of sensing due to depletion of the enzyme, while long-term robust real-time monitoring is one of the prerequisites for biosensors to be widely used by people, hence the need to include stability as one of the development indicators.

4.1.2.5. Flexibility

Flexibility enables biosensors to be stretchable, flexible and wearable, allowing them to be applied directly to the surface of the body or inside organs, which can enhance the detection performance and suitability.

4.1.2.6. Biocompatibility

Biocompatibility needs to be considered in environments where biosensing applications come into direct contact with the human body or are implanted in the human body.

4.2. Representative techniques

4.2.1. Electrochemical biosensor

Electrochemical biosensing is the most commonly used biosensing technology at present, with a wide range of electrochemical analysis techniques such as cyclic voltammetry (CV), differential pulse voltammetry (DPV), amperometric techniques, and electrochemical impedance spectroscopy (EIS).[214–216] Those analyses allow for a wide range of options for flexible electrochemical biosensing besides inherent rapid response time, high sensitivity, and compact size.[217,218] The electrochemical biosensor detects electrical or electrochemical signals derived from the reaction between target molecules and sensing probes.[219–222] Therefore, to enable its high flexibility and detection ability, a flexible and conductive substrate is the basis of the system, followed by high-performance fabrication techniques.

As shown in Figure 44, several materials including polymers, conductive polymers, conductive nanomaterials, and biomaterials etc have been developed recently.[223] Additionally, several highly effective fabrication techniques including sputtering, lithography, printing and electrodeposition, etc. are used to further construct high-performance flexible electrochemical biosensing systems.

4.2.2. Optical biosensor

Optical flexible biosensors are widely studied and used because of their non-invasive nature, ease of observation, and deep tissue monitoring capabilities.[224–226] Depending on the principle, they are classified as

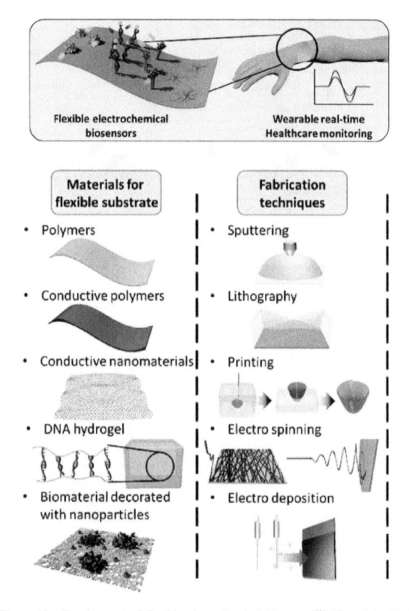

Figure 44. Development of flexible electrochemical biosensors[221] (Copyright 2017, American Association for the Advancement of Science).

surface-enhanced Raman scattering (SERS), localized surface plasmon resonance (LSPR), EL, colorimetry, fluorescence, and chemiluminescence (CL), as illustrated in Figure 45.[227]

Flexible Sensors 199

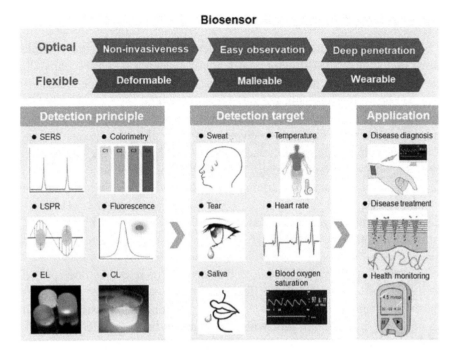

Figure 45. Development of flexible optical biosensors[227] (Copyright 2022, Elsevier Inc).

Regardless of the technology, the key point of research remains the development of flexible substrates and optical functionalization. For example, for SERS technology, substrates including flexible polymers, paper, nanowires etc. can provide not only flexibility but also good uniformity, bendability and optical transparency.[228,229] For LSPR, flexible substrates (e.g., polydimethylsiloxane, polyvinyl alcohol, polyethylene terephthalate and shape memory polyurethane) demonstrate excellent portability, dynamic stability, and customizability, showing attractive properties in terms of miniaturization, minimal disruption, and scalable production, a good combination with LSPR.[230,231] The construction of all flexible optical sensors requires consideration of the choice of flexible materials required for their optical function applications.

4.2.3. Field effect transistor biosensor

FET-based devices are semiconductor devices consisting of metal oxide semiconductor (MOS) structures. It features exceptional sensing capabilities with the benefits of flexibility, reduced complexity, and lightweight.[232]

As shown in Figure 46, when the metal potential changes, the electric field causes a corresponding bending of the energy band in the semiconductor channel.[233] This leads to changes in channel carrier concentration, such as accumulation, depletion, or inversion. Therefore, the gate potential can be controlled by other factors, such as the solution potential (e.g., pH) or the charge of a biomolecule, further influencing the state of channel carriers and shifting the current–voltage characteristics. This process is presented by the difference in threshold voltage. Therefore, the difference in threshold voltage (ΔV_{th}) can be calculated as an indicator of sensitivity.

To develop flexible FET-based biosensors, flexible substrates play a key role and can profoundly affect performance, and therefore require excellent mechanical properties to withstand bending and twisting, as well as excellent biocompatibility with their surrounding biological systems. The most commonly used substrates in flexible FET-based biosensors include PET, PI, PEN, Parylene C, PDMS, fiber (textiles), paper, etc.

4.3. Biomarker detection

Biomarkers distributed in various body fluids, including various glucose, uric acid, ions, metabolites, cancer markers, etc. are still the main targets of biosensors. Depending on the source of the test fluid, there are three main categories: invasive body fluid, minimally invasive body fluid and non-invasive. In the following sections the representative samples and commonly used body fluids are described in detail.

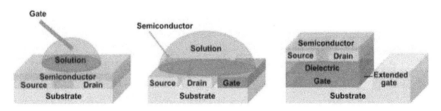

Figure 46. Different configurations of FET-based devices[233] (Copyright 2023, MDPI).

4.3.1. Invasive body fluid

Invasive body fluids such as blood and cerebrospinal fluid can be used to detect metabolites, proteins, and biomarkers that can be used for disease and health monitoring.

4.3.2. Blood

Testing for markers in the blood is by far the most commonly used clinical test, as some metabolites are evenly distributed in the blood and are easily detected, reflecting directly on the body's condition.[234] For example, glucose, in which diabetics need to have their blood glucose levels checked regularly to prevent their bodies from going too high or too low. A novel flexible graphene paper electrode based on 3D nanoporous gold (NPG)-supported graphene paper on binary platinum alloy NPs for glucose sensing is reported (Figure 47(a)).[235] The fabricated sensor showed a low detection limit of 5 μM and a wide detection range of 35 μM to 30 mM, which was also successfully applied to the detection of glucose in human serum samples.

4.3.3. Cerebrospinal fluid

Because of its location in the brain, cerebrospinal fluid contains many biomarkers (such as neuropeptides, proteins and neurotransmitters) associated with central nervous system disorders.[236] Abnormal concentrations of markers such as norepinephrine (NE) and 5-hydroxytryptamine (5-HT) may be directly related to several disorders such as anxiety, depression and Parkinson's.[237,238] Detection of these markers requires invasive sampling, necessitating *in situ* fabrication of a biocompatible flexible sensing device for cerebrospinal fluid analysis.

4.3.4. Minimally invasive body fluid

4.3.4.1. Interstitial Fluid (ISF)

The ISF is a body fluid that transports nutrients and metabolites between cells, blood and capillary lymphatics through capillary exchange, and the rich biomarkers in the ISF provide a variety of information about human health.[239] Because of the close correlation between ISF and blood,

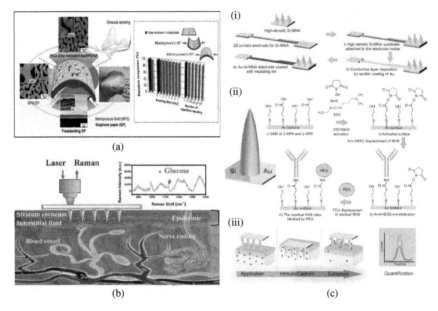

Figure 47. Biosensors for biomarker detection. (a) Left: a flexible PtCo/NPG/graphene paper (PtCo/NPG/GP) electrode for detecting glucose in human blood; Right: effects of bending states on the current responses of 10 mM glucose on flexible electrode[235] (Copyright 2016, Elsevier). (b) Preparation principle of the microneedle Ranman sensor used to detect glucose. (c) Schematic diagram of a wearable MNs sensor for breast cancer biomarker detection: (i) Preparation of MNs electrode; (ii) Au-Si-MNA electrode surface modification and immobilization of anti-HER2 monoclonal antibody; (iii) detection principle of MNs immunosensor. Reproduced with permission.

invasive monitoring of certain biomarker levels in ISF has the potential to reflect their true levels in the blood. The ISF is usually extracted by microneedle extraction and combined with other techniques such as optical, or electrochemical, detection of certain biomarkers. A novel SERS glucose microneedle sensor prepared from polymethyl methacrylate (PMMA) for direct *in situ* extraction and measurement of ISF is presented. The researchers coated the PMMA surface with silver nanoparticles (Ag NPs) to enhance the Raman surface signal (Figure 47(b)).[237] After modification with the glucose trapping agent 1-decanethiol (1-DT), the MNs can detect glucose *in situ* in the ISF of diabetic mice. there are five key points to the MNs sensor array: (i) PMMA MNs can successfully penetrate the skin; (ii) PMMA has high light transmission, which

Flexible Sensors 203

facilitates the transmission of Raman signals through the MNs; (iii) Ag NPs can enhance the Raman signal; (iv) 1-DT has the effect of enriching glucose at the tip of MNs; (v) the micropores left by the MNs array can self-heal within 10 min.

Through interstitial fluid, it is also possible to detect cancer markers to diagnose cancer, such as breast cancer. A wearable MNs device was developed for transdermal detection of the breast cancer biomarker ErbB2 (Figure 47(c)).[238] First, Si-MNs electrodes were prepared using UV lithography and deep reactive ion etching (DRIE), then the Si-MNs substrate was mounted on a 3D printed support and cured with an adhesive to obtain Si-MNs electrodes. Next, Au films were deposited and insulating inks were applied to the gold-plated support to obtain Au-Si-MNs. The electrodes were incubated overnight in an aqueous 3-mercaptopropionic acid solution to react thiol groups with the Au surface to form self-assembled monomolecular layers (SAMs). Next, the electrodes were immersed in a mixture of EDC and NHS, and then the NHS ester-functionalized Au-Si-MNs electrodes were immersed in an anti-HER2 antibody solution. Other NHS sites were combined with NH_2 functionalized PEG. Antibody-antigen binding was monitored by measuring the change in current intensity obtained before and after incubation with $ErbB_2$ artificial ISF.

4.3.4.2. Non-invasive

Non-invasive body fluid detection techniques focus on the detection of extracorporeal fluids such as sweat, tears, and saliva or non-invasive techniques to deeply reach inner body liquid.[240–242] The distribution of biomarkers in these fluids is relatively low and the challenge is that they are difficult to collect.

4.3.4.3. Sweat

High levels of biomarkers (e.g., glucose, lactate, uric acid, and ions) in sweat and non-invasive detection of biomarkers in sweat show attractive promise for establishing health monitoring devices.[243] Xiao *et al.* reported a wearable colorimetric optical biosensor that detects glucose concentration in sweat (Figure 48).[244] They integrated a microfluidic chip to collect the sweat from the human surface and a check valve on the sensor. Its flexible substrate enables its attaching to the surface well. The sensor can

204 Y. Guo et al.

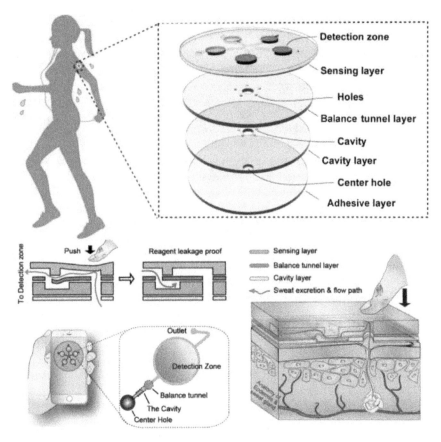

Figure 48. A wearable colorimetric sensor based on microfluidic chips that detects sweat glucose easily and quickly[244] (Copyright 2019, American Chemical Society).

perform five parallel tests at the same time and can detect subtle differences in sweat glucose with a detection limit of 30 mM. The results of human glucose concentration detection are visualized by the change in color, which allows accurate dosing of the corresponding drug to control blood glucose concentration. A wearable microfluidic platform was designed, which can provide accurate and multiplex colorimetric analysis of glucose, lactate, pH, chloride, and sweat in a wide variety of ambient lighting conditions.[245] This system has promising applications as it can be applied to the human body and detect multiple chemicals simultaneously. Higher levels of lactate, an important metabolite of sweat, reflect inadequate oxidative metabolism. Electrochemical lactate sensors based on

Flexible Sensors 205

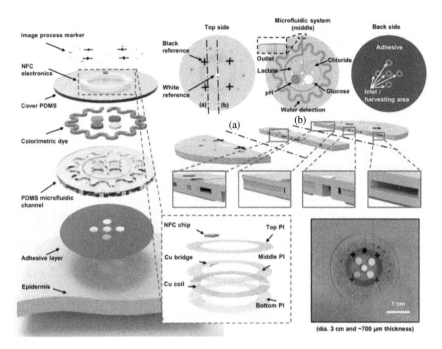

Figure 49. Schematic and optical image of an epidermal microfluidic biosensor for sweat monitoring, integrated with a flexible electronic device[245] (Copyright 2016, American Association for the Advancement of Science).

MIPs and Ag NWs modified flexible screen-printed electrodes were prepared by combining the advantages of high flexibility and conductivity of Ag NWs, where the MIPs were generated by the electropolymerisation of 3-APBA with lactate on Ag NWs and the subsequent washing of residual lactate to generate lactate anchor sites. The proposed flexible lactate sensor shows a linear concentration range of 10 μM–0.1 mM and an ultra-low detection limit of 0.22 μM. In addition, it allows for the non-invasive detection of sweat lactate on human skin during exercise (Figure 49).

4.3.4.4. Tears

Tears are often difficult to obtain but are always present in trace amounts in the eye, so flexible biosensors in the form of contact lenses have also been developed in great quantities.[246,247] As shown in Figure 50, a contact lens biosensing platform that combines multiple technologies and detects multiple biomarkers is already promising to build one of the future of wearables.[248, 249]

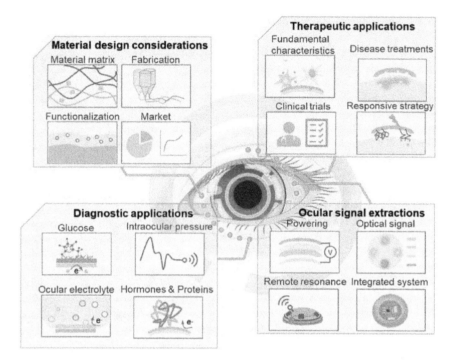

Figure 50. Lab-on-a-contact lens. A wearable SCL platform integrated with flexible power transfer systems is employed to wirelessly monitor physiological signals (intraocular pressure, corneal temperature, and pH), biomarkers (glucose, proteins, ions, and virus), and controlled drug release in diagnostic and therapeutic applications[248] (Copyright 2022, Wiley-VCH).

4.3.4.5. Others

Other body fluids such as urine and saliva have also received much attention and research because of their unique location and corresponding characteristic biomarkers to some diseases.

4.4. Biophysiological signal

4.4.1. Implantable in vivo signal

Many *in vivo* physiological signals are highly relevant for health monitoring and disease surveillance, especially EP signals such as electroencephalogram (EEG)/electrocorticography (ECoG) of the central nerve system (CNS), action potentials of peripheral nerve system (PNS).[250,251] The accurate and harmless detection of these signals in real-time is of great importance to the pharmaceutical and medical industries. Several flexible biological devices enhance the attachment of sensors to soft tissues such as blood vessels and the brain, allowing for obtaining more accurate and real-time data.

EEG and/or ECoG (known as intracranial electroencephalography, or iEEG) measures the voltage changes caused by ionic currents within neurons and has been an important method in cognitive neuroscience for understanding brain function and for the clinical study of neurological disorders such as epilepsy, Parkinson's disease and Alzheimer's disease.[252] This requires the ultimate in intracranial electrode biocompatibility and durability, where soluble silk protein membranes are often used as substrates. An ultra-thin electrode based on a soluble filamentous protein membrane is proposed (Figure 51(a)).[253] After the electrode has been immobilized in the brain region, its filament matrix is dissolved by saline, leaving only a 2.5 μm electrode network in highly conformal contact with the brain tissue, producing higher SNR and contact.

For some *in vivo* cardiac EP signal detection and electrotherapy, dynamic cardiac conformal integration and multifunctional mapping need to be considered. An advanced multifunctional mapping system that combines conformal high-density electrodes based on multiplexed circuits adhered to the surface of the heart with high spatial and temporal resolution is reported (Figure 51(b)).[254] Its representative ECG recordings for all 288 electrodes at 4 points in time exhibit that a wave of cardiac activation propagating from the left side of the array to the right side.

4.4.2. *Periphery signal*

Some physiological signals do not require implanted electrodes and provide stable and regular changes at the surface of the body. Flexible biomedical devices can therefore be attached to the skin and can measure biophysical signals from the body, such as temperature, strain, and pressure.[255] The development of peripheral flexible biosensors requires consideration of compatibility with allergies or infections from long-term wear, as well as conformal bonding to the skin epidermis.[256] Strain and pressure are covered in more detail in other chapters of this book, so this content will not be repeated in this section.

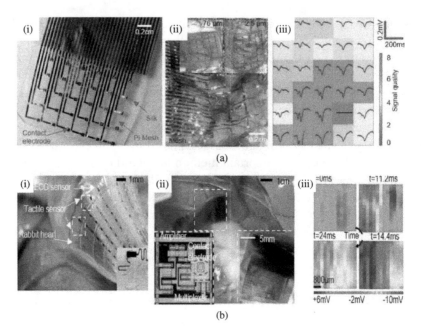

Figure 51. (a) (i) Completely conformal electrode array for the mapping of ECoGs; (ii) image of the electrode array (25 μm mesh) on a feline brain after the dissolution of the silk substrate. Inset, 76 and 2.5 μm electrode arrays; and (iii) the recorded ECoG with the color showing the SNR[253] (Copyright 2010, The Nature Publishing Group). (b) (i) Balloon catheter for cardiac electrophysiological mapping and ablation therapy. (ii) The image of the high-density electrodes on a porcine heart for the recording of the spatial-temporal cardiac electrical rhythms in real-time, the inset shows the circuit unit and the enlarged view of the electrode array and (iii) the recorded wave of cardiac activation propagating of the entire 288-electrodes[254] (Copyright 2011, The Nature Publishing Group).

The monitoring of human body temperature can provide an important indication for clinical diagnosis, therefore flexible temperature sensors have been widely developed and commonly used materials including metal, semiconductor, carbon group materials, polymer-carbon group materials composite and thermochromic liquid crystals and so on. A breathable and flexible temperature sensor with a porous structure is proposed, which not only can continuously monitor human body temperature, but also has extremely high breathability and flexibility (Figure 49(c)).[257] Its multilayer design allows perspiration to pass through and evaporate and also prevents impregnation and air from entering in contact with the skin.

Performance is maintained under prolonged wear and shower tests, and accuracy is almost identical to that of a mercury thermometer.

Blood flow monitoring is also an important indicator for clinical diagnosis, but has been a huge challenge for conventional medical monitoring.[258,259] It is less accurate because it is highly susceptible to changes in nature due to device exposure. The development of flexible biosensors offers new possibilities for blood flow monitoring as they can be controlled on the skin in such a mechanically invisible way that the signal can be obtained accurately and without interference. Webb et al. report an epi-device for continuous mapping of macro- and microvascular blood flow based on anisotropic heat transport phenomena around subsurface blood flow regions (Figure 49(d)).[260] The device consists of a metallic thermal actuator, which acts as a constant heat source to produce a well-controlled temperature rise, and 14 resistive sensors for sensing the spatio-temporal distribution of temperature generated by heating. During the measurement, the heat distribution is highly anisotropic and the deviations indicate the behavior of the blood flow. The device platform fits naturally onto the skin surface to eliminate relative motion between the actuator/detector and the blood, allowing accurate blood flow mapping as an indicator of vascular health, particularly in diseases with vascular-related pathologies.

5. Flexible Gas Sensor

5.1. Target gas sensing

As various gases are important indicators for various applications, the development of flexible sensors capable of detecting the presence of gases has been a research goal for many researchers. In particular, gaseous pollutants are generated in vehicles, industry, power generation units, agriculture/incineration of waste products, households, etc.[261–265] Exposure to gaseous pollutants increases health risks associated with heart disease, stroke, chronic obstructive pulmonary disease, lung cancer, acute respiratory infections, etc.[266–268] Therefore, flexible gas sensors that can monitor different pollutant gases in the surroundings is of great importance to improve human health and safety.[263,269,270] In this section, we discuss the latest practical articles on various gases and their responses, sensitivity, operating temperatures, and limits of detection (LOD), as well as ways to quickly detect these gases anywhere using gas sensors that can be on the body, in clothing, or embedded in wearable accessories.

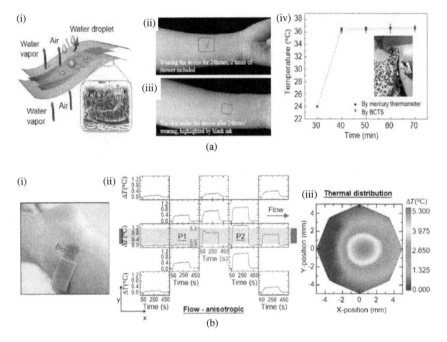

Figure 52. (a) (i) the illustration of waterproof and vapor permeable properties of the sensor; (ii) wearing the sensor for 24 h with two times of showers; (iii) the skin under the device after 24 h' of wearing, no sign of maceration or stimulation is observed; and (iv) in vitro test of the sensor and its comparison with the mercury thermometer. (b) (i) Photograph of a device on the skin; (ii) raw data from a device applied to an area above a large vessel; and (iii) spatial map of the temperature at $t = 300$ s.

5.1.1. NO_x sensors

Nitric Oxide (NO): Nitrous oxide is a colorless environmental pollutant usually produced by indoor household appliances such as gas stoves, gas, wood, etc., and causes a yellowish-brown smoke.[271,272] Long-term exposure may cause different chronic lung diseases and therefore requires special attention when manufacturing flexible gas sensors.[273]

Chen and his colleagues developed the flexible gas sensor based on micro-and nanostructured graphene/silver nanoparticle (G-Ag) thin film electrodes on polyimide (PI) substrates with combined coating and ultra-short pulse laser technology.[274] The prepared Ag NPs show nanomicrostructures retaining the properties of graphene (G). The irregularly distributed Ag NPs were deposited on the surface of G-sheets, which promoted the formation of stacked porous structures with a high specific surface area between Ag NPs and G-sheets. The G-Ag composite structure with high sensitivity and selectivity increases the gas detection of NO under temperature-assisted control. This is due to the high adsorption capacity and stability of Ag NPs on NO gas molecules, thus G-Ag films on flexible substrates are a potential gas detection sensor. Conventional methods that rely on indirect colorimetric measurements or NO detection using rigid and permanent material constructs do not provide continuous monitoring and/or require additional surgical retrieval of the implant, which can increase risk and hospital costs. A flexible, biodegradable, wirelessly operated electrochemical sensor was developed for real-time NO detection with a low detection limit (3.97 nmol), a wide sensing range (0.01–100 μM), and desirable anti-interference properties.[275] The device successfully captures NO evolution in cultured cells and organs with results comparable to those obtained by standard Griess assays (Figure 53(a)). The sensor platform, in combination with a wireless circuit, allows continuous detection of NO levels in live mammals over several days. This work could provide the necessary diagnostic and therapeutic information for health assessment, treatment optimization and postoperative monitoring.

Nitrogen dioxide (NO_2): Toxic NO_2 gas, an oxidized version of NO, is produced primarily by the combustion of fossil fuels and as a byproduct of the combustion of gasoline and fuel in automobile exhaust and is a major cause of haze, acid rain, and human health threats.[276,277] Exposure to low concentrations of nitrogen dioxide (1 ppm) may cause headaches, respiratory problems, and irritation of the eyes, nose, and throat; exposure to nitrogen dioxide above 20 ppm may be directly damaging, even fatal, as the gas accelerates the risk of respiratory disease in children and the elderly.[278] In addition, nitrogen oxides react with various volatile organic compounds to form ozone, which has several harmful effects on humans, such as damaging lung tissue and having a terrible effect on its function, mainly in asthmatic patients.[279] Therefore, there is an urgent need to develop nitrogen dioxide gas sensors that can be easily worn for everyday use. For wearable applications, the device platform needs to be flexible or stretchable to allow for easy integration into non-flat consumer products, on clothing, or on the body.

Cheng *et al.* developed an RT-operated gas sensing platform with fast response/recovery by exploiting the self-heating (Joule heating) effect of

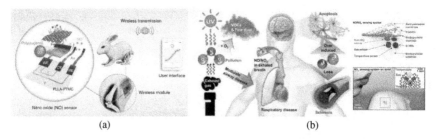

(a)   (b)

Figure 53. (a) Materials and designs for flexible and transient NO sensors composed of a bioresorbable substrate, Au nanomembrane electrode, and a poly(eugenol) thin film, and data transmission to a user interface through a customized wireless module[275] (Copyright 2020, The Nature Publishing Group). (b) Bioresorbable, flexible NO_x (NO/NO_2) sensing system including gas, temperature and humidity sensors on biodegradable polymer substrates, and optical micrograph of a representative sensor system laminated on a wrist, with a magnified view of the device in the inset[281] (Copyright 2020, The Nature Publishing Group).

the sensing interdigitated electrodes.[280] The highly porous laser-induced graphene (LIG) snake-patterned interconnect region was coated with Ag films to reduce the resistance, while the sensing region was decorated with molybdenum disulfide (MoS_2) NPs. The p–n junction formed at the rGO-MoS_2 composite interface enhances the sensitivity and selectivity of NO_2 gas. The sensing platform can withstand up to 20% tensile strain and can be used for epidermal electronics applications. A novel transient electronic technology, capable of complete dissolution or decomposition under certain conditions after a period of operation, offers unprecedented opportunities for medical implants, environmental sensors, and other applications. As shown in Figure 53(b), Hwang designed a biodegradable, flexible silicon-based electronic system that detects NO species with record sensitivity of 136 Rs (5 ppm, NO_2), is 100 times more selective for NO species than other substances, and has fast response (~30 s) and recovery (~60 s).[281] These exceptional capabilities depend not only on material, size and design layout, but also on temperature and electrical operation. Large-scale sensor arrays in mechanically flexible configurations exhibit negligible performance degradation under various modes of application loading, consistent with mechanical modeling. In vitro evaluations demonstrate the capability and stability of the integrated NO_x device in the harsh humid environment of biomedical applications.

5.1.2. CO_x sensors

CO_x (CO and CO_2) are toxic, odorless, colorless pollutants that are harmful to humans.[282,283-285] Low levels of exposure may cause only nausea, vomiting and dizziness.[286] However, higher consumption of these carbon gases can be very dangerous because carbon monoxide is toxic and carbon dioxide is an asphyxiant at high concentrations, increasing the risk of heart disease.[287] Therefore, to reduce injuries and accidents in the workplace, gas detectors are needed as a safety measure. In addition, carbon dioxide is the largest contributor to global warming.[288] Specifically, the radiative power of carbon dioxide is increasing every day, which leads to an imbalance of received and radiated sunlight.[289] As a result, CO_x sensors are needed for a variety of applications, including environmental monitoring and respiratory monitoring.

CO sensors: Currently, most commercial CO sensors are based on the semiconductor tin dioxide (SnO_2), as this technology offers high response and operational robustness.[290] However, bulk SnO_2 sensor media must typically be heated above 400°C to achieve adequate CO response. A process for fabrication of printed flexible CO sensor arrays based on reduced graphene oxide (rGO) sensor media modified with a printable semiconductor catalyst is demonstrated.[291] These sensors operate at room temperature with a fast response and are deposited on thin flexible substrates using high-throughput printing and coating methods. Nanoscale deposits of rGO sheets were decorated with SnO_2 by using a modified solvent thermal aerogel process. Ag finger fork electrodes were inkjet printed by nanoparticle ink on a 100 μm linewidth plastic substrate, then coated with SnO_2-rGO nanocomposites by inkjet or slot-mode coating, followed by heat treatment to further reduce rGO. In devices with slot-mode coated active layers, detection of 50 ppm CO in nitrogen was demonstrated. The response of these printed sensors was 15% with response time of 4.5 s and a recovery time of 12 s.

CO_2 sensors: Carbon dioxide is ubiquitous and is produced whenever carbonaceous materials are burned, and as a by-product of respiration, it will inevitably be produced as long as organisms are breathing.[292] Exceeding CO_2 levels may not only have adverse effects on human health, such as headaches, fatigue, eye diseases, and nasal or respiratory

disorders, but also contribute to the greenhouse effect and global warming.[293] However, carbon dioxide has neither color nor odor, so it is difficult for humans to detect its presence as well as concentration through the human sensory system.[294] Three-dimensional electron-conductive hydrogel has a high specific surface area, which helps to facilitate the adsorption of the gas and to directly sense gaseous CO_2. Wu et al developed a hydrogel material that can directly react with CO_2.[295] They prepared 3D chemically functionalized reduced graphite oxide hydrogel (FRGOH) by a simple one-step hydrothermal method using hydroquinone molecules to achieve high-performance sensing of carbon dioxide. Compared to the unmodified RGOH sensor, the FRGOH device is almost twice as responsive to CO_2 (1.65% response to 1,000 ppm CO_2) with a lower detection limit (20 ppm). When CO_2 is adsorbed on the hydrogel surface, the highest occupied molecular orbital (HOMO) is higher than the Fermi energy level of the RGO. The charge is transferred from CO_2 to RGO by mixing (hybridization) with RGO orbitals, leading to an increase in the charge (hole) concentration in the RGO and a decrease in the resistance of the hydrogel. Therefore, a hydrogel device with a higher initial resistance will have a larger response space and thus produce a higher response.

5.1.3. Ammonia (NH_3) sensors

Ammonia (NH_3), a colorless gas with a strong odor that has been used as a refrigeration gas for centuries, is one of the toxic environmental pollutants that occurs naturally through human waste and industry.[262,296,297] A large number of people are exposed to ammonia by breathing the vapors of many clean products. The main drawback of ammonia is that it can be flammable at high concentrations. In addition, because ammonia has a suffocating odor, inhalation of low concentrations can cause coughing and nose pain, and above a certain concentration, a burning sensation in the nose and throat.[298] NH_3 produces a sharp, irritating odor that is readily detectable at concentrations above 50 ppm.[299] Therefore, there is a high risk of unknown low-level ammonia exposure that can cause serious damage to human health. NH_3 gas at concentrations above 35 ppm can damage the eyes, throat, skin, and respiratory system.[300] In addition, NH_3 gas reacts exothermically with water and ammonia is corrosive to many metals such as Zn and Cu and their alloys.[301] Therefore, the development of a highly selective and sensitive NH_3 sensor is necessary for environmental and health monitoring.

By encapsulating citric acid (CA) into cross-linked polyethylene glycol diacrylate (PEGDA) hydrogels using thiol-ene photochemistry, Zhang et al. prepared a highly sensitive and selective CA/PEGDA hydrogel-based NH_3 gas sensor operating at high humidity (80% RH).[302] The sensor exhibited a high response to 20 ppm NH_3 at room temperature, as well as a low LOD (50 ppb). This is due to the equilibrium water content of the hydrogel composite and the moderate acid dissociation constant of the acid group. This sensor exhibits good chemical stability, while the stable hydrogel structure facilitates the detection of NH_3 in humid environments. Specifically, CA has an extremely high hydrophilicity and NH_3 adsorption capacity. At 80% RH, the large amount of adsorbed water enhances the activity of the carboxyl group, which causes the carboxyl group to react with the large amount of adsorbed NH_3 to form an ammonium salt. The ammonium salt has stronger hydrophilicity than the carboxyl group, and the adsorption of water increases, leading to CA membrane deliquescence. Unlike CA, PEGDA exhibits relatively poor hydrophilicity, and the amount of NH_3 adsorbed is relatively small, preventing the formation of carboxylate salts even under higher mild environments. Compounding CA with PEGDA allows the hydrogel to maintain great NH_3 adsorption capacity while still possessing proper hydrophilicity. Thus, CA/PEGDA has both a high NH_3 response and good reproducibility. In addition, the carboxyl group in the CA structure can provide more adsorption sites for NH_3, and therefore, the CA/PEGDA composites have good response properties to NH_3 at 80% RH. They also tested the presence of NH_3 in human exhalation, and the sensor can accurately identify different concentrations of NH_3 in exhalation, demonstrating the potential of the sensor for clinical applications. Similarly, they developed an environmentally friendly and non-toxic biomass hydrogel-based NH_3 sensor using poly-l-glutamic acid and l-glutamic acid (PAA and GA) as sensing materials based on a moisture activation mechanism at room temperature.[303] The sensor has a NH_3 response from 9.2 to 50 ppm at room temperature and 80% relative humidity. The PAA/GA hydrogel has better sensing characteristics for NH_3 in high humidity environments, especially its selectivity. In addition, they did tests on human exhalation and the sensor was able to accurately identify the presence of 5 ppm NH_3 in the exhaled gas, which means that the sensor could potentially be used in healthcare.

Graphene is a typical 2D material that contains a monolayer of sp2-hybridized carbon atom sheets with zero band gap energy and excellent electrical conductivity. It is commonly described as a p-type semiconductor, mainly due to oxygen- containing functional groups and defects.[304] The edges of the graphene-based composite form a porous structure, which increases the specific surface area and leads to better sensor response. Due to its large surface area, the material shows excellent electrical properties, possessing high electrical conductivity and ultra-fast electron mobility, with a specific surface area approximately twice that of CNTs. Furthermore, combining gas sensing materials with wireless sensors effectively improves the sensors for wearable, health monitoring and environmental monitoring applications. RGO modified with AgNPs is proposed as a sensing material for NH_3, which is a simple, environmentally friendly and cost-effective method.[305] AgNPs enhance the electron adsorption capacity of RGO for NH_3. Wireless gas sensors were constructed by combining the materials with commercial NFC tags (Figure 54(a)). The NH_3 gas can be monitored at room temperature (5°C) at low concentrations (25 ppm), and the impedance of the wireless sensor circuit changes when the NH_3 molecules are absorbed by the Ag NPs-decorated RGO. In addition to conventional sensors that can detect a single stimulus, some advanced sensors for simultaneous monitoring of multiple stimuli (multifunctional sensors) are actively investigated. PANI is considered one of the most promising and widely used sensing

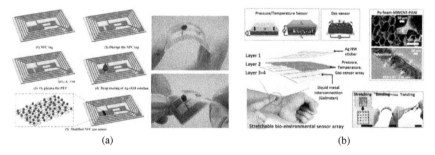

Figure 54. (a) The preparation process of the NFC tag sensor, and this sensor tag under twisting[305] (Copyright 2019, The Nature Publishing Group). (b) Schematic illustration of the skin-attached multi-functional sensor array with pressure/temperature and NH_3 gas sensors, integrated using directly printed liquid metal interconnections[307] (Copyright 2017, The Nature Publishing Group).

materials because of its low production cost, environmental stability and acceptable conductivity as well as its unique NH_3 gas sensing capability.[306] As shown in Figure 54(b), based on a single sensing material, polyurethane foam coated with multi-walled carbon nanotubes (MWCNTs)/PANI nanocomposites, a skin-like stretchable multifunctional sensor array was designed and fabricated that is capable of detecting body temperature, wrist pulse, and NH_3 gas simultaneously.[307] These sensors offer high sensitivity, fast response, and excellent durability. In addition, the fabricated sensor arrays exhibit stable performance at up to 50% biaxial stretching and attachment to the skin due to the use of directly printed Galinstan liquid metal interconnects.

5.1.4. Hydrogen sulfide (H_2S) sensors

Hydrogen sulfide, a colorless, corrosive, highly flammable, explosive, and extremely toxic gas, is generated from different industries (e.g., natural gas plants, petroleum refineries, wastewater treatment plants, tanneries, paper mills, etc.) and produces odor-like rotten eggs.[308–310] Depending upon exposure concentration and duration, H_2S may either cause corneal erosion or even cause death.[311,312] Hence, the detection of H_2S gas is necessary for health and safety.

Asad et al. fabricated a flexible H_2S sensor by spin-coating CuNP-decorated SWCNTs as a sensing layer on a flexible PET substrate.[313] CuNP–SWCNTs-based sensors exhibited a remarkably selective n-type response to H_2S gas at RT and showed a linearly increasing response upon exposure to H_2S from 5 to 150 ppm. Later, Asad et al. also prepared CuO–SWCNTs nanostructures for low concentration (100 ppb) detection of H_2S and showed a wide range of operation from 100 ppb to 50 ppm concentration of H_2S gas.[314] The group further modified a commercial 13.56 MHz radio RFID tag by immobilizing CuO–SWCNTs nanostructures onto the antenna of the tag and wirelessly monitored the change in reflectance upon H_2S gas exposure. Upon H_2S exposure, due to the electron donation ability of H_2S, the resistance value of the sensing elements was increased and further shifted the resonance frequency of the RFID tag antenna and decreased the reflection value. They also attached the RFID tag to a human hand and demonstrated wearable gas-sensing ability. Choi et al. fabricated a self-powered wristband for H_2S detection (5–20 ppm) by integrating an rGO-based chemo-resistive gas sensor on a flexible printed circuit board (FPCB) with a wireless Bluetooth communication module.

5.1.5. Methane (CH_4) sensors

Ambient methane (CH_4) is a well-mixed ozone-depleting gas that is second only to carbon dioxide in terms of incremental radiative constraints.[315–317] The main sources of CH_4 gas are natural gas production and combustion of agricultural biomass.[318] CH_4 is a colorless and odorless gas that can explode when mixed with oxygen in the environment.[319] Therefore, its presence in even small concentrations is a significant threat and needs to be detected. Conventional MOS gas sensors have limited applications in wearable devices because of their inflexibility and high power consumption due to substantial heat loss. Doped Si/SiO_2 flexible fibers were prepared to fabricate MOS gas sensors by thermal drawing method as a substrate.[320] As shown in Figure 55(a), CH_4 gas sensors were fabricated by *in situ* synthesis of Co-doped ZnO nanorods on the fiber surface. The doped silicon core acts as a heat source via Joule heat, which conducts heat to the sensing material and reduces heat loss; the SiO_2 cladding is an insulating substrate. The gas sensor is integrated into the mineral cloth as a wearable device, and CH_4 is monitored in real time by light-emitting diodes of different colors. The feasibility of using Si/SiO_2 doped optical fiber as a substrate for fabricating wearable MOS gas sensors is demonstrated, where the sensors have substantial advantages over conventional sensors in terms of flexibility, thermal utilization, etc.

5.1.6. Hydrogen (H_2) sensors

Hydrogen (H_2) is a non-toxic, non-radioactive, non-polluting gas that does not produce dangerous combustion products but can be very dangerous under certain conditions, such as mixing 18–59% in air, given that it requires low ignition energy to become highly explosive.[321–323] In addition, H_2 is a colorless and odorless gas and therefore, given that free hydrogen is highly reactive, it is impossible to detect it without any detection equipment.[324] It is 10 times more flammable than gasoline, and even twenty times more explosive.[325] In addition, the flame of a fire caused by H_2 is invisible, which makes it a serious hazard at work.[326] In addition, the explosion limit of H_2 depends on the confinement, and it can generate explosive waves that destroy nearby buildings and injure people.[327] Therefore, a large

Flexible Sensors 219

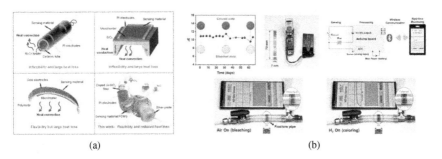

Figure 55. (a) MOS gas sensors on different substrates: ceramic tube, MEMS, polyimide, and doped Si/SiO$_2$ fibers[320] (Copyright 2023, American Association for the Advancement of Science, AAAS). (b) Long-term stability of the sensor for over two months at 4% H$_2$, and the wireless sensor module when the hybrid H$_2$ gas sensors wrapped around a pipe[329] (Copyright 2023, Elsevier Inc).

number of investigations of hydrogen stations have focused on their ignition, deflagration and detonation. Flexible and highly concentrated H$_2$ sensing was developed by "dipping-drying" a 3D porous PU foam integrated with graphene oxide (GO-PU).[328] The multilayered honeycomb structure of graphene oxide appears to be tightly adhered to the favorite PU. Thanks to the numerous adsorption sites and abundant surface functional groups of the GO "double honeycomb" structure, the GO-PU foam exhibits a unique response and linearity for 2–100 v/v% H$_2$ and shows excellent light weight, customizability and flexibility. Notably, the foam has excellent inductive stability against 0–180° bending and low 0–20% strain, as well as excellent H$_2$ detectable performance even after being pressed by 200 g of weight, immersed in water, and frozen at –10.8°C in a refrigerator. Room temperature, wireless, wrapable, gas-chromic H$_2$ platforms with chemoresistive and naked eye readable dual signals were developed using solution-based low temperature annealed SnO$_2$ sensors/WO$_3$ films on polyimide substrates.[329] The sensor can be wrapped around any point source, such as a flange, to provide early detection and precise localization of H$_2$ leaks. An optimized spray-coated gasochromic WO$_3$ film achieves a color change value of 11 with a 6 s response time at 4% H$_2$ and low eye detectable limit of 0.2%. Also, because the patterned sprayed vapor discoloration layer does not interfere with the effective area of SnO$_2$ based on a chemiresistive sensor, the sensor's resistance value varies by more than 11,000 times at 1% H$_2$ with a wide concentration range from 0.005% to 2% of H$_2$ at room temperature. As shown in Figure 55(b), this

220   Y. Guo et al.

sensor can be easily applied to a variety of non-planar surfaces, while Bluetooth system integration allows wireless monitoring via smartphone.

5.2. Volatile organic compounds

With the increasing use of organic chemicals in various products, many VOCs evaporate at RT. Inhalation of VOCs can affect health.[330–335] Thus, there is a pressing need for monitoring of VOCs not only in industries but also inside homes.[336–338] Thus, to gather localized information about the presence of VOCs, an easy-to-carry, easy-to-wear, and comfortable sensor is needed. In this part of the review, we discuss potential target VOCs and their detection technologies.

5.2.1. Methanol

Methanol is a common volatile organic compound that typically occurs as a solvent in breweries, the biotechnology industry, clinical areas, and the chemical industry.[339–341] Because it readily evaporates from aqueous solutions at RT, inhalation, consumption, or dermal absorption may lead to irreversible damage to tissues, eyes, or nervous system, and further lead to blindness, organ failure, and even death.[342] Methanol may form toxic formic acid and formaldehyde after metabolism in the human body.[343,344] Therefore, rapid identification and quantification of methanol in the working environment is necessary.

Jiang et al. demonstrated an electrochemical sensor for flexible methanol sensing applications.[345] Ag/AgCl RE, carbon CE and WE were screen printed on PET, T-shirts (for cotton-based sensors) and nitrile gloves. Carbon WE was modified with a mixture of 5% Pt, Nafion, and ethanol. In addition, the three-electrode system was covered with a solid electrolyte of Nafion film. Pt is the electrocatalyst for methanol oxidation on WE. The current flow from WE to CE increases linearly with increasing methanol concentration from 2% to 20%. The wearable sensor also showed high selectivity, reproducibility and good stability to methanol vapor. Recently, Ma et al. developed a wearable smart electrochemical sensor for methanol quantification.[346] A three-electrode system was screen-printed on PET or nitrile glove substrates using Ag/AgCl ink (for RE) and carbon ink (for WE and CE). In addition, WE was modified by Pt/rGO composites. During the CV process, methanol was decomposed

Flexible Sensors 221

by Pt electrocatalyst at WE and oxidized to form CO_2 through an intermediate partial oxidation process, which produced two anodic peaks of 0.6 and 0.21 V during forward and reverse scanning, respectively. The amperometric current increases with increasing methanol concentration and is linear with 1–10% of methanol concentration. The flexible sensor exhibited stable sensing behavior even over a wide range of humidity (15–85%) and temperature (22–42°C) and good mechanical stability under bending (200 cycles, 180°) and stretching. They also demonstrated the use of Bluetooth technology to transfer measurement data to a smartphone for real-time wireless remote monitoring of methanol.

Toluene
Toluene is a commonly available indoor air pollutant that causes health hazards as it is used in preparation of paints, varnish, metal cleaners, plastics, and detergents. Toluene is an exhaled breath biomarker for lung cancer.[347–349] Conventionally, detection of exhaled breath biomarkers usually needs a bulky instrument and expertise, limiting real-time monitoring. On the other hand, noninvasive and real-time monitoring of toluene above a certain threshold in exhaled breath could be a lifesaver.[350–352] Jin et al. designed ionic chemiresistor skin (ICS) for sensing VOCs.[353] The sensing channel of the chemiresistor skin was formed by depositing i-TPU. The i-TPU solid-state polymer electrolyte sensing layer was formed by noncovalent mixing of [EMIM][TFSI] IL and intrinsically stretchable thermoplastic PU (TPU). The ICS showed ultra-stable sensing capabilities for five VOCs (toluene, hexane, propanol, ethanol, and acetone), with responses based on intermolecular interaction. The ICS showed a greater response to toluene than the other VOCs tested. During exposure to VOCs, the IL in the i-TPU channel and the VOC interact and increase the ionic conductivity of the channel, decreasing channel resistance. With varying toluene concentrations (1–1000 ppm), the sensor showed a gradually decreasing response. The ICS showed excellent operation ability even after 85% humidity exposure or to exposure of 100°C temperature and demonstrated outstanding mechanical stability after applying 100% stretching strain. The transparent, intrinsically stretchable, skin-attachable chemiresistor showed exceptional sensing behavior for different VOCs and thus could be a good candidate for a human-adaptive, skin-attachable sensing platform for daily use and early diagnosis of VOCs present either in the environment or in exhaled breath.

5.2.2. Acetic acid

Acetic acid, a common chemical used for many purposes in industries, can be found in common household products.[354] Prolonged inhalation of acetic acid vapor at 10 ppm could irritate eyes, throat, and nose and may cause irreversible damage to eyes, skin, or lungs at exposure to as high as 100 ppm.[355] Thus, the detection of acetic acid vapor beyond the threshold is important for personal safety and health. Jun et al. fabricated a flexible and wearable sensor composed of C-PPy NPs.[356] The sensor showed excellent sensing performance for acetic acid at concentrations from 1 to 100 ppm. The conductivity of the C-PPy NPs was influenced by negatively charged CH_3HOO^- ions; thus, the resistance of the sensor gradually decreased with increasing concentration of acetic acid. Also, the group further modified the commercially available RFID tag with C-PPy NPs to form a passive wireless sensor. The reflection coefficient at the resonance frequency of the RFID tag increased with increasing concentration of acetic acid due to the decrease of resistance.

5.2.3. Gas in breath

A large variety of volatile organic and inorganic compounds can be found in human exhaled breath, along with major gaseous constituents of N_2 (78.04%), O_2 (16%), CO_2 (4–5%), and water vapor. VOCs are usually generated as a byproduct of various cellular metabolism processes or oxidative stress; thus, the concentration of such byproducts (in the ppt to ppm range) usually depends on diet, environmental exposure, and disease states of the patient. Extensive clinical trials have linked various critical diseases including cancer, Alzheimer's, Parkinson's, diabetes, asthma, etc. to exhaled volatile molecules from human breath. Thus, analysis of biomarkers in human exhaled breath is an alternate, inexpensive, fast, nonhazardous, noninvasive route for disease detection and health monitoring. But the existing breath analysis methods involve preconcentration of exhaled breath and thermal desorption of concentrated gas, followed by conventional gas sensing. Those traditional methods require expensive equipment and high levels of expertise and therefore are not suitable for real-time health monitoring. In addition to VOCs and other inorganic compounds, the exhaled breath also contains water vapor. By sensing the presence of breath vapor, the breath profile could be monitored and also could predict sleep apnea, asthma, anemia, and other chronic pulmonary diseases. In this part of the review, detection techniques of breath biomarkers along with breathing patterns analysis for noninvasive disease diagnosis are discussed in brief.

5.2.4. Gas in food

Consumption of spoiled food can cause different types of foodborne illnesses, and this has become a rising social problem.[357,358] Therefore, it is important to develop economically feasible and convenient methods to inspect or monitor food for spoilage and contamination, as well as preservation to ensure food safety and hygiene. Spoiled/deteriorated foods can usually be easily identified by visual observation or by normal olfaction. These classical methods are not suitable for early detection of food spoilage, i.e., before spoilage or at the onset of decay, which can inhibit food poisoning. NH_3 gas is produced during the decomposition of foods containing high protein (e.g., eggs, dairy products, meat, etc.).[359,360] Freshness can be monitored by monitoring the presence of this gas in/surrounding the food. In order to monitor the presence or production of NH_3 gas around food products, a low-cost, environmentally friendly sensing system with high sensitivity, low LOD and RT operability is needed. On the other hand, for packaged foods, freshness is determined based on the expiration date printed on the package. However, the printed date does not reflect the actual freshness of the food inside, since the quality depends mainly on external conditions, such as preservation conditions and packaging quality.[361] Therefore, relying on expiration dates for food quality determination may lead to the consumption of inedible food or the disposal of unwanted good food. To reduce food waste and food poisoning, disposable sensors need to be integrated into smart food quality monitoring. The presence of pesticide residues on fresh vegetables is a further risk to human health. Therefore, it is necessary to evaluate food products using low-cost and reliable sensors. In this section, we discuss the development of flexible and portable food safety-related gas sensors for the detection of food spoilage marker gases (e.g., NH_3, biogenic amines (BAs), oxygen, etc.) and pesticide residues.

Barandun *et al.* developed a simple, low-cost, environmentally friendly, paper-based chemical gas sensor to monitor the freshness of food products wirelessly, as shown in Figure 56(a).[362] Since paper is usually made of very hydrophilic cellulose fibers, it readily absorbs moisture from the environment, even if it appears to be dry. Therefore, in the presence of water-soluble gases (e.g., NH_3, CO_2, CO, H_2S, etc.), the surface conductivity of the paper can change depending on the chemical nature and concentration of the gas.[363] The group innovatively combined the water solubility of NH_3 gas and the hygroscopicity of paper to fabricate a near-zero cost NH_3 sensor without any additional sensing material. The electrical properties of the paper (i.e., surface conductivity) are monitored by printing interlocking graphite electrodes on the surface of the paper. In the presence of NH_3, the electrical conductivity of the paper increases as the ion content of water-soluble NH_3 gas increases after dissociation in the surface-bound water film. Therefore, by monitoring the conductivity of the paper, the presence of NH_3 can be detected precisely. Due to the better water solubility of NH_3 gas than other water-soluble gases, the paper-based sensor shows high sensitivity and selectivity for NH_3 gas, with detection of NH_3 concentrations ranging from 200 ppb to 10 ppm. the high-performance NH_3 sensor is further integrated with commercially available NFC tags. To predict food spoilage, the presence of NH_3 in food containers was continuously measured by wirelessly monitoring the conductivity of the paper-based sensor using a smartphone.

In another work, a flexible nanowire-based chemiresistive NH_3 sensor was developed for real-time food freshness detection in smartphones.[364] The authors fabricated highly aligned sub-100 nm CP PEDOT:PSS nanowires on a flexible PET substrate using a convenient and low-cost capillary-fill-based soft lithography technique. Using a copper shadow mask, gold-top electrodes were deposited on the PET substrate to

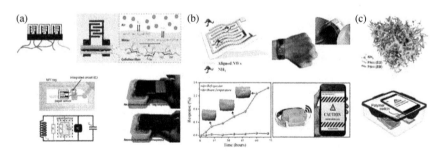

Figure 56. Flexible gas sensors for food quality and safety monitoring. (a) Paper-based gas sensor for monitoring the freshness of packaged foods[362] (Copyright 2021, Springer Inc). (b) A Bluetooth-enabled wearable sensor for food spoilage detection[364] (Copyright 2019, American Chemical Society). (c) NFC-enabled, conductive, polymer-based gas sensors for food spoilage detection[365] (Copyright 2018, American Chemical Society).

fabricate the final functional devices. The flexible nanowire sensor exhibits good mechanical flexibility. Since NH_3 is an electron-donating gas, its presence consumes holes in the valence band of the sensing material (PEDOT:PSS nanowires), increasing the overall resistance of the gas sensor. This flexible sensor shows excellent sensing performance in the range of 750 ppb to 6 ppm NH_3. In addition, it showed a sensitivity of 0.2524% ppm^{-1} with good linearity and LOD down to 100 ppb. The researchers integrated the flexible nanowire sensor onto an FPCB with a Bluetooth module to fabricate a wristband-style portable electronic system (Figure 56(b)). They further demonstrated the usability of this portable sensor by wirelessly monitoring NH_3 emissions during the spoilage of salmon stored in the refrigerator and RT. This smartphone system can monitor food quality in real-time and can send spoilage-related warnings to the user. Similarly, Ma et al. demonstrated an NFC-based food spoilage detector during food spoilage, which generates BAs and causes malodor due to the decarboxylation of amino acids after interaction with microorganisms.[365] This group modified the NFC label by printing a p-toluenesulfonic acid hexahydrate (PTS)-doped PANI sensing layer, as shown in Figure 56(c). The nanostructured PTS-PANI conductive polymer showed very high sensitivity to various total volatile base nitrogen (TVBN) compounds, including BA (e.g., putrescine and cadaverine) and ammonia, which are important markers of food spoilage. The conductivity of the PTS-PANI nanostructures decreases upon interaction with amine gases due to dephasing. As a result, the resistance of the sensing layer increases with increasing NH_3 concentration. The sensor shows 225%, 46% and 17% response near 5 ppm concentration of NH_3, putrescine and cadaverine, respectively. In addition, the sensor showed excellent dynamic response to NH_3 and BAs over a wide concentration range of 5–200 and 5–40 ppm, respectively. The detection of food spoilage with PTS-PANi-modified NFC tags depends on the threshold switching, which is largely dependent on the reflection coefficient (S11). Typically, the power emitted by the smartphone is transmitted wirelessly to the NFC tag and a voltage is generated across the antenna coil. Then, the NFC tag resonates and reflects the signal to the mobile device. The readability of the NFC tag depends on the reflection coefficient, which depends on the impedance matching between the NFC tag and the reader. The reflection coefficient depends on the resistance of the PTS-PANI component, which is connected in parallel to the NFC coil. In the absence of amine gas, the S11 value drops due to the low resistance of the PTS-PANI assembly and the

NFC tag cannot be read by the smartphone. With exposure to amine gas, the resistance of the PTS-PANI assembly increases. When the resistance exceeds a certain threshold, the S11 value is sufficient to reflect the signal to the smartphone, and the NFC tag becomes readable and indicates the beginning of food spoilage. The spoiled meat can be detected by the smartphone using the NFC badge.

6. Flexible Magnetic Sensor

6.1. *Magnetic smart composites*

The flexible magnetic sensors are capable of detecting and responding to magnetic fields while maintaining conformability to varied surface profiles.[366-370] Their development is rapidly evolving in response to the growing demands of emerging fields like wearable electronics, healthcare, soft robotics, and the IoT.[371-374] One of the key advancements is the employment of magnetoelectric composites, which simultaneously exhibit both magnetic and electric properties, showing an immense potential for the development of highly sensitive and adaptable magnetic sensors.[375,376]

Using magnetic stimulus-responsive polymers allows smart composites to have the ability to sense stimuli in the environment including touch, vibration, and fluid sensing, and the stress response in the developments of magnetic field.[377] Magnetic tactile sensors have been widely studied, and highly sensitive and low-power sensors have been developed using different physical principles. Su *et al.* developed a sea-anemone-inspired soft robot capable of sensing and morphing underwater.[378] The robot consisted of two parts: a top soft body tentacle for sensing external stimuli and a bottom magnetically stimulated constrictor for receiving signals. When the water velocity increases to more than 1.1 m/s, the top tentacle can convert the mechanical stimulus into an electrical signal, and the bottom magnetically stimulated contraction body receives the signal and contracts and deforms to avoid being swept away by water. The contraction ability of a robot can be adjusted using parameters such as the magnetic particle content, body thickness, and folding depth.

To implement the mobility of flexible sensors, self-powered sensing systems are highly desirable because they do not need an external power source. Gong *et al.* fabricated a flexible self-powered magnetism/pressure dual-mode sensor based on magnetorheological plastomer (MRP) with high sensitivity of a slight pressure (1.3 kPa) and a small-angle bending

(27°), and good stability and the response of human joints motion and the movement of the object (Figure 57(a)).[379] Under an external magnetic field, the micro-scale carbonyl iron (CI) particles in the MRP electrode aggregated into chain-like and cluster-like structures, which enhanced the electrochemical activity of ions in the electrolyte of the electrode materials and formed the conductive network. Specifically, a smart magnetic sensing writing board assembled by the flexible dual-mode magnetism/pressure sensor array showed an application in the field of contactless electronics. Furthermore, Xuan et al. prepared shape-deformable flexible multifunctional sensors based on a liquid-metal-filled magnetorheological plastomer to enable magnetic switching control by dispersing liquid metal and magnetic CI particles into a soft low-modulus polyurethane matrix, which exhibited highly sensitive feedback to external signals, including magnetic fields, temperature, and pressure.[380] The potential of smart electronics such as intelligent control, environmental monitoring, motion recognition was demonstrated. As shown in Figure 57(b), a dual-mode sensor with two electrodes and tubular mechanically heterogeneous structure enabling simultaneous sensing of strain and temperature without cross-talk is demonstrated. The structure consists of a thermocouple coiled around an elastic strain-to-magnetic induction conversion unit, revealing a giant magnetoelastic effect, and accommodating a magnetic amorphous wire. The thermocouple provides access to temperature and its coil structure allows to measure impedance changes caused by the applied strain. The proposed sensing platform is relevant for multifunctional wearable electronics for applications in early disease prevention, health monitoring, and human–machine interaction.

In addition to sensing-only capabilities, tactile displays are pertinent applications for magnetic soft materials. The ability of magnetic soft materials to deform in a magnetic field can be exploited to develop multifunctional devices, such as the development of tactile displays for palpation by surgeons. Zhao et al. used 3D printing technology to design soft electronic devices that could deform into two different shapes under an applied magnetic field, with each transformation mode producing two different electronic functions (Figure 58(a)).[381] Diverse functions derived from complex shape changes including reconfigurable soft electronics were further demonstrate, and a mechanical metamaterial that can jump and a soft robot that crawls, rolls, catches fast-moving objects and transports a pharmaceutical dose. Huang et al. integrated electronic devices such as antennas, RF power harvesters, and light-emitting diode arrays

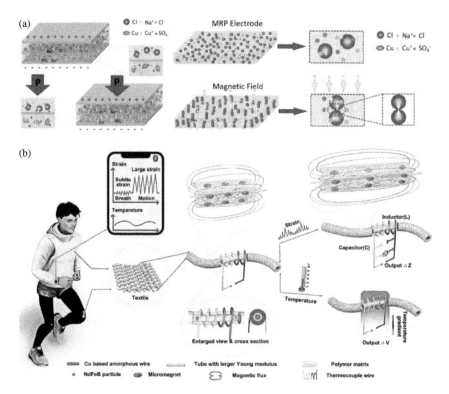

Figure 57. (a) Schematic diagram of the microscopic change of the dual-mode sensor under the pressure and magnetic field[379] (Copyright 2019, Elsevier Inc). (b) Schematics revealing a possibility to use STDMS for cross-talk free assessment of strain and temperature information for the specific example of smart wearables[380] (Copyright 2023, Wiley-VCH).

on origami-based magnetic films and reconfigured them by deforming the origami film, which is programmable with magnetic polarity, to achieve reconfigurable electronic devices (Figure 58(b)).[382] The origami magnetic membranes can be combined to produce more complex patterns and magnetic polarities, leading to innovative applications in surgical robots, tunable antennas, and a variety of reconfigurable flexible electronics.

6.2. Microstructure fabrication by magnetic field regulation

Recently, the magnetic field regulation method has been widely studied for its controllability, and microstructures are fabricated by this effective

Flexible Sensors 229

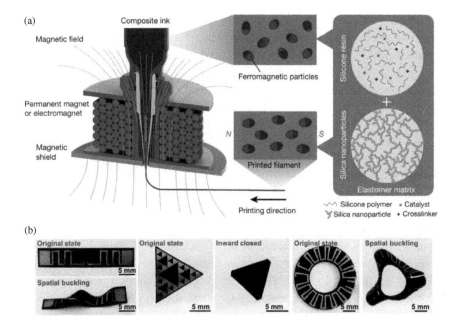

Figure 58. (a) Schematics of the printing process and the material composition. The ferromagnetic particles embedded in the composite ink are reoriented by the applied magnetic field generated by a permanent magnet or an electromagnet placed around the dispensing nozzle[381] (Copyright 2018, The Nature Publishing Group). (b) Images of three kinds of flexible antennas on rectangular, triangular, and annular membranes, a dynamic process to fold the LED array circuit, and various kinds of free-standing 3D structures built by three basic membranes[382] (Copyright 2021, Wiley-VCH).

method for use in various applications.[383,384] Typically, magnetic particles are mixed with the polymer matrix and then placed in the magnetic field to modulate the stacking direction and size of metal particles to obtain diverse and novel 3D microstructures.[385,386] For example, in nature, many animals possess cilia with mechanosensory functions that have inspired the advancement of artificial ciliary sensors.[387] Alfadhel and Kosel reported a haptic sensing method integrating highly elastic permanent magnetic nanocomposite artificial cilia with multilayer giant magnetic impedance (GMI) sensors. Using the permanent magnetic properties of iron nanowires, no additional magnetic-field magnetization is required to allow remote operation, thus simplifying the system integration and minimizing power consumption. When the system is operating, external forces cause the average magnetic field value of the GMI sensor to change, thereby changing its impedance. Human skin, as the external interface between the body and the environment, consists of the epidermis, dermis and hypodermis. As shown in Figure 59, the hair roots from the hair follicle in the dermis through the epidermis to the external environment, where it sensitively transmits environmental signals changes to the brain with the help of mechanoreceptors and sensory nerve fibers.[388] Inspired by the hairy skin, Zhou et al. developed a capacitive pressure sensor with high sensitivity and wide detection range by employing micro-cilia arrays as dielectric layers. The carbonyl iron particles (CIPs), PDMS solution, and curing agent were mixed evenly and then spin-coated onto the cured PDMS film. When placed in a permanent magnet, the PDMS/CIP layer was magnetized and lined into chains which were then eventually converted into the pattern of micro-cilia arrays by the assistance of magnetic field. The fabrication of micro-cilia arrays is due to the material's gravity, applied magnetic force, and surface tension. The ferromagnetism of CIP can attract the PDMS/CIP mixture by the magnetic force of permanent magnet and then concentrate along the direction of magnetic field. Moreover, since the mixture is not cured, PDMS/CIP will show a trend to restore to the original state, but the viscous PDMS gel has a strong surface tension to bind the movement of particles. The magnetic sensor based on the micro cilia array-dielectric layer exhibits a broad detection range, high sensitivities, a low detectable pressure limit, fast response time, and excellent recoverability and robustness. Additionally, it can be applied to pulse-sensing, voice recognition, gas-flow monitoring, and loading mapping, as well as high pressure demonstrations. This group further proposes a magnetic sensor based on tilted flexible micromagnet array (t-FMA) as the interaction interface (Figure 60).[389] Using the bidirectional bending capability of the t-FMA, the sensor can identify the magnitude and direction of the magnetic field in real time by a non-overlapping capacitive signal. The optimized sensor exhibits high sensitivity, low detection limits and good durability. As a proof of concept, the sensor has been successfully demonstrated for convenient, efficient and programmable interactive systems, e.g., contactless Morse code and Braille communication. The identifiability of magnetic field direction and magnitude further enables the sensor unit to be a high-capacity transmitter for cryptographic message interaction (e.g., coded ID identification) and multiple control command output. Similarly, Li et al. presented a flexible pressure sensor based on the magnetically grown dielectric interface (Figure 61).[390] The electrodes were prepared by pouring a curable magnetorheological fluid on the bottom electrode, which was homogeneously mixed with silver-coated nickel

Flexible Sensors 231

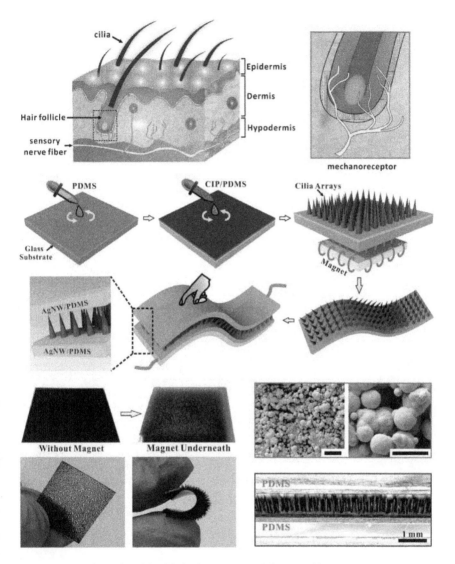

Figure 59. Schematic of the fabrication process of the capacitive pressure sensor using the micro cilia array as the dielectric layer[388] (Copyright 2019, The Royal Society of Chemistry).

particles and PDMS solution. When the electrode was placed in a permanent magnets, the magnetorheological fluid forms a vertically aligned array of microneedles. Microneedles were arranged along the magnetic field direction and rotate with the direction of the external perpendicular

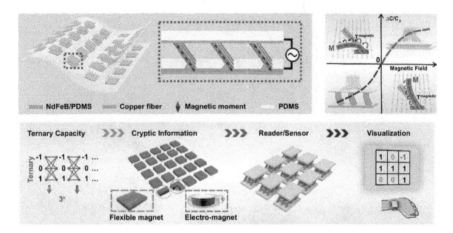

Figure 60. Structure, magnetic property and working mechanism of the flexible magnetic field sensor based on t-FMA, and the touchless and cryptic information recognition system of encoding, decoding and extraction of the targeted information[389] (Copyright 2021, Springer Inc.).

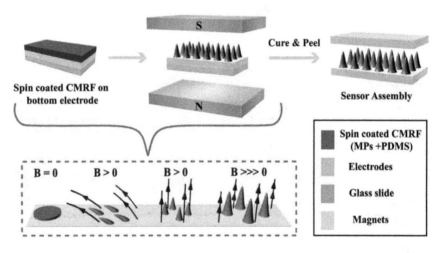

Figure 61. Illustration for the fabrication of magnetically grown dielectric interface-incorporated pressure sensor[390] (Copyright 2020, Wiley-VCH).

curing magnetic field. This approach provides a fast and simple route to mass production of FMPSs, and it is easy to control the fabrication parameters of sensors with different properties. The resulting sensors can detect low pressures with fast response times and good recyclability, making it a

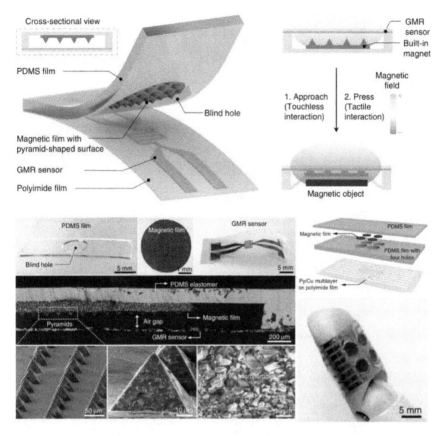

Figure 62. Schematic structure of the m-MEMS platform and mechanisms of the touchless (proximity) and tactile (pressure) sensing modes[393] (Copyright 2019, The Nature Publishing Group).

wise choice for the applications of motion detection, health monitoring, human-machine interaction, and soft robotics.

Inspired by the automatic recognition of stimulated signals in human skin, as shown in Figure 62, Makarov *et al.* implemented a bifunctional electronic skin with tactile and touchless perception by integrating a magnetic microelectromechanical system (m-MEMS).[391] The equipped m-MEMS system can transmit both mechanical pressure (tactile signals) and magnetic field signals (touchless signals) and process them into two distinct regions. The e-skin can clearly distinguish between these two modalities and offer real-time complicated interactions with physical objects enhanced by augmented reality, robotics technology, and virtual

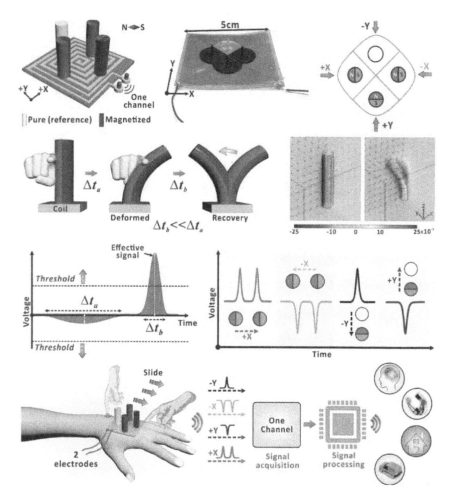

Figure 63. Diagram of the wearable interface that consists of four micropillars and one coil substrate for force direction perception and HMI applications[392] (Copyright 2023, Elsevier Inc).

content data, which broadens their potential applications in the medical field. In contrast to pressure and strain, sliding is a directional operation that requires the interface to recognize the direction of the applied force in order to achieve accurate interaction. Previous explorations of "directional forces" in HMI systems have been based on sensor arrays, which have raised concerns about complex electrode design and multiple communication channels to prevent crosstalk. As indicated in Figure 63, a self-powered and wearable human-machine interface was developed that can distinguish the axial direction of in-plane forces according to Faraday's law.[392] Based on Faraday's law of induction. The interface consists of well-oriented magnetized micro-pillars and a conductive coil. The conductive coil, collects and transmits electrical signals during the interaction. When sliding forces are applied in the plane, differentiable signals are generated to reflect the different axial orientations depending on the polarity and number of voltage peaks. With this unique behavior, the HMI process can be accomplished in an interference-free manner with two electrodes and one communication channel. By The intrinsic oscillation of the microcolumn through the optimization of the system leads to a significantly enhanced signal with high accuracy and reliability in practical applications. Interfaces based on magnetized micropillars are successfully established for HMI platforms such as intelligent robot control and Morse code communication.

Therefore, the technique of preparing microstructures by magnetic field conditioning is attractive, and the induction properties can be easily tuned within the desired pressure range by conveniently adjusting the fabrication parameters of the microstructures.[394] It will be a powerful alternative to the general fabrication method of flexible magnetic sensors.

References

1. Amoli, V., Kim, J.S., Kim, S.Y., Koo, J., Chung, Y.S., Choi, H. and Kim, D.H. (2020). Ionic tactile sensors for emerging human-interactive technologies: A review of recent progress, *Advanced Materials, 30*, 1904532.
2. Xu, K., Lu, Y. and Takei, K. (2019). Multifunctional skin-inspired flexible sensor systems for wearable electronics, *Advanced Electronic Materials, 4*, 1800628.
3. Shu, Q., Pang, Y., Li, Q., Gu, Y., Liu, Z., Liu, B., Li, J. and Li, Y. (2024). Flexible resistive tactile pressure sensors, *Journal of Materials Chemistry A, 12*, 9296–9321.
4. Wu, G., Li, X., Bao, R. and Pan, C. (2024). Innovations in tactile sensing: Microstructural designs for superior flexible sensor performance, *Advanced Functional Materials*, 2405722.
5. Wang, L., Liu, J., Qi, X., Zhang, X., Wang, H., Tian, M. and Qu, L. (2024). Flexible micro/nanopatterned pressure tactile sensors: Technologies, morphology and applications, *Journal of Materials Chemistry A, 12*, 8065–8099.

6. Guo, Y., Zhang, X., Jiang, F., Tian, X., Wu, J. and Yan, J. (2024). Large-scale synthesis of flexible cermet interdigital electrodes with stable ceramic-metal contact for fire-resistant pressure tactile sensors, *Advanced Functional Materials*, *34*, 2313645.
7. Huang, X., Ma, Z., Xia, W., Hao, L., Wu, Y., Lu, S., Luo, Y., Qin, L. and Dong, G. (2024). A high-sensitivity flexible piezoelectric tactile sensor utilizing an innovative rigid-in-soft structure, *Nano Energy*, *129*, 110019.
8. Mhanna, V., Bashour, H., Lê Quý, K., Barennes, P., Rawat, P., Greiff, V. and Mariotti-Ferrandiz, E. (2024). Adaptive immune receptor repertoire analysis, *Nature Reviews Methods Primers*, *4*, 6.
9. Carroll, S.L., Pasare, C. and Barton, G.M. (2024). Control of adaptive immunity by pattern recognition receptors, *Immunity*, *57*, 632–648.
10. Wang, C., Dong, L., Peng, D. and Pan, C. (2019). Tactile sensors for advanced intelligent systems, *Advanced Intelligent Systems*, *1*, 1900090.
11. Ha, K.-H., Li, Z., Kim, S., Huh, H., Wang, Z., Shi, H., Block, C., Bhattacharya, S. and Lu, N. (2024). Stretchable hybrid response pressure sensors, *Matter*, *7*, 1895–1908.
12. Wang, Y., Adam, M.L., Zhao, Y., Zheng, W., Gao, L., Yin, Z. and Zhao, H. (2023). Machine learning-enhanced flexible mechanical sensing, *Nano-Micro Letters*, *15*, 55.
13. Pyo, S., Lee, J., Bae, K., Sim, S. and Kim, J. (2021). Recent progress in flexible tactile sensors for human-interactive systems: From sensors to advanced applications, *Advanced Materials*, *33*, 2005902.
14. Shi, Z., Meng, L., Shi, X., Li, H., Zhang, J., Sun, Q., Liu, X., Chen, J. and Liu, S. (2022). Morphological engineering of sensing materials for flexible pressure sensors and artificial intelligence applications, *Nano-Micro Letters*, *14*, 141.
15. Yang, J.C., Kim, J.-O., Oh, J., Kwon, S.Y., Sim, J.Y., Kim, D.W., Choi, H.B. and Park, S. (2019). Microstructured porous pyramid-based ultrahigh sensitive pressure sensor insensitive to strain and temperature, *ACS Applied Materials & Interfaces*, *11*, 19472–19480.
16. Gao, Y., Yan, C., Huang, H., Yang, T., Tian, G., Xiong, D., Chen, N., Chu, X., Zhong, S., Deng, W., et al. (2020). Microchannel-confined MXene based flexible piezoresistive multifunctional micro-force sensor, *Advanced Materials*, *30*, 1909603.
17. Fu, X., Wang, L., Zhao, L., Yuan, Z., Zhang, Y., Wang, D., Wang, D., Li, J., Li, D., Shulga, V., et al. (2021). Controlled assembly of MXene nanosheets as an electrode and active layer for high-performance electronic skin, *Advanced Materials*, *31*, 2010533.
18. Yi, Q., Pei, X., Das, P., Qin, H., Lee, S.W. and Esfandyarpour, R. (2022). A self-powered triboelectric MXene-based 3D-printed wearable physiological biosignal sensing system for on-demand, wireless, and real-time health monitoring, *Nano Energy*, *101*, 107511.

19. Gao, Y., Yu, L., Yeo, J.C. and Lim, C.T. (2020). Flexible hybrid sensors for health monitoring: Materials and mechanisms to render wearability, *Advanced Materials*, *32*, 1902133.
20. Li, J., Yao, Z., Meng, X., Zhang, X., Wang, Z., Wang, J., Ma, G., Liu, L., Zhang, J., Niu, S., et al. (2024). High-fidelity, Low-hysteresis bionic flexible strain sensors for Soft machines, *ACS Nano*, *18*, 2520–2530.
21. Lu, J., Su, L., Zhang, Z., Song, W., Hu, S., Wang, J., Li, X., Huang, Y., He, Z., Lei, M., et al. (2024). A flexible silver-nanoparticle/polyacrylonitrile biomimetic strain sensor by patterned UV reduction for artificial intelligence flexible electronics, *Journal of Materials Chemistry A*, *12*, 11895–11906.
22. Zhao, C., Wang, Y., Tang, G., Ru, J., Zhu, Z., Li, B., Guo, C.F., Li, L. and Zhu, D. (2022). Ionic flexible sensors: Mechanisms, materials, structures, and applications, *Advanced Materials*, *32*, 2110417.
23. Amoli, V., Kim, S.Y., Kim, J.S., Choi, H., Koo, J. and Kim, D.H. (2019). Biomimetics for high-performance flexible tactile sensors and advanced artificial sensory systems, *Journal of Materials Chemistry C*, *7*, 14816–14844.
24. Li, Y., Zhang, W., Zhao, C., Li, W., Dong, E., Xu, M., Huang, H., Yang, Y., Li, L., Zheng, L., et al. (2024). Breaking the saturation of sensitivity for ultrawide range flexible pressure sensors by soft-strain effect, *Advanced Materials*, 2405405.
25. Wang, X., Zhao, X., Yu, Y., Zhai, W., Yue, X., Dai, K., Liu, C. and Shen, C. (2024). Design of flexible micro-porous fiber with double conductive network synergy for high-performance strain sensor, *Chemical Engineering Journal*, *495*, 153641.
26. Lu, Y., Yue, Y., Ding, Q., Mei, C., Xu, X., Jiang, S., He, S., Wu, Q., Xiao, H. and Han, J. (2023). Environment-tolerant ionic hydrogel–elastomer hybrids with robust interfaces, high transparence, and biocompatibility for a mechanical–thermal multimode sensor, *Innovation*, *5*, e12409.
27. Fu, H., Wang, B., Li, J., Cao, D., Zhang, W., Xu, J., Li, J., Zeng, J., Gao, W. and Chen, K. (2024). Ultra-strong, nonfreezing, and flexible strain sensors enabled by biomass-based hydrogels through triple dynamic bond design, *Materials Horizons*, *11*, 1588–1596.
28. Cui, L., Wang, W., Zheng, J., Hu, C., Zhu, Z. and Liu, B. (2024). Wide-humidity, anti-freezing and stretchable multifunctional conductive carboxymethyl cellulose-based hydrogels for flexible wearable strain sensors and arrays, *Carbohydrate Polymers*, *342*, 122406.
29. Luo, J., Gao, S., Luo, H., Wang, L., Huang, X., Guo, Z., Lai, X., Lin, L., Li, R.K. and Gao, J. (2021). Superhydrophobic and breathable smart MXene-based textile for multifunctional wearable sensing electronics, *Chemical Engineering Journal*, *406*, 126898.
30. Guan, L., Liu, H., Ren, X., Wang, T., Zhu, W., Zhao, Y., Feng, Y., Shen, C., Zvyagin, A.V. and Fang, L. (2022). Balloon inspired conductive hydrogel

strain sensor for reducing radiation damage in peritumoral organs during brachytherapy, *Advanced Functional Materials*, *32*, 2112281.
31. Chen, Y.-W., Pancham, P.P., Mukherjee, A., Martincic, E. and Lo, C.-Y. (2022). Recent advances in flexible force sensors and their applications: A review, *Flexible and Printed Electronics*, *7*, 033002.
32. Chen, Z.-J., Shen, T.-Y., Zhang, M.-H., Xiao, X., Wang, H.-Q., Lu, Q.-R., Luo, Y.-L., Jin, Z. and Li, C.-H. (2024). Tough, anti-fatigue, self-adhesive, and anti-freezing hydrogel electrolytes for dendrite-free flexible zinc ion batteries and strain sensors, *Advanced Functional Materials*, *34*, 2314864.
33. Medina, I.I., Arana, G., Atoche, A.C.C., López, J.J.E., Castillo, J.V., Avilés, F. and Atoche, A.A.C. (2024). Adhesion testing system based on convolutional neural networks for quality inspection of flexible strain sensors, *IEEE Transactions on Industrial Informatics*, *20*, 9235–9243.
34. Jiang, F., Lee, G.-B., Tai, Y.-C. and Ho, C.-M. (2000). A flexible micromachine-based shear-stress sensor array and its application to separation-point detection, *Sensors and Actuators A: Physical*, *79*, 194–203.
35. Yu, H., Ai, L., Rouhanizadeh, M., Patel, D., Kim, E.S. and Hsiai, T.K. (2008). Flexible polymer sensors for *in vivo* intravascular shear stress analysis, *Journal of Micromechanics and Microengineering*, *17*, 1178–1186.
36. Missinne, J., Bosman, E., Van Hoe, B., Van Steenberge, G., Kalathimekkad, S., Van Daele, P. and Vanfleteren, J. (2011). Flexible shear sensor based on embedded optoelectronic components, *IEEE Photonics Technology Letters*, *23*, 771–773.
37. Mu, C., Song, Y., Huang, W., Ran, A., Sun, R., Xie, W. and Zhang, H. (2018). Flexible normal-tangential force sensor with opposite resistance responding for highly sensitive artificial skin, *Advanced Functional Materials*, *28*, 1707503.
38. Xie, M., Zhang, Y., Kraśny, M.J., Bowen, C., Khanbareh, H. and Gathercole, N. (2018). Flexible and active self-powered pressure, shear sensors based on freeze casting ceramic–polymer composites, *Energy & Environmental Science*, *11*, 2919–2927.
39. Huang, X., Bu, T., Zheng, Q., Liu, S., Li, Y., Fang, H., Qiu, Y., Xie, B., Yin, Z. and Wu, H. (2024). Flexible sensors with zero Poisson's ratio, *National Science Review*, *11*, nwae027.
40. Pang, C., Lee, G.-Y., Kim, T.-i., Kim, S.M., Kim, H.N., Ahn, S.-H. and Suh, K.-Y. (2012). A flexible and highly sensitive strain-gauge sensor using reversible interlocking of nanofibres, *Nature Materials*, *11*, 795–801.
41. Choi, D., Jang, S., Kim, J.S., Kim, H.J., Kim, D.H. and Kwon, J.Y. (2019). A highly sensitive tactile sensor using a pyramid-plug structure for detecting pressure, shear force, and torsion, *Advanced Materials Technologies*, *4*, 1800284.

42. Gao, Y.-H., Jen, Y.-H., Chen, R., Aw, K., Yamane, D. and Lo, C.-Y. (2019). Five-fold sensitivity enhancement in a capacitive tactile sensor by reducing material and structural rigidity, *Sensors and Actuators A: Physical*, *293*, 167–177.
43. Kim, K.K., Ha, I., Won, P., Seo, D.-G., Cho, K.-J. and Ko, S.H. (2019). Transparent wearable three-dimensional touch by self-generated multiscale structure, *Nature Communications*, *10*, 2582.
44. Lu, Y., Xu, K., Zhang, L., Deguchi, M., Shishido, H., Arie, T., Pan, R., Hayashi, A., Shen, L., Akita, S. and Takei, K. (2020). Multimodal plant healthcare flexible sensor system, *ACS Nano*, *14*, 10966.
45. You, I., Mackanic, D.G., Matsuhisa, N., Kang, J., Kwon, J., Beker, L., Mun, J., Suh, W., Kim, T.Y. and Tok, J.B.-H. (2020). Artificial multimodal receptors based on ion relaxation dynamics, *Science*, *370*, 961–965.
46. Cai, M., Jiao, Z., Nie, S., Wang, C., Zou, J. and Song, J. (2021). A multifunctional electronic skin based on patterned metal films for tactile sensing with a broad linear response range, *Science Advances*, *7*, eabl8313.
47. Yang, R., Zhang, W., Tiwari, N., Yan, H., Li, T. and Cheng, H. (2022). Multimodal sensors with decoupled sensing mechanisms, *Advanced Science*, *9*, 2202470.
48. Sundaram, S., Kellnhofer, P., Li, Y., Zhu, J.-Y., Torralba, A. and Matusik, W. (2019). Learning the signatures of the human grasp using a scalable tactile glove, *Nature*, *569*, 698–702.
49. Chun, K.Y., Son, Y.J., Jeon, E.S., Lee, S. and Han, C.S. (2018). A self-powered sensor mimicking slow-and fast-adapting cutaneous mechanoreceptors, *Advanced Materials*, *30*, 1706299.
50. Chun, S., Son, W., Kim, H., Lim, S.K., Pang, C. and Choi, C. (2019). Self-powered pressure-and vibration-sensitive tactile sensors for learning technique-based neural finger skin, *Nano Letters*, *19*, 3305–3312.
51. Zhang, Y., Hu, Y., Zhu, P., Han, F., Zhu, Y., Sun, R. and Wong, C.-P. (2017). Flexible and highly sensitive pressure sensor based on microdome-patterned PDMS forming with assistance of colloid self-assembly and replica technique for wearable electronics, *ACS Applied Materials & Interfaces*, *9*, 35968–35976.
52. Lee, Y.H., Kweon, O.Y., Kim, H., Yoo, J.H., Han, S.G. and Oh, J.H. (2018). Recent advances in organic sensors for health self-monitoring systems, *Journal of Materials Chemistry C*, *6*, 8569–8612.
53. Li, J., Zhao, Y., Zhai, W., Zhao, X., Dai, K., Liu, C. and Shen, C. (2024). Highly aligned electrospun film with wave-like structure for multidirectional strain and visual sensing, *Chemical Engineering Journal*, *485*, 149952.
54. Shan, X., Ding, L., Wang, D., Wen, S., Shi, J., Chen, C., Wang, Y., Zhu, H., Huang, Z., Wang, S.S., *et al.* (2024). Sub-femtonewton force sensing in solution by super-resolved photonic force microscopy, *Nature Photonics*, *18*, 913–921.

55. Madani, Z., Silva, P.E.S., Baniasadi, H., Vaara, M., Das, S., Arias, J.C., Seppälä, J., Sun, Z. and Vapaavuori, J. (2024). Light-driven multidirectional bending in artificial muscles, *Advanced Materials*, 2405917.
56. Zhang, X., He, W., Bai, J., Billinghurst, M., Liu, D., Dong, J., Qin, Y., Liu, T. and Wang, Z. (2024). Integrate augmented reality and force sensing devices to assist blind area assembly, *Journal of Manufacturing Systems*, 74, 594–605.
57. Wu, G., Li, X., Bao, R. and Pan, C. (2024). Innovations in tactile sensing: Microstructural designs for superior flexible sensor performance, *Advanced Functional Materials*, 2405722.
58. Tan, Y., Liu, X., Tang, W., Chen, J., Zhu, Z., Li, L., Zhou, N., Kang, X., Xu, D., Wang, L., et al. (2022). Flexible pressure sensors based on bionic microstructures: From plants to animals, *Advanced Materials Interfaces*, 9, 2101312.
59. Mishra, R.B., El-Atab, N., Hussain, A.M. and Hussain, M.M. (2021). Recent progress on flexible capacitive pressure sensors: From design and materials to applications, *Advanced Materials Technologies*, 6, 2001023.
60. Ma, Y., Li, H., Chen, S., Liu, Y., Meng, Y., Cheng, J. and Feng, X. (2021). Skin-like electronics for perception and interaction: Materials, structural designs, and applications, *Advanced Intelligent Systems*, 3, 2000108.
61. Nur, R., Matsuhisa, N., Jiang, Z., Nayeem, M.O.G., Yokota, T. and Someya, T. (2018). A highly sensitive capacitive-type strain sensor using wrinkled ultrathin gold films, *Nano Letters*, 18, 5610–5617.
62. Mu, J., Hou, C., Wang, G., Wang, X., Zhang, Q., Li, Y., Wang, H. and Zhu, M. (2016). An elastic transparent conductor based on hierarchically wrinkled reduced graphene oxide for artificial muscles and sensors, *Advanced Materials*, 28, 9491–9497.
63. Cheng, Y., Wang, R., Chan, K.H., Lu, X., Sun, J. and Ho, G.W. (2018). A biomimetic conductive tendril for ultrastretchable and integratable electronics, muscles, and sensors, *ACS Nano*, 12, 3898–3907.
64. Li, X., Ye, B., Jiang, L., Li, X., Zhao, Y., Qu, L., Yi, P., Li, T., Li, M., Li, L., et al. (2024). Helical micropillar processed by one-step 3D printing for solar thermal conversion, *Small*, 2400569.
65. Deshmukh, A.P., Zheng, W., Chuang, C., Bailey, A.D., Williams, J.A., Sletten, E.M., Egelman, E.H. and Caram, J.R. (2024). Near-atomic-resolution structure of J-aggregated helical light-harvesting nanotubes, *Nature Chemistry*, 16, 800–808.
66. Zhao, Y., Yang, W., Tan, Y.J., Li, S., Zeng, X., Liu, Z. and Tee, B.C.K. (2019). Highly conductive 3D metal-rubber composites for stretchable electronic applications, *APL Materials*, 7, 031508.
67. Xu, S., Yan, Z., Jang, K.-I., Huang, W., Fu, H., Kim, J., Wei, Z., Flavin, M., McCracken, J., Wang, R., et al. (2015). Assembly of micro/nanomaterials

into complex, three-dimensional architectures by compressive buckling, *Science*, *347*, 154–159.
68. Jang, K.-I., Li, K., Chung, H.U., Xu, S., Jung, H.N., Yang, Y., Kwak, J.W., Jung, H.H., Song, J., Yang, C., *et al.* (2017). Self-assembled three dimensional network designs for soft electronics, *Nature Communications*, *8*, 15894.
69. Zhao, H., Li, K., Han, M., Zhu, F., Vázquez-Guardado, A., Guo, P., Xie, Z., Park, Y., Chen, L., Wang, X., *et al.* (2019). Buckling and twisting of advanced materials into morphable 3D mesostructures, *Proceedings of the National Academy of Sciences*, *116*, 13239–13248.
70. Fu, H., Nan, K., Bai, W., Huang, W., Bai, K., Lu, L., Zhou, C., Liu, Y., Liu, F., Wang, J., *et al.* (2018). Morphable 3D mesostructures and microelectronic devices by multistable buckling mechanics, *Nature Materials*, *17*, 268–276.
71. Won, S.M., Wang, H., Kim, B.H., Lee, K., Jang, H., Kwon, K., Han, M., Crawford, K.E., Li, H., Lee, Y., *et al.* (2019). Multimodal sensing with a three-dimensional piezoresistive structure, *ACS Nano*, *13*, 10972–10979.
72. Ma, L., Liu, Q., Wu, R., Meng, Z., Patil, A., Yu, R., Yang, Y., Zhu, S., Fan, X., Hou, C., *et al.* (2020). From molecular reconstruction of mesoscopic functional conductive silk fibrous Materials to remote respiration monitoring, *Small*, *16*, e2000203.
73. Qi, M., Yang, R., Wang, Z., Liu, Y., Zhang, Q., He, B., Li, K., Yang, Q., Wei, L., Pan, C., *et al.* (2023). Bioinspired self-healing soft electronics, *Advanced Functional Materials*, *33*, 2214479.
74. Liu, Y., Wang, F., Hu, Z., Li, M., Ouyang, S., Wu, Y., Wang, S., Li, Z., Qian, J., Wang, L., *et al.* (2024). Applications of cellulose-based flexible self-healing sensors for human health monitoring, *Nano Energy*, *127*, 109790.
75. Luo, X., Wu, H., Wang, C., Jin, Q., Luo, C., Ma, G., Guo, W. and Long, Y. (2024). 3D printing of self-healing and degradable conductive ionoelastomers for customized flexible sensors, *Chemical Engineering Journal*, *483*, 149330.
76. Ni, Y., Chen, J. and Chen, K. (2024). Flexible vanillin-polyacrylate/chitosan/mesoporous nanosilica-MXene composite film with self-healing ability towards dual-mode sensors, *Carbohydrate Polymers*, *335*, 122042.
77. Cheng, Y., Xiao, X., Pan, K. and Pang, H. (2020). Development and application of self-healing materials in smart batteries and supercapacitors, *Chemical Engineering Journal*, *380*, 122565.
78. Zhang, Y., Yuan, Y., Yu, H., Cai, C., Sun, J. and Tian, X. (2024). A stretchable conductive elastomer sensor with self-healing and highly linear strain for human movement detection and pressure response, *Materials Horizons*, *11*, 3911–3920.
79. Hu, F., Dong, B., Yu, D., Zhao, R., Chen, W., Song, Z., Lu, P., Zhang, F., Wang, Z., Liu, X., *et al.* (2024). Highly stretchable, self-healing,

antibacterial, conductive, and amylopectin-enhanced hydrogels with gallium droplets loading as strain sensors, *Carbohydrate Polymers*, *342*, 122357.
80. Jiang, Q., Wan, Y., Qin, Y., Qu, X., Zhou, M., Huo, S., Wang, X., Yu, Z. and He, H. (2024). Durable and wearable self-powered temperature sensor based on self-healing thermoelectric fiber by coaxial wet spinning strategy for fire safety of firefighting clothing, *Advanced Fiber Materials*, *6*, 1387–1401.
81. Cao, Y., Tan, Y.J., Li, S., Lee, W.W., Guo, H., Cai, Y., Wang, C. and Tee, B.C.K. (2019). Self-healing electronic skins for aquatic environments, *Nature Electronics*, *2*, 75–82.
82. Son, D., Kang, J., Vardoulis, O., Kim, Y., Matsuhisa, N., Oh, J.Y., To, J.W.F., Mun, J., Katsumata, T., Liu, Y., *et al.* (2018). An integrated self-healable electronic skin system fabricated via dynamic reconstruction of a nanostructured conducting network, *Nature Nanotechnology*, *13*, 1057–1065.
83. Ge, G., Lu, Y., Qu, X., Zhao, W., Ren, Y., Wang, W., Wang, Q., Huang, W. and Dong, X. (2020). Muscle-inspired self-healing hydrogels for strain and temperature sensor, *ACS Nano*, *14*, 218–228.
84. Ying, W.B., Wang, G., Kong, Z., Yao, C.K., Wang, Y., Hu, H., Li, F., Chen, C., Tian, Y., Zhang, J., *et al.* (2021). A biologically muscle-inspired polyurethane with super-tough, thermal reparable and self-healing capabilities for stretchable electronics, *Advanced Functional Materials*, *31*, 2009869.
85. Wang, Y., Yu, Y., Guo, J., Zhang, Z., Zhang, X. and Zhao, Y. (2020). Bio-inspired stretchable, adhesive, and conductive structural color film for visually flexible electronics, *Advanced Functional Materials*, *30*, 2000151.
86. Bi, S., Jin, W., Han, X., Metts, J., Ostrosky, A.D., Lehotsky, J., He, Z., Jiang, C. and Asare-Yeboah, K. (2023). Flexible pressure visualization equipment for human-computer interaction, *Materials Today Sustainability*, *21*, 100318.
87. Shao, B., Zhang, S., Hu, Y., Zheng, Z., Zhu, H., Wang, L., Zhao, L., Xu, F., Wang, L., Li, M., *et al.* (2024). Color-shifting iontronic skin for on-site, nonpixelated pressure mapping visualization, *Nano Letters*, *24*, 4741–4748.
88. Kim, M.-S., Timilsina, S., Jang, S.-M., Kim, J.-S. and Park, S.-S. (2024). A mechanoluminescent ZnS:Cu/PDMS and biocompatible piezoelectric silk fibroin/PDMS hybrid sensor for self-powered sensing and artificial intelligence control, *Advanced Materials Technologies*, *9*, 2400255.
89. Guo, X., Li, Y., Hong, W., Yan, Z., Duan, Z., Zhang, A., Zhang, X., Jin, C., Liu, T., Li, X., *et al.* (2024). Bamboo-inspired, environmental friendly PDMS/Plant fiber composites-based capacitive flexible pressure sensors by origami for human–machine interaction, *ACS Sustainable Chemistry & Engineering*, *12*, 4835–4845.

90. Pyo, S., Lee, J., Bae, K., Sim, S. and Kim, J. (2021). Recent progress in flexible tactile sensors for human-interactive systems: From sensors to advanced applications, *Advanced Materials*, *33*, 2005902.
91. Lee, B., Oh, J.-Y., Cho, H., Joo, C.W., Yoon, H., Jeong, S., Oh, E., Byun, J., Kim, H., Lee, S., et al. (2020). Ultraflexible and transparent electroluminescent skin for real-time and super-resolution imaging of pressure distribution, *Nature Communications*, *11*, 663.
92. Pan, C., Dong, L., Zhu, G., Niu, S., Yu, R., Yang, Q., Liu, Y. and Wang, Z.L. (2013). High-resolution electroluminescent imaging of pressure distribution using a piezoelectric nanowire LED array, *Nature Photonics*, *7*, 752–758.
93. Xu, S., Bao, C., Hasan, M.A.M., Zhang, X., Li, C. and Yang, Y. (2024). Low-frequency porous composite foam catalyst for efficient dye wastewater degradation via coupling photoelectric and piezoelectric effects, *Nano Energy*, *127*, 109720.
94. Li, M., Lu, J., Wan, P., Jiang, M., Mo, Y. and Pan, C. (2024). An Ultrasensitive perovskite single-model plasmonic strain sensor based on piezoelectric effect, *Advanced Functional Materials*, 2403840.
95. Park, K.-H., Lee, M.-K., Kim, B.-H., Baek, C. and Lee, G.-J. (2024). Lead-free hybrid piezoelectric ceramic stack for both potent and temperature-stable piezoelectricity, *Journal of Materials Chemistry A*, *12*, 22299–22309.
96. Wei, X.Y., Wang, X., Kuang, S.Y., Su, L., Li, H.Y., Wang, Y., Pan, C., Wang, Z.L. and Zhu, G. (2016). Dynamic triboelectrification-induced electroluminescence and its Use in visualized sensing, *Advanced Materials*, *28*, 6656–6664.
97. Kim, J., Shim, H.J., Yang, J., Choi, M.K., Kim, D.C., Kim, J., Hyeon, T. and Kim, D.-H. (2017). Flexible displays: Ultrathin quantum dot display integrated with wearable electronics (Adv. Mater. 38/2017), *Advanced Materials*, *29*, 201770278.
98. Koo, J., Amoli, V., Kim, S.Y., Lee, C., Kim, J., Park, S.-M., Kim, J., Ahn, J.M., Jung, K.J. and Kim, D.H. (2020). Low-power, deformable, dynamic multicolor electrochromic skin, *Nano Energy*, *78*, 105199.
99. Guo, Q., Zhao, X., Li, Z., Wang, D. and Nie, G. (2020). A novel solid-state electrochromic supercapacitor with high energy storage capacity and cycle stability based on poly(5-formylindole)/WO3 honeycombed porous nanocomposites, *Chemical Engineering Journal*, *384*, 123370.
100. Du, J., Wang, T., Li, Y., Xiong, P., Xiao, Y., Feng, A., Wang, X., Jiang, K. and Lin, H. (2024). Efficient synthesis of quaternary piezo-photonic materials for pressure visualization and E-signature anti-counterfeiting, *Chemical Engineering Journal*, *494*, 152989.
101. Luo, Z., Chen, W., Lai, M., Shi, S., Chen, P., Yang, X., Chen, Z., Wang, B., Zhang, Y. and Zhou, X. (2024). Fully printable and reconfigurable

hufu-type electroluminescent devices for visualized encryption, *Advanced Materials*, *36*, 2313909.
102. Li, R., Ma, X., Li, J., Cao, J., Gao, H., Li, T., Zhang, X., Wang, L., Zhang, Q., Wang, G., et al. (2021). Flexible and high-performance electrochromic devices enabled by self-assembled 2D TiO2/MXene heterostructures, *Nature Communications*, *12*, 1587.
103. Bi, S., Jin, W., Han, X., Cao, X., He, Z., Asare-Yeboah, K. and Jiang, C. (2022). Ultra-fast-responsivity with sharp contrast integrated flexible piezo electrochromic based tactile sensing display, *Nano Energy*, *102*, 107629.
104. Kokubo, A., Kuwabara, M., Tomitani, N., Yamashita, S., Shiga, T. and Kario, K. (2024). Development of beat-by-beat blood pressure monitoring device and nocturnal sec-surge detection algorithm, *Hypertension Research*, *47*, 1576–1587.
105. Chou, H.-H., Nguyen, A., Chortos, A., To, J.W.F., Lu, C., Mei, J., Kurosawa, T., Bae, W.-G., Tok, J.B.H. and Bao, Z. (2015). A chameleon-inspired stretchable electronic skin with interactive colour changing controlled by tactile sensing, *Nature Communications*, *6*, 8011.
106. Park, H., Kim, D.S., Hong, S.Y., Kim, C., Yun, J.Y., Oh, S.Y., Jin, S.W., Jeong, Y.R., Kim, G.T. and Ha, J.S. (2017). A skin-integrated transparent and stretchable strain sensor with interactive color-changing electrochromic displays, *Nanoscale*, *9*, 7631–7640.
107. Cho, S.H., Lee, S.W., Yu, S., Kim, H., Chang, S., Kang, D., Hwang, I., Kang, H.S., Jeong, B., Kim, E.H., et al. (2017). Micropatterned pyramidal ionic gels for sensing broad-range pressures with high sensitivity, *ACS Applied Materials & Interfaces*, *9*, 10128–10135.
108. Ahn, J., Lee, Y., Kim, J., Yoon, S., Jeong, Y.-C. and Cho, K.Y. (2022). Thiol-ene UV-curable sponge electrolyte for low-voltage color changing wearable tactile device, *Polymer*, *250*, 124898.
109. Han, X., Du, W., Chen, M., Wang, X., Zhang, X., Li, X., Li, J., Peng, Z., Pan, C. and Wang, Z.L. (2017). Visualization recording and storage of pressure distribution through a smart matrix based on the piezotronic effect, *Advanced Materials*, *29*, 1701253.
110. Amoli, V., Kim, J.S., Kim, S.Y., Koo, J., Chung, Y.S., Choi, H. and Kim, D.H. (2019). Ionic tactile sensors for emerging human-interactive technologies: A review of recent progress, *Advanced Functional Materials*, *20*, 1904532.
111. Giltrap, A.M., Yuan, Y. and Davis, B.G. (2024). Late-stage functionalization of living organisms: Rethinking selectivity in biology, *Chemical Reviews*, *124*, 889–928.
112. Lei, Z.-C., Wang, X., Yang, L., Qu, H., Sun, Y., Yang, Y., Li, W., Zhang, W.-B., Cao, X.-Y., Fan, C., et al. (2024). What can molecular assembly learn from catalysed assembly in living organisms?, *Chemical Society Reviews*, *53*, 1892–1914.

113. Xu, K., Lu, Y. and Takei, K. (2021). Flexible hybrid sensor systems with feedback functions, *Advanced Functional Materials*, *31*, 2007436.
114. Kim, K.K., Kim, M., Pyun, K., Kim, J., Min, J., Koh, S., Root, S.E., Kim, J., Nguyen, B.-N.T., Nishio, Y., et al. (2023). A substrate-less nanomesh receptor with meta-learning for rapid hand task recognition, *Nature Electronics*, *6*, 64–75.
115. Hwang, S.-W., Tao, H., Kim, D.-H., Cheng, H., Song, J.-K., Rill, E., Brenckle, M.A., Panilaitis, B., Won, S.M., Kim, Y.-S., et al. (2012). A physically transient form of silicon electronics, *Science*, *337*, 1640–1644.
116. Kang, S.-K., Koo, J., Lee, Y.K. and Rogers, J.A. (2018). Advanced materials and devices for bioresorbable electronics, *Accounts of Chemical Research*, *51*, 988–998.
117. Asghari, F., Samiei, M., Adibkia, K., Akbarzadeh, A. and Davaran, S. (2017). Biodegradable and biocompatible polymers for tissue engineering application: A review, *Artificial Cells, Nanomedicine, and Biotechnology*, *45*, 185–192.
118. Guo, Y., Zhong, M., Fang, Z., Wan, P. and Yu, G. (2019). A wearable transient pressure sensor made with MXene nanosheets for sensitive broad-range human–machine interfacing, *Nano Letters*, *19*, 1143–1150.
119. Qi, D., Liu, Z., Yu, M., Liu, Y., Tang, Y., Lv, J., Li, Y., Wei, J., Liedberg, B., Yu, Z., et al. (2015). Highly stretchable gold nanobelts with sinusoidal structures for recording electrocorticograms, *Advanced Materials*, *27*, 3145–3151.
120. Tybrandt, K., Khodagholy, D., Dielacher, B., Stauffer, F., Renz, A.F., Buzsáki, G. and Vörös, J. (2018). High-density stretchable electrode grids for chronic neural recording, *Advanced Materials*, *30*, 1706520.
121. Ye, C., Lukas, H., Wang, M., Lee, Y. and Gao, W. (2024). Nucleic acid-based wearable and implantable electrochemical sensors, *Chemical Society Reviews*, *53*, 7960–7982.
122. Kim, J., Hong, J., Park, K., Lee, S., Hoang, A.T., Pak, S., Zhao, H., Ji, S., Yang, S., Chung, C.K., et al. (2024). Injectable 2D material-based sensor array for minimally invasive neural implants, *Advanced Materials*, 2400261.
123. Khodagholy, D., Gelinas, J.N., Thesen, T., Doyle, W., Devinsky, O., Malliaras, G.G. and Buzsáki, G. (2015). NeuroGrid: Recording action potentials from the surface of the brain, *Nature Neuroscience*, *18*, 310–315.
124. Ouyang, H., Li, Z., Gu, M., Hu, Y., Xu, L., Jiang, D., Cheng, S., Zou, Y., Deng, Y., Shi, B., et al. (2021). A bioresorbable dynamic pressure sensor for cardiovascular postoperative care, *Advanced Materials*, *33*, 2102302.
125. Zhao, B., Eid, L., Zhang, Y., Sun, X. and Cui, W. (2023). Implantable sensors for post-surgical monitoring of vascular complications, *Advanced Sensor Research*, *2*, 2200095.

126. Lee, J., Ihle, S.J., Pellegrino, G.S., Kim, H., Yea, J., Jeon, C.-Y., Son, H.-C., Jin, C., Eberli, D., Schmid, F., et al. (2021). Stretchable and suturable fibre sensors for wireless monitoring of connective tissue strain, *Nature Electronics*, *4*, 291–301.
127. Herbert, R., Lim, H.-R., Rigo, B. and Yeo, W.-H. (2022). Fully implantable wireless batteryless vascular electronics with printed soft sensors for multiplex sensing of hemodynamics, *Science Advances*, *8*, eabm1175.
128. Peng, J. and Snyder, G.J. (2019). A figure of merit for flexibility, *Science*, *366*, 690–691.
129. Rogers, J.A., Someya, T. and Huang, Y. (2010). Materials and mechanics for stretchable electronics, *Science*, *327*, 1603–1607.
130. Rogers, J.A., Chen, X. and Feng, X. (2020). Flexible hybrid electronics, *Advanced Materials*, *32*, 1905590.
131. Ren, Q., Zhu, C., Ma, S., Wang, Z., Yan, J., Wan, T., Yan, W. and Chai, Y. (2024). optoelectronic devices for in-sensor computing, *Advanced Materials*, 2407476.
132. Lee, S.Y., Pakela, J.M., Na, K., Shi, J., McKenna, B.J., Simeone, D.M., Yoon, E., Scheiman, J.M. and Mycek, M.-A. (2020). Needle-compatible miniaturized optoelectronic sensor for pancreatic cancer detection, *Science Advances*, *6*, eabc1746.
133. Konstantatos, G. (2018). Current status and technological prospect of photodetectors based on two-dimensional materials, *Nature Communications*, *9*, 5266.
134. Yin, R., Wang, D., Zhao, S., Lou, Z. and Shen, G. (2021). Wearable sensors-enabled human–machine interaction systems: From design to application, *Advanced Functional Materials*, *31*, 2008936.
135. Zhang, Z., Wang, W., Jiang, Y., Wang, Y.-X., Wu, Y., Lai, J.-C., Niu, S., Xu, C., Shih, C.-C., Wang, C., et al. (2022). High-brightness all-polymer stretchable LED with charge-trapping dilution, *Nature*, *603*, 624–630.
136. Simone, G., Dyson, M.J., Meskers, S.C.J., Janssen, R.A.J. and Gelinck, G.H. (2020). Organic photodetectors and their application in large area and flexible image sensors: The role of dark current, *Advanced Functional Materials*, *30*, 1904205.
137. Wang, Z. and Shen, G. (2023). Flexible optoelectronic sensors: Status and prospects, *Materials Chemistry Frontiers*, *7*, 1496–1519.
138. Wang, W., He, Z., Di, C.-a. and Zhu, D. (2023). Advances in organic transistors for artificial perception applications, *Materials Today Electronics*, *3*, 100028.
139. Zhang, J.-B., Tian, Y.-B., Gu, Z.-G. and Zhang, J. (2024). Metal–organic framework-based photodetectors, *Nano-Micro Letters*, *16*, 253.
140. Akhavan, S., Najafabadi, A.T., Mignuzzi, S., Jalebi, M.A., Ruocco, A., Paradisanos, I., Balci, O., Andaji-Garmaroudi, Z., Goykhman, I., Occhipinti,

L.G., et al. (2024). Graphene-Perovskite Fibre Photodetectors, *Advanced Materials*, 2400703.
141. Weng, Y., Yu, Z., Wu, T., Liang, L. and Liu, S. (2023). Recent progress in stretchable organic field-effect transistors: Key materials, fabrication and applications, *New Journal of Chemistry*, *47*, 5086–5109.
142. Li, H., Ma, Y. and Huang, Y. (2021). Material innovation and mechanics design for substrates and encapsulation of flexible electronics: A review, *Materials Horizons*, *8*, 383–400.
143. Qiu, T., Akinoglu, E.M., Luo, B., Konarova, M., Yun, J.-H., Gentle, I.R. and Wang, L. (2022). Nanosphere lithography: A versatile approach to develop transparent conductive films for optoelectronic applications, *Advanced Materials*, *34*, 2103842.
144. Gao, W.-H. and Chen, C. (2024). Perovskites and their constructed near-infrared photodetectors, *Nano Energy*, *128*, 109904.
145. Li, D., Lai, W.-Y., Zhang, Y.-Z. and Huang, W. (2018). Printable transparent conductive films for flexible electronics, *Advanced Materials*, *30*, 1704738.
146. Wang, C., Xia, K., Wang, H., Liang, X., Yin, Z. and Zhang, Y. (2019). Advanced carbon for flexible and wearable electronics, *Advanced Materials*, *31*, 1801072.
147. Fan, X., Nie, W., Tsai, H., Wang, N., Huang, H., Cheng, Y., Wen, R., Ma, L., Yan, F. and Xia, Y. (2019). PEDOT:PSS for flexible and stretchable electronics: Modifications, strategies, and applications, *Advanced Science*, *6*, 1900813.
148. Liu, J., Wang, M., Wang, P., Hou, F., Meng, C., Hashimoto, K. and Guo, S. (2020). Cost-efficient flexible supercapacitive tactile sensor with superior sensitivity and high spatial resolution for human-robot interaction, *IEEE Access*, *8*, 64836–64845.
149. Xiang, Z.-L., Tan, Q.-H., Zhu, T., Yang, P.-Z., Liu, Y.-P., Liu, Y.-K. and Wang, Q.-J. (2024). High-performance 1D CsPbBr3/CdS photodetectors, *Rare Metals*, *43*, 5932–5942.
150. Pan, T., Liu, S., Zhang, L. and Xie, W. (2022). Flexible organic optoelectronic devices on paper, *iScience*, *25*, 103782.
151. Huang, Q., Zhang, K., Yang, Y., Ren, J., Sun, R., Huang, F. and Wang, X. (2019). Highly smooth, stable and reflective Ag-paper electrode enabled by silver mirror reaction for organic optoelectronics, *Chemical Engineering Journal*, *370*, 1048–1056.
152. Chen, Z., Li, L., Xiao, X., Zhang, Y., Zhang, J., Jiang, Q., Hu, X. and Wang, Y. (2024). Used tissue paper as a 3D substrate for non-enzyme glucose sensors, *Green Chemistry*, *26*, 3801–3813.
153. He, Z., Wang, L., Liu, G.-S., Xu, Y., Qiu, Z., Zhong, M., Li, X., Gui, X., Lin, Y.-S., Qin, Z., et al. (2020). Constructing electrophoretic displays on

foldable paper-based electrodes by a facile transferring method, *ACS Applied Electronic Materials*, *2*, 1335–1342.
154. Jiang, Y., Wang, Z., Liu, X., Yang, Q., Huang, Q., Wang, L., Dai, Y., Qin, C. and Wang, S. (2020). Highly transparent, UV-shielding, and water-resistant lignocellulose nanopaper from agro-industrial waste for green optoelectronics, *ACS Sustainable Chemistry & Engineering*, *8*, 17508–17519.
155. Oh, J.Y., Rondeau-Gagné, S., Chiu, Y.-C., Chortos, A., Lissel, F., Wang, G.-J.N., Schroeder, B.C., Kurosawa, T., Lopez, J., Katsumata, T., *et al.* (2016). Intrinsically stretchable and healable semiconducting polymer for organic transistors, *Nature*, *539*, 411–415.
156. Khang, D.-Y., Jiang, H., Huang, Y. and Rogers, J.A. (2006). A stretchable form of single-crystal silicon for high-performance electronics on rubber substrates, *Science*, *311*, 208–212.
157. Ko, H.C., Stoykovich, M.P., Song, J., Malyarchuk, V., Choi, W.M., Yu, C.-J., Geddes III, J.B., Xiao, J., Wang, S., Huang, Y., *et al.* (2008). A hemispherical electronic eye camera based on compressible silicon optoelectronics, *Nature*, *454*, 748–753.
158. Konstantatos, G., Howard, I., Fischer, A., Hoogland, S., Clifford, J., Klem, E., Levina, L. and Sargent, E.H. (2006). Ultrasensitive solution-cast quantum dot photodetectors, *Nature*, *442*, 180–183.
159. Jia, C., Lin, Z., Huang, Y. and Duan, X. (2019). Nanowire electronics: From nanoscale to macroscale, *Chemical Reviews*, *119*, 9074–9135.
160. Deng, T., Zhang, Z., Liu, Y., Wang, Y., Su, F., Li, S., Zhang, Y., Li, H., Chen, H., Zhao, Z., *et al.* (2019). Three-dimensional graphene field-effect transistors as high-performance photodetectors, *Nano Letters*, *19*, 1494–1503.
161. Lim, K.-G., Han, T.-H. and Lee, T.-W. (2021). Engineering electrodes and metal halide perovskite materials for flexible/stretchable perovskite solar cells and light-emitting diodes, *Energy & Environmental Science*, *14*, 2009–2035.
162. Kim, D.-H., Lu, N., Ma, R., Kim, Y.-S., Kim, R.-H., Wang, S., Wu, J., Won, S.M., Tao, H., Islam, A., *et al.* (2011). Epidermal electronics, *Science*, *333*, 838–843.
163. Sheng, X., Bower, C.A., Bonafede, S., Wilson, J.W., Fisher, B., Meitl, M., Yuen, H., Wang, S., Shen, L., Banks, A.R., *et al.* (2014). Printing-based assembly of quadruple-junction four-terminal microscale solar cells and their use in high-efficiency modules, *Nature Materials*, *13*, 593–598.
164. Wang, Z., Xiao, X., Wu, W., Zhang, X. and Pang, Y. (2024). Ultra-conformal epidermal antenna for multifunctional motion artifact-free sensing and point-of-care monitoring, *Biosensors and Bioelectronics*, *253*, 116150.
165. Zhao, S., Zhou, Y., Xia, M., Zhang, Y., Yang, S., Tuan Hoang, A., Cao, D., Gao, Y. and Lai, Y. (2024). Highly stretchable, sensitive and healable epidermal electronics enabled by dynamic hard domains and multidimensional

conductive fillers for human motion and biopotential sensing, *Chemical Engineering Journal*, *489*, 151192.
166. Gao, W., Emaminejad, S., Nyein, H.Y.Y., Challa, S., Chen, K., Peck, A., Fahad, H.M., Ota, H., Shiraki, H., Kiriya, D., *et al.* (2016). Fully integrated wearable sensor arrays for multiplexed in situ perspiration analysis, *Nature*, 529, 509–514.
167. Chen, Y., Wan, X., Li, G., Ye, J., Gao, J. and Wen, D. (2024). Metal hydrogel-based integrated wearable biofuel cell for self-powered epidermal sweat biomarker monitoring, *Advanced Functional Materials*, 2404329.
168. Lochner, C.M., Khan, Y., Pierre, A. and Arias, A.C. (2014). All-organic optoelectronic sensor for pulse oximetry, *Nature Communications*, 5, 5745.
169. Kim, J., Salvatore, G.A., Araki, H., Chiarelli, A.M., Xie, Z., Banks, A., Sheng, X., Liu, Y., Lee, J.W., Jang, K.-I., *et al.* (2016). Battery-free, stretchable optoelectronic systems for wireless optical characterization of the skin, *Science Advances*, 2, e1600418.
170. Bansal, A.K., Hou, S., Kulyk, O., Bowman, E.M. and Samuel, I.D.W. (2015). Wearable Organic optoelectronic sensors for medicine, *Advanced Materials*, 27, 7638–7644.
171. Yokota, T., Zalar, P., Kaltenbrunner, M., Jinno, H., Matsuhisa, N., Kitanosako, H., Tachibana, Y., Yukita, W., Koizumi, M. and Someya, T. (2016). Ultraflexible organic photonic skin, *Science Advances*, 2, e1501856.
172. Mathieson, K., Loudin, J., Goetz, G., Huie, P., Wang, L., Kamins, T.I., Galambos, L., Smith, R., Harris, J.S., Sher, A., *et al.* (2012). Photovoltaic retinal prosthesis with high pixel density, *Nature Photonics*, 6, 391–397.
173. Choi, M., Choi, J.W., Kim, S., Nizamoglu, S., Hahn, S.K. and Yun, S.H. (2013). Light-guiding hydrogels for cell-based sensing and optogenetic synthesis *in vivo*, *Nature Photonics*, 7, 987–994.
174. Yun, S.H. and Kwok, S.J.J. (2017). Light in diagnosis, therapy and surgery, *Nature Biomedical Engineering*, 1, 0008.
175. Boppart, S.A. and Richards-Kortum, R. (2014). Point-of-care and point-of-procedure optical imaging technologies for primary care and global health, *Science Translational Medicine*, 6, 253rv252–253rv252.
176. Lu, C., Froriep, U.P., Koppes, R.A., Canales, A., Caggiano, V., Selvidge, J., Bizzi, E. and Anikeeva, P. (2014). Polymer fiber probes enable optical control of spinal cord and muscle function *in vivo*, *Advanced Functional Materials*, 24, 6594–6600.
177. Canales, A., Jia, X., Froriep, U.P., Koppes, R.A., Tringides, C.M., Selvidge, J., Lu, C., Hou, C., Wei, L., Fink, Y., *et al.* (2015). Multifunctional fibers for simultaneous optical, electrical and chemical interrogation of neural circuits *in vivo*, *Nature Biotechnology*, 33, 277–284.
178. Lu, C., Park, S., Richner, T.J., Derry, A., Brown, I., Hou, C., Rao, S., Kang, J., Moritz, C.T., Fink, Y., *et al.* (2017). Flexible and stretchable

nanowire-coated fibers for optoelectronic probing of spinal cord circuits, *Science Advances*, *3*, e1600955.
179. Xu, H., Yin, L., Liu, C., Sheng, X. and Zhao, N. (2018). Recent advances in biointegrated optoelectronic devices, *Advanced Materials*, *30*, 1800156.
180. Kim, T.-i., McCall, J.G., Jung, Y.H., Huang, X., Siuda, E.R., Li, Y., Song, J., Song, Y.M., Pao, H.A., Kim, R.-H., et al. (2013). Injectable, cellular-scale optoelectronics with applications for wireless optogenetics, *Science*, *340*, 211–216.
181. Jeong, J.-W., McCall, J.G., Shin, G., Zhang, Y., Al-Hasani, R., Kim, M., Li, S., Sim, J.Y., Jang, K.-I., Shi, Y., et al. (2015). Wireless optofluidic systems for programmable in vivo pharmacology and optogenetics, *Cell*, *162*, 662–674.
182. Wu, F., Stark, E., Ku, P.-C., Wise, K.D., Buzsáki, G. and Yoon, E. (2015). Monolithically integrated μLEDs on silicon neural probes for high-resolution optogenetic studies in behaving animals, *Neuron*, *88*, 1136–1148.
183. Ohta, J., Ohta, Y., Takehara, H., Noda, T., Sasagawa, K., Tokuda, T., Haruta, M., Kobayashi, T., Akay, Y.M. and Akay, M. (2017). Implantable microimaging device for observing brain activities of rodents, *Proceedings of the IEEE*, *105*, 158–166.
184. Song, Y.M., Xie, Y., Malyarchuk, V., Xiao, J., Jung, I., Choi, K.-J., Liu, Z., Park, H., Lu, C., Kim, R.-H., et al. (2013). Digital cameras with designs inspired by the arthropod eye, *Nature*, *497*, 95–99.
185. Choi, C., Choi, M.K., Liu, S., Kim, M., Park, O.K., Im, C., Kim, J., Qin, X., Lee, G.J., Cho, K.W., et al. (2017). Human eye-inspired soft optoelectronic device using high-density MoS2-graphene curved image sensor array, *Nature Communications*, *8*, 1664.
186. Wang, W., Zhou, H., Xu, Z., Li, Z., Zhang, L. and Wan, P. (2024). Flexible conformally bioadhesive MXene hydrogel electronics for machine learning-facilitated human-interactive sensing, *Advanced Materials*, 2401035.
187. Feng, Y., Wu, C., Chen, M., Sun, H., Vellaisamy, A.L.R., Daoud, W.A., Yu, X., Zhang, G. and Li, W.J. (2024). Amoeba-inspired self-healing electronic slime for adaptable, durable epidermal wearable electronics, *Advanced Functional Materials*, 2402393.
188. Zhang, K., Jung, Y.H., Mikael, S., Seo, J.-H., Kim, M., Mi, H., Zhou, H., Xia, Z., Zhou, W., Gong, S., et al. (2017). Origami silicon optoelectronics for hemispherical electronic eye systems, *Nature Communications*, *8*, 1782.
189. Yu, C., Li, Y., Zhang, X., Huang, X., Malyarchuk, V., Wang, S., Shi, Y., Gao, L., Su, Y., Zhang, Y., et al. (2014). Adaptive optoelectronic camouflage systems with designs inspired by cephalopod skins, *Proceedings of the National Academy of Sciences*, *111*, 12998–13003.

190. Wang, H., Liu, H., Zhao, Q., Ni, Z., Zou, Y., Yang, J., Wang, L., Sun, Y., Guo, Y., Hu, W., et al. (2017). A retina-like dual band organic photosensor array for filter-free near-infrared-to-memory operations, *Advanced Materials*, 29, 1701772.
191. Gu, L., Poddar, S., Lin, Y., Long, Z., Zhang, D., Zhang, Q., Shu, L., Qiu, X., Kam, M., Javey, A., et al. (2020). A biomimetic eye with a hemispherical perovskite nanowire array retina, *Nature*, 581, 278–282.
192. Chen, S., Lou, Z., Chen, D. and Shen, G. (2018). An artificial flexible visual memory system based on an UV-motivated memristor, *Advanced Materials*, 30, 1705400.
193. Lee, J.-H., Cho, K.H. and Cho, K. (2023). Emerging trends in soft electronics: Integrating machine intelligence with soft acoustic/vibration sensors, *Advanced Materials*, 35, 2209673.
194. Ji, Z., Zhou, J., Guo, Y., Xia, Y., Abkar, A., Liang, D. and Fu, Y. (2024). Achieving consistency of flexible surface acoustic wave sensors with artificial intelligence, *Microsystems & Nanoengineering*, 10, 94.
195. Zhang, Y., Zhou, X., Zhang, N., Zhu, J., Bai, N., Hou, X., Sun, T., Li, G., Zhao, L., Chen, Y., et al. (2024). Ultrafast piezocapacitive soft pressure sensors with over 10 kHz bandwidth via bonded microstructured interfaces, *Nature Communications*, 15, 3048.
196. Mallegni, N., Molinari, G., Ricci, C., Lazzeri, A., La Rosa, D., Crivello, A. and Milazzo, M. (2022). Sensing devices for detecting and processing acoustic signals in healthcare. *Biosensors*, 12, 835.
197. Jung, Y.H., Hong, S.K., Wang, H.S., Han, J.H., Pham, T.X., Park, H., Kim, J., Kang, S., Yoo, C.D. and Lee, K.J. (2020). Flexible piezoelectric acoustic sensors and machine learning for speech processing, *Advanced Materials*, 32, 1904020.
198. Ding, H., Shu, X., Jin, Y., Fan, T. and Zhang, H. (2019). Recent advances in nanomaterial-enabled acoustic devices for audible sound generation and detection, *Nanoscale*, 11, 5839–5860.
199. Guo, H., Pu, X., Chen, J., Meng, Y., Yeh, M.-H., Liu, G., Tang, Q., Chen, B., Liu, D., Qi, S., et al. (2018). A highly sensitive, self-powered triboelectric auditory sensor for social robotics and hearing aids, *Science Robotics*, 3, eaat2516.
200. Lin, Z., Zhang, G., Xiao, X., Au, C., Zhou, Y., Sun, C., Zhou, Z., Yan, R., Fan, E., Si, S., et al. (2022). A personalized acoustic interface for wearable human–machine interaction, *Advanced Functional Materials*, 32, 2109430.
201. Lee, S., Kim, J., Yun, I., Bae, G.Y., Kim, D., Park, S., Yi, I.-M., Moon, W., Chung, Y. and Cho, K. (2019). An ultrathin conformable vibration-responsive electronic skin for quantitative vocal recognition, *Nature Communications*, 10, 2468.

202. Jiang, Y., Zhang, Y., Ning, C., Ji, Q., Peng, X., Dong, K. and Wang, Z.L. (2022). Ultrathin eardrum-inspired self-powered acoustic sensor for vocal synchronization recognition with the assistance of machine learning, *Small*, *18*, 2106960.
203. Wang, H.S., Hong, S.K., Han, J.H., Jung, Y.H., Jeong, H.K., Im, T.H., Jeong, C.K., Lee, B.-Y., Kim, G., Yoo, C.D., et al. (2021). Biomimetic and flexible piezoelectric mobile acoustic sensors with multiresonant ultrathin structures for machine learning biometrics, *Science Advances*, *7*, eabe5683.
204. Lee, K., Ni, X., Lee, J.Y., Arafa, H., Pe, D.J., Xu, S., Avila, R., Irie, M., Lee, J.H., Easterlin, R.L., et al. (2020). Mechano-acoustic sensing of physiological processes and body motions via a soft wireless device placed at the suprasternal notch, *Nature Biomedical Engineering*, *4*, 148–158.
205. Lee, S., Roh, H., Kim, J., Chung, S., Seo, D., Moon, W. and Cho, K. (2022). An electret-powered skin-attachable auditory sensor that functions in harsh acoustic environments, *Advanced Materials*, *34*, 2205537.
206. Ding, H., Zeng, Z., Wang, Z., Li, X., Yildirim, T., Xie, Q., Zhang, H., Wageh, S., Al-Ghamdi, A.A., Zhang, X., et al. (2022). Deep Learning-Enabled MXene/PEDOT:PSS Acoustic sensor for speech recognition and skin-vibration detection, *Advanced Intelligent Systems*, *4*, 2200140.
207. Lee, S., Kim, J., Roh, H., Kim, W., Chung, S., Moon, W. and Cho, K. (2022). A High-Fidelity Skin-attachable acoustic sensor for realizing auditory electronic skin, *Advanced Materials*, *34*, 2270161.
208. Jeerapan, I., Sempionatto, J.R. and Wang, J. (2020). On-body bioelectronics: Wearable biofuel cells for bioenergy harvesting and self-powered biosensing, *Advanced Functional Materials*, *30*, 1906243.
209. Yu, Y., Nyein, H.Y.Y., Gao, W. and Javey, A. (2020). Flexible electrochemical bioelectronics: The rise of *in situ* bioanalysis, *Advanced Materials*, *32*, 1902083.
210. Gao, H., Wen, L., Tian, J., Wu, Y., Liu, F., Lin, Y., Hua, W. and Wu, G. (2019). A portable electrochemical immunosensor for highly sensitive point-of-care testing of genetically modified crops, *Biosensors and Bioelectronics*, *142*, 111504.
211. Liu, Z., Zhang, Y., Li, B., Ren, X., Ma, H. and Wei, Q. (2020). Novel ratiometric electrochemical sensor for no-wash detection of fluorene-9-bisphenol based on combining CoN nanoarrays with molecularly imprinted polymers, *Analytica Chimica Acta*, *145*, 3320–3328.
212. Zhou, S., Jia, C., Shu, G., Guan, Z., Wu, H., Li, J. and Ou-Yang, W. (2024). Recent advances in TENGs collecting acoustic energy: From low-frequency sound to ultrasound, *Nano Energy*, *129*, 109951.
213. Bhalla, P. and Singh, N. (2016). Generalized drude scattering rate from the memory function formalism: An independent verification of the sharapov-carbotte result, *The European Physical Journal B*, *89*, 1–8.

214. Kucherenko, D.Y., Kucherenko, I., Soldatkin, O., Topolnikova, Y.V., Dzyadevych, S. and Soldatkin, A. (2019). A highly selective amperometric biosensor array for the simultaneous determination of glutamate, glucose, choline, acetylcholine, lactate and pyruvate, *Biosensors, 128*, 100–108.
215. Park, J., Lee, W., Kim, I., Kim, M., Jo, S., Kim, W., Park, H., Lee, G., Choi, W., Yoon, D.S., *et al.* (2019). Ultrasensitive detection of fibrinogen using erythrocyte membrane-draped electrochemical impedance biosensor, *Sensors and Actuators B: Chemical, 293*, 296–303.
216. Kim, J., Jeong, J. and Ko, S.H. (2024). Electrochemical biosensors for point-of-care testing, *Bio-Design and Manufacturing, 7*, 548–565.
217. Zhang, Y. and Chen, X. (2019). Nanotechnology and nanomaterial-based no-wash electrochemical biosensors: From design to application, *Nanoscale, 11*, 19105–19118.
218. Jeerapan, I. and Poorahong, S. (2020). Flexible and stretchable electrochemical sensing systems: Materials, energy sources, and integrations, *Journal of The Electrochemical Society, 167*, 037573.
219. Park, Y.M., Lim, S.Y., Jeong, S.W., Song, Y., Bae, N.H., Hong, S.B., Choi, B.G., Lee, S.J. and Lee, K.G. (2018). Flexible nanopillar-based electrochemical sensors for genetic detection of foodborne pathogens, *Nanoscale Horizons, 5*, 1–8.
220. Noviana, E., McCord, C.P., Clark, K.M., Jang, I. and Henry, C.S. (2020). Electrochemical paper-based devices: Sensing approaches and progress toward practical applications, *Lab on a Chip, 20*, 9–34.
221. Chen, Y., Lu, S., Zhang, S., Li, Y., Qu, Z., Chen, Y., Lu, B., Wang, X. and Feng, X. (2017). Skin-like biosensor system via electrochemical channels for noninvasive blood glucose monitoring, *Science Advances, 3*, e1701629.
222. Zhao, Y., Han, J., Huang, J., Huang, Q., Tao, Y., Gu, R., Li, H.-Y., Zhang, Y., Zhang, H. and Liu, H. (2024). A miniprotein receptor electrochemical biosensor chip based on quantum dots, *Lab on a Chip, 24*, 1875–1886.
223. Yoon, J., Cho, H.-Y., Shin, M., Choi, H.K., Lee, T. and Choi, J.-W. (2020). Flexible electrochemical biosensors for healthcare monitoring, *Journal of Materials Chemistry B, 8*, 7303–7318.
224. Deng, M., Song, G., Zhong, K., Wang, Z., Xia, X. and Tian, Y. (2022). Wearable fluorescent contact lenses for monitoring glucose via a smartphone, *Sensors and Actuators B: Chemical, 352*, 131067.
225. Joseph, J., Wießner, S., König, T.A. and Fery, A. (2021). Exploring plasmonic resonances toward "Large-Scale" flexible optical sensors with deformation stability. *Advanced Functional Materials, 31*, 2101959.
226. Thamilselvan, A. and Kim, M.I. (2024). Recent advances on nanozyme-based electrochemical biosensors for cancer biomarker detection, *TrAC Trends in Analytical Chemistry, 177*, 117815.
227. Wu, W., Wang, L., Yang, Y., Du, W., Ji, W., Fang, Z., Hou, X., Wu, Q., Zhang, C., Li, L., *et al.* (2022). Optical flexible biosensors: From detection

principles to biomedical applications, *Biosensors and Bioelectronics*, *210*, 114328.
228. Shao, M., Liu, D., Lu, J., Zhao, X., Yu, J., Zhang, C., Man, B., Pan, H. and Li, Z. (2023). Giant enhancement of the initial SERS activity for plasmonic nanostructures via pyroelectric PMN-PT, *Nanoscale Horizons*, *8*, 948–957.
229. Koh, E.H., Lee, W.-C., Choi, Y.-J., Moon, J.-I., Jang, J., Park, S.-G., Choo, J., Kim, D.-H. and Jung, H.S. (2021). A Wearable surface-enhanced raman scattering sensor for label-free molecular detection, *ACS Applied Materials & Interfaces*, *13*, 3024–3032.
230. Tang, Z., Wu, J., Yu, X., Hong, R., Zu, X., Lin, X., Luo, H., Lin, W. and Yi, G. (2021). Fabrication of au nanoparticle arrays on flexible substrate for tunable localized surface plasmon resonance, *ACS Applied Materials & Interfaces*, *13*, 9281–9288.
231. Ghosh, A.K., Sarkar, S., Nebel, L.J., Aftenieva, O., Gupta, V., Sander, O., Das, A., Joseph, J., Wießner, S. and König, T.A. (2021). Exploring plasmonic resonances toward "Large-Scale" flexible optical sensors with deformation stability, *Advanced Functional Materials*, *31*, 2101959.
232. Zhang, Z. and Yates Jr, J.T. (2012). Band bending in semiconductors: Chemical and physical consequences at surfaces and interfaces, *Chemical Reviews*, *112*, 5520–5551.
233. Hao, R., Liu, L., Yuan, J., Wu, L. and Lei, S. (2023). Recent advances in field effect transistor biosensors: Designing strategies and applications for sensitive assay, *Biosensors*, *13*, 426.
234. Lin, T., Xu, Y., Zhao, A., He, W. and Xiao, F. (2022). Flexible electrochemical sensors integrated with nanomaterials for *in situ* determination of small molecules in biological samples: A review, *Analytica Chimica Acta*, *1207*, 339461.
235. Zhao, A., Zhang, Z., Zhang, P., Xiao, S., Wang, L., Dong, Y., Yuan, H., Li, P., Sun, Y. and Jiang, X. (2016). 3D nanoporous gold scaffold supported on graphene paper: Freestanding and flexible electrode with high loading of ultrafine PtCo alloy nanoparticles for electrochemical glucose sensing, *Analytica Chimica Acta*, *938*, 63–71.
236. Fame, R.M. and Lehtinen, M.K. (2020). Emergence and developmental roles of the cerebrospinal fluid system, *Developmental Cell*, *52*, 261–275.
237. Ju, J., Hsieh, C.-M., Tian, Y., Kang, J., Chia, R., Chang, H., Bai, Y., Xu, C., Wang, X. and Liu, Q. (2020). Surface enhanced raman spectroscopy based biosensor with a microneedle array for minimally invasive *in vivo* glucose measurements, *ACS Sensors*, *5*, 1777–1785.
238. Dervisevic, M., Alba, M., Adams, T.E., Prieto-Simon, B. and Voelcker, N.H. (2021). Electrochemical immunosensor for breast cancer biomarker detection using high-density silicon microneedle array, *Biosensors and Bioelectronics*, *192*, 113496.

239. Xu, N., Xu, W., Zhang, M., Yu, J., Ling, G. and Zhang, P. (2022). Microneedle-based technology: Toward minimally invasive disease diagnostics, *Advanced Materials Technologies*, *7*, 2101595.
240. Kim, J., Campbell, A.S., Ávila, B.E.-F.d. and Wang, J. (2019). Wearable biosensors for healthcare monitoring, *Nature Biotechnology*, *37*, 389–406.
241. Ma, X., Ahadian, S., Liu, S., Zhang, J., Liu, S., Cao, T., Lin, W., Wu, D., de Barros, N.R., Zare, M.R., *et al.* (2021). Smart Contact Lenses for Biosensing Applications. *Advanced Intelligent Systems*, 2000263.
242. Moreddu, R., Nasrollahi, V., Kassanos, P., Dimov, S., Vigolo, D. and Yetisen, A.K. (2021). Lab-on-a-Contact lens platforms fabricated by multi-axis femtosecond laser ablation, *Small*, *17*, 2102008.
243. Khor, S.M., Choi, J., Won, P. and Ko, S.H. (2022). Challenges and strategies in developing an enzymatic wearable sweat glucose biosensor as a practical point-of-care monitoring tool for type II diabetes, *Nanoscale*, *12*, 221.
244. Xiao, J., Liu, Y., Su, L., Zhao, D., Zhao, L. and Zhang, X. (2019). Microfluidic chip-based wearable colorimetric sensor for simple and facile detection of sweat glucose, *Analytical Chemistry*, *91*, 14803–14807.
245. Koh, A., Kang, D., Xue, Y., Lee, S., Pielak, R.M., Kim, J., Hwang, T., Min, S., Banks, A. and Bastien, P. (2016). A soft, wearable microfluidic device for the capture, storage, and colorimetric sensing of sweat, *Science Translational Medicine*, *8*, 366ra165.
246. Kim, J., Park, J., Park, Y.-G., Cha, E., Ku, M., An, H.S., Lee, K.-P., Huh, M.-I., Kim, J. and Kim, T.-S. (2021). A soft and transparent contact lens for the wireless quantitative monitoring of intraocular pressure, *Nature Biomedical Engineering*, *5*, 772–782.
247. Lu, X., Zhou, X., Song, B., Zhang, H., Cheng, M., Zhu, X., Wu, Y., Shi, H., Chu, B., He, Y., *et al.* (2024). Framework nucleic acids combined with 3D hybridization chain reaction amplifiers for monitoring multiple human tear cytokines, *Advanced Materials*, *36*, 2400622.
248. Zhu, Y., Li, S., Li, J., Falcone, N., Cui, Q., Shah, S., Hartel, M.C., Yu, N., Young, P. and de Barros, N.R. (2022). Lab-on-a-contact lens: Recent advances and future opportunities in diagnostics and therapeutics, *Advanced Materials*, *34*, 2108389.
249. Wang, Z., Dong, Y., Sui, X., Shao, X., Li, K., Zhang, H., Xu, Z. and Zhang, D. (2024). An artificial intelligence-assisted microfluidic colorimetric wearable sensor system for monitoring of key tear biomarkers, *npj Flexible Electronics*, *8*, 35.
250. Chen, Y., Zhang, Y., Liang, Z., Cao, Y., Han, Z. and Feng, X. (2020). Flexible inorganic bioelectronics, *npj Flexible Electronics*, *4*, 20.
251. Kwon, S.H., Zhang, C., Jiang, Z. and Dong, L. (2024). Textured nanofibers inspired by nature for harvesting biomechanical energy and sensing biophysiological signals, *Nano Energy*, *122*, 109334.

252. Jin, Chen, Xingyi, Huang, Bin, Sun, Pingkai (2019). Highly thermally conductive yet electrically insulating polymer/boron nitride nanosheets nanocomposite films for improved thermal management capability, *ACS Nano, 13*(1), 337–345.
253. Kim, D.H., Viventi, J., Amsden, J.J., Xiao, J., Vigeland, L., Kim, Y.S., Blanco, J.A., Panilaitis, B., Frechette, E.S. and Contreras, D. Dissolvable films of silk fibroin for ultrathin conformal bio-integrated electronics, *Nature Materials, 9,* 511–517.
254. Kim, D.H., Lu, N., Ghaffari, R., Kim, Y.S., Lee, S.P., Xu, L., Wu, J., Kim, R.H., Song, J. and Liu, Z. (2011). Materials for multifunctional balloon catheters with capabilities in cardiac electrophysiological mapping and ablation therapy, *Nature Materials, 10,* 316–323.
255. Zhang, Y., Zheng, N., Cao, Y., Wang, F. and Feng, X. (2019). Climbing-inspired twining electrodes using shape memory for peripheral nerve stimulation and recording, *Science Advances, 5,* eaaw1066.
256. Heikenfeld, J., Jajack, A., Feldman, B., Granger, S.W. and Katchman, B.A. (2019). Accessing analytes in biofluids for peripheral biochemical monitoring, *Nature Biotechnology, 37,* 407–419.
257. Chen, Y., Lu, B., Chen, Y. and Feng, X. (2015). Breathable and stretchable temperature sensors inspired by Skin, *Scientific Reports, 5,* 11505.
258. Y., Liu, J., J., S., Norton, R., Qazi, Z. and Z. (2016). Epidermal mechano-acoustic sensing electronics for cardiovascular diagnostics and human-machine interfaces, *Science Advances, 2,* e1601185.
259. Wang, F., Dong, W.U., Jin, P., Zhang, Y., Yang, Y., Yinji, M.A., Yang, A., Ji, F.U. and Feng, X. (2019). A flexible skin-mounted wireless acoustic device for bowel sounds monitoring and evaluation. *Science China Information Sciences, 62,* 202402.
260. Webb, R.C., Ma, Y., Krishnan, S., Li, Y., Yoon, S., Guo, X., Feng, X., Shi, Y., Seidel, M. and Cho, N.H. (2015). Epidermal devices for noninvasive, precise, and continuous mapping of macrovascular and microvascular blood flow, *Science Advances, 1,* e1500701.
261. Dincer, C., Bruch, R., Costa-Rama, E., Fernández-Abedul, M.T., Merkoçi, A., Manz, A., Urban, G.A. and Güder, F. (2019). Disposable sensors in diagnostics, food, and environmental monitoring, *Advanced Materials, 31,* 1806739.
262. Bag, A., Moon, D.-B., Park, K.-H., Cho, C.-Y. and Lee, N.-E. (2019). Room-temperature-operated fast and reversible vertical-heterostructure-diode gas sensor composed of reduced graphene oxide and AlGaN/GaN, *Sensors and Actuators B: Chemical, 296,* 126684.
263. Lee, M.Y., Lee, H.R., Park, C.H., Han, S.G. and Oh, J.H. (2018). Organic transistor-based chemical sensors for wearable bioelectronics, *Accounts of Chemical Research, 51,* 2829–2838.

264. Kou, Y., Hua, L., Chen, W.-J., Xu, X., Song, L., Yu, S. and Lu, Z. (2024). Material design and application progress of flexible chemiresistive gas sensors, *Journal of Materials Chemistry A*, *12*, 21583–21604.
265. Men, Y., Qin, Z., Yang, Z., Zhang, P., Li, M., Wang, Q., Zeng, D., Yin, X. and Ji, H. (2024). Antibacterial defective-ZIF-8/PPY/BC-based flexible electronics as stress-strain and NO_2 gas sensors, *Advanced Functional Materials*, *34*, 2316633.
266. Bag, A. and Lee, N.-E. (2019). Gas sensing with heterostructures based on two-dimensional nanostructured materials: A review, *Journal of Materials Chemistry C*, *7*, 13367–13383.
267. Zhang, X., Ye, T., Meng, X., Tian, Z., Pang, L., Han, Y., Li, H., Lu, G., Xiu, F. and Yu, H.-D. (2020). Sustainable and transparent fish gelatin films for flexible electroluminescent devices, *ACS Nano*, *14*, 3876–3884.
268. Wang, P., Tang, C., Song, H., Zhang, L., Lu, Y. and Huang, F. (2024). 1D/2D heterostructured WS2@PANI composite for highly sensitive, flexible, and room temperature ammonia gas sensor, *ACS Applied Materials & Interfaces*, *16*, 14082–14092.
269. Kim, S.-J., Choi, S.-J., Jang, J.-S., Cho, H.-J. and Kim, I.-D. (2017). Innovative nanosensor for disease diagnosis, *Accounts of Chemical Research*, *50*, 1587–1596.
270. Wang, C., Xia, K., Wang, H., Liang, X., Yin, Z. and Zhang, Y. (2019). Advanced carbon for flexible and wearable electronics, *Advanced Materials*, *31*, 1801072.
271. Smith, M.K. and Mirica, K.A. (2017). Self-organized frameworks on textiles (SOFT): Conductive fabrics for simultaneous sensing, capture, and filtration of gases, *Journal of the American Chemical Society*, *139*, 16759–16767.
272. He, H., Guo, J., Zhao, J., Xu, J., Zhao, C., Gao, Z. and Song, Y.-Y. (2022). Engineering CuMOF in TiO_2 nanochannels as flexible gas sensor for high-performance NO detection at room temperature, *Advanced Science*, *7*, 2750–2758.
273. Wang, L., Chen, X., Yi, Z., Xu, R., Dong, J., Wang, S., Zhao, Y. and Liu, Y. (2022). Facile Synthesis of Conductive Metal–Organic Frameworks Nanotubes for Ultrahigh-Performance Flexible NO Sensors, *Small Methods*, *6*, 2200581.
274. Chen, Z.-C., Chang, T.-L., Chen, C.-H., Liou, D.-S., Han, T.-Y. and Wu, Q.-X. (2021). Flexible NO gas sensor fabricated using graphene/silver nanoparticles stacked electrode structures, *Materials Letters*, *295*, 129826.
275. Li, R., Qi, H., Ma, Y., Deng, Y., Liu, S., Jie, Y., Jing, J., He, J., Zhang, X. and Wheatley, L. (2020). A flexible and physically transient electrochemical sensor for real-time wireless nitric oxide monitoring, *Nature Communications*, *11*, 1–11.

276. Li, W., Chen, R., Qi, W., Cai, L., Sun, Y., Sun, M., Li, C., Yang, X., Xiang, L. and Xie, D. (2019). Reduced graphene oxide/mesoporous ZnO NSs hybrid fibers for flexible, stretchable, twisted, and wearable NO_2 E-textile gas sensor, *Advanced Science*, *4*, 2809–2818.
277. Han, C., Li, X., Liu, Y., Tang, Y., Liu, M., Li, X., Shao, C., Ma, J. and Liu, Y. (2021). Flexible all-inorganic room-temperature chemiresistors based on fibrous ceramic substrate and visible-light-powered semiconductor sensing layer, *Advanced Science*, *8*, 2102471.
278. Wang, B., Thukral, A., Xie, Z., Liu, L., Zhang, X., Huang, W., Yu, X., Yu, C., Marks, T.J. and Facchetti, A. (2020). Flexible and stretchable metal oxide nanofiber networks for multimodal and monolithically integrated wearable electronics, *Nature Communications*, *11*, 2405.
279. Wang, X.-X., Li, H.-Y. and Guo, X. (2020). Flexible and transparent sensors for ultra-low NO2 detection at room temperature under visible light illumination, *Journal of Materials Chemistry A*, *8*, 14482–14490.
280. Yang, L., Yi, N., Zhu, J., Cheng, Z., Yin, X., Zhang, X., Zhu, H. and Cheng, H. (2020). Novel gas sensing platform based on a stretchable laser-induced graphene pattern with self-heating capabilities, *Journal of Materials Chemistry A*, *8*, 6487–6500.
281. Ko, G.-J., Han, S.D., Kim, J.-K., Zhu, J., Han, W.B., Chung, J., Yang, S.M., Cheng, H., Kim, D.-H. and Kang, C.-Y. (2020). Biodegradable, flexible silicon nanomembrane-based NOx gas sensor system with record-high performance for transient environmental monitors and medical implants, *NPG Asia Materials*, *12*, 71.
282. Shahrbabaki, Z., Farajikhah, S., Ghasemian, M.B., Oveissi, F., Rath, R.J., Yun, J., Dehghani, F. and Naficy, S. (2023). A Flexible and polymer-based chemiresistive CO2 Gas sensor at room temperature, *Advanced Materials Technologies*, *8*, 2201510.
283. Liu, X., Zheng, W., Kumar, R., Kumar, M. and Zhang, J. (2022). Conducting polymer-based nanostructures for gas sensors, *Coordination Chemistry Reviews*, *462*, 214517.
284. Rath, R.J., Naficy, S., Giaretta, J., Oveissi, F., Yun, J., Dehghani, F. and Farajikhah, S. (2024). Chemiresistive sensor for enhanced CO_2 gas monitoring, *ACS Sensors*, *9*, 1735–1742.
285. Soltabayev, B., Raiymbekov, Y., Nuftolla, A., Turlybekuly, A., Yergaliuly, G. and Mentbayeva, A. (2024). Sensitivity enhancement of CO_2 sensors at room temperature based on the CZO nanorod architecture, *ACS Sensors*, *9*, 1227–1238.
286. Kim, J.-H., Mirzaei, A., Kim, J.-Y., Yang, D.-H., Kim, S.S. and Kim, H.W. (2022). Selective CO gas sensing by Au-decorated WS2-SnO2 core-shell nanosheets on flexible substrates in self-heating mode, *Sensors and Actuators B: Chemical*, *353*, 131197.

287. Cho, J. and Shin, G. (2022). Fabrication of a flexible, wireless micro-heater on elastomer for wearable gas sensor applications, *Polymers*, *14*, 1557.
288. Kim, J.-H., Mirzaei, A., Kim, H.W. and Kim, S.S. (2020). Flexible and low power CO gas sensor with Au-functionalized 2D WS2 nanoflakes, *Sensors and Actuators B: Chemical*, *313*, 128040.
289. Kim, J.-H., Kim, J.-Y., Mirzaei, A., Kim, H.W. and Kim, S.S. (2021). Synergistic effects of SnO_2 and Au nanoparticles decorated on WS2 nanosheets for flexible, room-temperature CO gas sensing, *Sensors and Actuators B: Chemical*, *332*, 129493.
290. Harrison, P. and Willett, M. (1988). The mechanism of operation of tin (IV) oxide carbon monoxide sensors, *Nature*, *332*, 337–339.
291. Zuo, J., Tavakoli, S., Mathavakrishnan, D., Ma, T., Lim, M., Rotondo, B., Pauzauskie, P., Pavinatto, F. and MacKenzie, D. (2020). Additive manufacturing of a flexible carbon monoxide sensor based on a SnO_2-graphene nanoink, *Communications Materials*, *8*, 36.
292. Kokila, V., Prasanna, R., Kumar, A., Nishanth, S., Shukla, J., Gulia, U., Nain, L., Shivay, Y.S. and Singh, A.K. (2022). Cyanobacterial inoculation in elevated CO_2 environment stimulates soil C enrichment and plant growth of tomato, *Environmental Technology & Innovation*, *26*, 102234.
293. Wang, R., Zhang, M., Guan, Y., Chen, M. and Zhang, Y. (2019). A CO_2-responsive hydrogel film for optical sensing of dissolved CO_2, *Soft Matter*, *15*, 6107–6115.
294. Hafiz, S.M., Ritikos, R., Whitcher, T.J., Razib, N.M., Bien, D.C.S., Chanlek, N., Nakajima, H., Saisopa, T., Songsiriritthigul, P., Huang, N.M., *et al.* (2014). A practical carbon dioxide gas sensor using room-temperature hydrogen plasma reduced graphene oxide, *Sensors and Actuators A: Physical*, *193*, 692–700.
295. Wu, J., Tao, K., Zhang, J., Guo, Y., Miao, J. and Norford, L.K. (2016). Chemically functionalized 3D graphene hydrogel for high performance gas sensing, *Journal of Materials Chemistry A*, *4*, 8130–8140.
296. Hong, S.Y., Oh, J.H., Park, H., Yun, J.Y., Jin, S.W., Sun, L., Zi, G. and Ha, J.S. (2017). Polyurethane foam coated with a multi-walled carbon nanotube/polyaniline nanocomposite for a skin-like stretchable array of multifunctional sensors, *NPG Asia Materials*, *9*, e448.
297. Jaishi, L.R., Yu, J., Ding, W., Tsow, F. and Xian, X. (2024). A novel colorimetric tuning fork sensor for ammonia monitoring, *Sensors and Actuators B: Chemical*, *405*, 135342.
298. Ghosh, R., Nayak, A.K., Santra, S., Pradhan, D. and Guha, P.K. (2015). Enhanced ammonia sensing at room temperature with reduced graphene oxide/tin oxide hybrid films, *RSC Advances*, *5*, 50165–50173.
299. Konvalina, G. and Haick, H. (2014). Sensors for breath testing: From nanomaterials to comprehensive disease detection, *Accounts of Chemical Research*, *47*, 66–76.

300. Luo, M. Huang, X. Xiong, E. Cai, S. Li, S. Jia, Z. and Gao, Z. (2024). Fast response/recovery and sub-ppm ammonia gas sensors based on a novel V2CTx@MoS2 composite. *Journal of Materials Chemistry A, 12,* 12225–12236.
301. Liu, L., Fei, T., Guan, X., Zhao, H. and Zhang, T. (2021). Highly sensitive and chemically stable NH3 sensors based on an organic acid-sensitized cross-linked hydrogel for exhaled breath analysis, *Biosensors and Bioelectronics, 191,* 113459.
302. Liu, L., Fei, T., Guan, X., Zhao, H. and Zhang, T. (2021). Humidity-activated ammonia sensor with excellent selectivity for exhaled breath analysis, *Sensors and Actuators B: Chemical, 334,* 129625.
303. Nayak, S.P., Ramamurthy, S.S. and Kiran Kumar, J.K. (2020). Green synthesis of silver nanoparticles decorated reduced graphene oxide nanocomposite as an electrocatalytic platform for the simultaneous detection of dopamine and uric acid, *Materials Chemistry and Physics, 252,* 123302.
304. Zhang, L., Tan, Q., Kou, H., Wu, D., Zhang, W. and Xiong, J. (2019). Highly sensitive NH3 wireless sensor based on Ag-RGO composite operated at room-temperature, *Scientific Reports, 9,* 9942.
305. Park, H., Jeong, Y.R., Yun, J., Hong, S.Y., Jin, S., Lee, S.-J., Zi, G. and Ha, J.S. (2015). Stretchable array of highly sensitive pressure sensors consisting of polyaniline nanofibers and Au-coated polydimethylsiloxane micropillars, *ACS Nano, 9,* 9974–9985.
306. Hong, S., Oh, J., Park, H., Yun, J., Jin, S., Sun, L., Zi, G. and Ha, J. (2017). Polyurethane foam coated with a multi-walled carbon nanotube/polyaniline nanocomposite for a skin-like stretchable array of multi-functional sensors, *NPG Asia Materials, 9,* 1–10.
307. Xue, D., Zhou, R., Lin, X., Duan, X., Li, Q. and Wang, T. (2019). A highly selective and sensitive H2S sensor at low temperatures based on Cr-doped α-Fe2O3 nanoparticles, *RSC Advances, 9,* 4150–4156.
308. Song, Z., Wei, Z., Wang, B., Luo, Z., Xu, S., Zhang, W., Yu, H., Li, M., Huang, Z. and Zang, J. (2016). Sensitive room-temperature H_2S gas sensors employing SnO_2 quantum wire/reduced graphene oxide nanocomposites, *Chemistry of Materials, 28,* 1205–1212.
309. Cui, G., Zhang, P., Chen, L., Wang, X., Li, J., shi, C. and Wang, D. (2017). Highly sensitive H_2S sensors based on Cu_2O/Co_3O_4 nano/microstructure heteroarrays at and below room temperature, *Scientific Reports, 7,* 43887.
310. Park, K.-R., Cho, H.-B., Lee, J., Song, Y., Kim, W.-B. and Choa, Y.-H. (2020). Design of highly porous SnO_2-CuO nanotubes for enhancing H_2S gas sensor performance, *Sensors and Actuators B: Chemical, 302,* 127179.
311. Sui, L., Yu, T., Zhao, D., Cheng, X., Zhang, X., Wang, P., Xu, Y., Gao, S., Zhao, H. and Gao, Y. (2020). *In situ* deposited hierarchical CuO/NiO

nanowall arrays film sensor with enhanced gas sensing performance to H_2S, *Journal of Hazardous Materials*, *385*, 121570.
312. Asad, M., Sheikhi, M.H., Pourfath, M. and Moradi, M. (2015). High sensitive and selective flexible H2S gas sensors based on Cu nanoparticle decorated SWCNTs, *Sensors and Actuators B: Chemical*, *210*, 1–8.
313. Asad, M. and Sheikhi, M.H. (2016). Highly sensitive wireless H2S gas sensors at room temperature based on CuO-SWCNT hybrid nanomaterials, *Sensors and Actuators B: Chemical*, *231*, 474–483.
314. Shaalan, N., Morsy, A.E., Abdel-Rahim, M. and Rashad, M. (2021). Simple preparation of Ni/CuO nanocomposites with superior sensing activity toward the detection of methane gas, *Applied Physics A*, *127*, 455.
315. Peng, Z., Huang, Y., Zheng, K., Min, Y., Zhao, H., Pi, M., Song, F., Zheng, C., Zhang, Y., Chang, Z., et al. (2024). Slow-light-enhanced polarization division multiplexing infrared absorption spectroscopy for on-chip wideband multigas detection in a 1D photonic crystal waveguide, *Analytical Chemistry*, *96*, 3445–3453.
316. Zhang, X., Zhang, J., Li, C., Zhang, X., Yun, J. and Cao, D. (2024). A review on nanofiber-based composites for toxic and flammable gas sensing, *Advanced Composites and Hybrid Materials*, *7*, 108.
317. Chimowa, G., Tshabalala, Z.P., Akande, A.A., Bepete, G., Mwakikunga, B., Ray, S.S. and Benecha, E.M. (2017). Improving methane gas sensing properties of multi-walled carbon nanotubes by vanadium oxide filling, *Sensors and Actuators B: Chemical*, *247*, 11–18.
318. Xia, Y., Wang, J., Xu, L., Li, X. and Huang, S. (2020). A room-temperature methane sensor based on Pd-decorated ZnO/rGO hybrids enhanced by visible light photocatalysis, *Sensors and Actuators B: Chemical*, *304*, 127334.
319. Niu, F., Zhou, F., Wang, Z., Wei, L., Hu, J., Dong, L., Ma, Y., Wang, M., Jia, S. and Chen, X. (2023). Synthesizing metal oxide semiconductors on doped Si/SiO_2 flexible fiber substrates for wearable gas sensing, *Results*, *6*, 0100.
320. Luo, Y., Zhang, C., Zheng, B., Geng, X. and Debliquy, M. (2017). Hydrogen sensors based on noble metal doped metal-oxide semiconductor: A review, *International Journal of Hydrogen Energy*, *42*, 20386–20397.
321. Lee, J.S., Oh, J., Jun, J. and Jang, J. (2015). Wireless hydrogen smart sensor based on Pt/graphene-immobilized radio-frequency identification tag, *ACS Nano*, *9*, 7783–7790.
322. Cho, S.H., Suh, J.M., Jeong, B., Lee, T.H., Choi, K.S., Eom, T.H., Choi, S.W., Nam, G.B., Kim, Y.J. and Jang, H.W. (2024). Substantially accelerated response and recovery in Pd-decorated WO_3 nanorods gasochromic hydrogen sensor, *Small*, 2309744.
323. Rashid, T.R., Phan, D.T. and Chung, G.S. (2014). Effect of Ga-modified layer on flexible hydrogen sensor using ZnO nanorods decorated by Pd catalysts, *Sensors and Actuators B: Chemical*, *193*, 869–876.

324. Punetha, D., Kar, M. and Pandey, S.K. (2020). A new type low-cost, flexible and wearable tertiary nanocomposite sensor for room temperature hydrogen gas sensing, *Scientific Reports*, *10*, 2151.
325. Rashid, T.-R., Phan, D.-T. and Chung, G.-S. (2013). A flexible hydrogen sensor based on Pd nanoparticles decorated ZnO nanorods grown on polyimide tape, *Sensors and Actuators B: Chemical*, *185*, 777–784.
326. Gu, H., Wang, Z. and Hu, Y. (2012). Hydrogen gas sensors based on semiconductor oxide nanostructures, *Sensors*, *12*, 5517–5550.
327. Wang, C., Du, L., Xing, X., Feng, D. and Yang, D. (2022). Lightweight porous polyurethane foam integrated with graphene oxide for flexible and high-concentration hydrogen sensing, *ACS Sensors*, *7*, 2420–2428.
328. Girma, H.G., Lee, H.M., Kim, Y., Ryu, G.-S., Jeon, S., Kim, J.Y., Jung, S.-H., Kim, S.H., Noh, Y.-Y. and Lim, B. (2023). Highly sensitive and wrappable room temperature wireless gasochromic and chemiresistive dual-response H2 sensors using spray coating, *Nano Energy*, *113*, 108551.
329. Konvalina, G. and Haick, H. (2013). Sensors for breath testing: From nanomaterials to comprehensive disease detection, *Accounts of Chemical Research*, *47*, 66–76.
330. Zeng, G., Wu, C., Chang, Y., Zhou, C., Chen, B., Zhang, M., Li, J., Duan, X., Yang, Q. and Pang, W. (2019). Detection and discrimination of volatile organic compounds using a single film bulk acoustic wave resonator with temperature modulation as a multiparameter virtual sensor array, *ACS Sensors*, *4*, 1524–1533.
331. Das, S. and Pal, M. (2020). Non-invasive monitoring of human health by exhaled breath analysis: A comprehensive review, *Journal of The Electrochemical Society*, *167*, 037562.
332. Qu, X., Hu, Y., Xu, C., Li, Y., Zhang, L., Huang, Q., Sadat Moshirian-Farahi, S., Zhang, J., Xu, X., Liao, M., *et al.* (2024). Optical sensors of volatile organic compounds for non-invasive diagnosis of diseases, *Chemical Engineering Journal*, *485*, 149804.
333. Allegretto, J.A. and Dostalek, J. (2024). Metal–organic frameworks in surface enhanced raman spectroscopy-based analysis of volatile organic compounds, *Advanced Science*, 2401437.
334. Mirzaei, A., Leonardi, S. and Neri, G. (2016). Detection of hazardous volatile organic compounds (VOCs) by metal oxide nanostructures-based gas sensors: A review, *Ceramics International*, *42*, 15119–15141.
335. Liu, Z., Yang, T., Dong, Y. and Wang, X. (2018). A room temperature VOCs gas sensor based on a layer by layer multi-walled carbon nanotubes/polyethylene glycol composite, *Sensors*, *18*, 3113.
336. Zhang, F., Hu, H., Islam, M., Peng, S., Wu, S., Lim, S., Zhou, Y. and Wang, C.-H. (2020). Multi-modal strain and temperature sensor by hybridizing

reduced graphene oxide and PEDOT:PSS, *Composites Science and Technology*, *187*, 107959.
337. Pal, R., Goyal, S.L., Rawal, I. and Sharma, S. (2020). Efficient room temperature methanol sensors based on polyaniline/graphene micro/nanocomposites, *Indian Physics Journal*, *29*, 591–603.
338. van den Broek, J., Abegg, S., Pratsinis, S.E. and Güntner, A.T. (2019). Highly selective detection of methanol over ethanol by a handheld gas sensor, *Nature Communications*, *10*, 4220.
339. Sivakumar, G., Kumar, R., Yadav, V., Gupta, V. and Balaraman, E. (2024). Correction to "Multi-functionality of methanol in sustainable catalysis: Beyond methanol economy", *ACS Catalysis*, *14*, 6016–6020.
340. Yang, C., Su, X., Wang, J., Cao, X., Wang, S. and Zhang, L. (2013). Facile microwave-assisted hydrothermal synthesis of varied-shaped CuO nanoparticles and their gas sensing properties, *Sensors and Actuators B: Chemical*, *185*, 159–165.
341. Liu, B., Yang, H., Zhao, H., An, L., Zhang, L., Shi, R., Wang, L., Bao, L. and Chen, Y. (2011). Synthesis and enhanced gas-sensing properties of ultralong NiO nanowires assembled with NiO nanocrystals, *Sensors and Actuators B: Chemical*, *156*, 251–262.
342. Zhang, Y., Mi, J., Wu, W., Fei, J., Lv, B., Yu, X., Wen, K., Shen, J. and Wang, Z. (2024). Investigation of antibody tolerance in methanol for analytical purposes: Methanol effect patterns and molecular mechanisms, *Advanced Science*, 2402050.
343. Jiang, Y., Ma, J., Lv, J., Ma, H., Xia, H., Wang, J., Yang, C., Xue, M., Li, G. and Zhu, N. (2019). Facile wearable vapor/liquid amphibious methanol sensor, *ACS Sensors*, *4*, 152–160.
344. Zhu, P., Wang, Y., Wang, Y., Mao, H., Zhang, Q. and Deng, Y. (2020). Flexible 3D architectured piezo/thermoelectric bimodal tactile sensor array for E-skin application, *Advanced Electronic Materials*, *10*, 2001945.
345. Kim, N.-H., Choi, S.-J., Yang, D.-J., Bae, J., Park, J. and Kim, I.-D. (2014). Highly sensitive and selective hydrogen sulfide and toluene sensors using Pd functionalized WO_3 nanofibers for potential diagnosis of halitosis and lung cancer, *Sensors and Actuators B: Chemical*, *193*, 574–581.
346. Minitha, C.R., Anithaa, V.S., Subramaniam, V. and Rajendra Kumar, R.T. (2018). Impact of oxygen functional groups on reduced graphene oxide-based sensors for ammonia and toluene detection at room temperature, *ACS Omega*, *3*, 4105–4112.
347. Li, Y., Chen, B., Liu, L., Zhu, B. and Zhang, D. (2024). Water-resistance-based S-scheme heterojunction for deep mineralization of toluene, *Angewandte Chemie International Edition*, *63*, e202319432.
348. Ma, H., Xu, Y., Rong, Z., Cheng, X., Gao, S., Zhang, X., Zhao, H. and Huo, L. (2012). Highly toluene sensing performance based on

monodispersed Cr2O3 porous microspheres, *Sensors and Actuators B: Chemical, 174*, 325–331.
349. Seekaew, Y., Wisitsoraat, A., Phokharatkul, D. and Wongchoosuk, C. (2019). Room temperature toluene gas sensor based on TiO_2 nanoparticles decorated 3D graphene-carbon nanotube nanostructures, *Sensors and Actuators B: Chemical, 279*, 69–78.
350. Mao, Y., Yang, L., Liu, S., Song, Y., Luo, M. and Guo, Y. (2024). A theoretical study on toluene oxidization by OH radical, *BMC Chemistry, 18*, 72.
351. Jin, M.L., Park, S., Kim, J.S., Kwon, S.H., Zhang, S., Yoo, M.S., Jang, S., Koh, H.J., Cho, S.Y. and Kim, S.Y. (2018). An ultrastable ionic chemiresistor skin with an intrinsically stretchable polymer electrolyte, *Advanced Materials, 30*, 1706851.
352. Ernstgård, L., Iregren, A., Sjögren, B. and Johanson, G. (2006). Acute effects of exposure to vapours of acetic acid in humans, *Toxicology Letters, 165*, 22–30.
353. Bag, A. and Lee, N.E. (2021). Recent advancements in development of wearable gas sensors, *Advanced Materials Technologies, 6*, 2000883.
354. Jun, J., Oh, J., Shin, D.H., Kim, S.G., Lee, J.S., Kim, W. and Jang, J. (2016). Wireless, room temperature volatile organic compound sensor based on polypyrrole nanoparticle immobilized ultrahigh frequency radio frequency identification tag, *ACS Applied Materials & Interfaces, 8*, 33139–33147.
355. Schroeder, V., Savagatrup, S., He, M., Lin, S. and Swager, T.M. (2018). Carbon nanotube chemical sensors, *Chemical Reviews, 119*, 599–663.
356. Li, S., Chen, S., Zhuo, B., Li, Q., Liu, W. and Guo, X. (2017). Flexible ammonia sensor based on PEDOT:PSS/silver nanowire composite film for meat freshness monitoring, *IEEE Electron Device Letters, 38*, 975–978.
357. Nguyen, L.H., Naficy, S., McConchie, R., Dehghani, F. and Chandrawati, R. (2019). Polydiacetylene-based sensors to detect food spoilage at low temperatures, *Journal of Materials Chemistry C, 7*, 1919–1926.
358. Shaalan, N.M., Ahmed, F., Kumar, S., Melaibari, A., Hasan, P.M. and Aljaafari, A. (2020). Monitoring food spoilage based on a defect-induced multiwall carbon nanotube sensor at room temperature: Preventing food waste, *ACS Omega, 5*, 30531–30537.
359. Barreca, D., Gasparotto, A., Gri, F., Comini, E. and Maccato, C. (2018). Plasma-assisted growth of β-MnO2 nanosystems as gas sensors for safety and food industry applications, *Advanced Materials Interfaces, 5*, 1800792.
360. Trul, A., Agina, E. and Ponomarenko, S. (2021). Gas sensors based on conjugated oligomers and polymers as promising sensitive elements for toxic gases monitoring in the atmosphere, *Polymer Science, Series B, 63*, 443–458.
361. Queiroz, E.L., Araújo, G.S., Almeida, T.B., Martinez, E.A. and Souza, S.M.A.d. (2021). Propriedades químicas e mecânicas de filme bioativo de

amido de mandioca com adição de extrato de jamelão (Syzygium cumini L.), *Brazilian Journal of Food Technology*, *24*, e2020216.
362. Tang, N., Zhou, C., Xu, L., Jiang, Y., Qu, H. and Duan, X. (2019). A fully integrated wireless flexible ammonia sensor fabricated by soft nanolithography, *Advanced Science*, *4*, 726–732.
363. Ma, Z., Chen, P., Cheng, W., Yan, K., Pan, L., Shi, Y. and Yu, G. (2018). Highly sensitive, printable nanostructured conductive polymer wireless sensor for food spoilage detection, *Nano Letters*, *18*, 4570–4575.
364. Pan, L., Xie, Y., Yang, H., Li, M., Bao, X., Shang, J. and Li, R.-W. (2023). Flexible magnetic sensors, *Sensors*, *23*, 4083.
365. Xu, J., Pang, H., Gong, X., Pei, L. and Xuan, S. (2021). Shape-deformable liquid metal-filled magnetorheological plastomer sensor with a magnetic field "On-off" Switch, *iScience*, *6*, 2021, 102549.
366. Wu, L., Zhang, Y., Zhao, F., Sheng, J. and Gu, N. (2024). Magnetic characterization of human intrinsically magnetic monocytes through a novel optical tracking-based magnetic sensor, *Small*, *20*, 2307306.
367. Esat, T., Borodin, D., Oh, J., Heinrich, A.J., Tautz, F.S., Bae, Y. and Temirov, R. (2024). A quantum sensor for atomic-scale electric and magnetic fields, *Nature Nanotechnology*, *19*, 1466–1471.
368. Ma, Z., Wu, Y., Lu, S., Li, J., Liu, J., Huang, X., Zhang, X., Zhang, Y., Dong, G., Qin, L., *et al.* (2024). Magnetically assisted 3D printing of ultraantiwear flexible sensor, *Advanced Functional Materials*, 2406108.
369. Ha, M., Cañón Bermúdez, G.S., Liu, J.A.-C., Oliveros Mata, E.S., Evans, B.A., Tracy, J.B. and Makarov, D. (2021). Reconfigurable magnetic origami actuators with on-board sensing for guided assembly, *Advanced Materials*, *33*, 2008751.
370. Kim, Y., Parada, G.A., Liu, S. and Zhao, X. (2019). Ferromagnetic soft continuum robots, *Science Robotics*, *4*, eaax7329.
371. Man, J., Jin, Z. and Chen, J. (2024). Magnetic tactile sensor with bionic hair array for sliding sensing and object recognition, *Advanced Science*, *11*, 2306832.
372. Wang, W., Luo, L., Li, Y., Hong, B., Ma, Y., Kang, K. and Wang, J. (2024). Detection of SARS-CoV-2 using machine learning-enabled paper-assisted ratiometric fluorescent sensors based on target-induced magnetic DNAzyme, *Biosensors and Bioelectronics*, *255*, 116272.
373. Kim, Y., Genevriere, E., Harker, P., Choe, J., Balicki, M., Regenhardt, R.W., Vranic, J.E., Dmytriw, A.A., Patel, A.B. and Zhao, X. (2022). telerobotic neurovascular interventions with magnetic manipulation, *Science Robotics*, *7*, eabg9907.
374. Shasha, C. and Krishnan, K.M. (2021). Nonequilibrium dynamics of magnetic nanoparticles with applications in biomedicine, *Advanced Materials*, *33*, 1904131.

375. Ma, C., Wu, S., Ze, Q., Kuang, X., Zhang, R., Qi, H.J. and Zhao, R. (2020). Magnetic multimaterial printing for multimodal shape transformation with tunable properties and shiftable mechanical behaviors, *ACS Applied Materials & Interfaces*, *13*, 12639–12648.
376. Wang, Q., Wu, Z., Huang, J., Du, Z., Yue, Y., Chen, D., Li, D. and Su, B. (2021). Integration of sensing and shape-deforming capabilities for a bioinspired soft robot, *Composites Part B: Engineering*, *223*, 109116.
377. Xu, J., Pei, L., Li, J., Pang, H., Li, Z., Li, B., Xuan, S. and Gong, X. (2019). Flexible, self-powered, magnetism/pressure dual-mode sensor based on magnetorheological plastomer, *Composites Science and Technology*, *183*, 107820.
378. Xiao, H., Li, S., He, Z., Wu, Y., Gao, Z., Hu, C., Hu, S., Wang, S., Liu, C. and Shang, J. (2023). Dual mode strain–temperature sensor with high stimuli discriminability and resolution for smart wearables, *Advanced Functional Materials*, *33*, 2214907.
379. Kim, Y., Yuk, H., Zhao, R., Chester, S.A. and Zhao, X. (2018). Printing ferromagnetic domains for untethered fast-transforming soft materials, *Nature*, *558*, 274–279.
380. Qi, Z., Zhou, M., Li, Y., Xia, Z., Huo, W. and Huang, X. (2021). Reconfigurable flexible electronics driven by origami magnetic membranes, *Advanced Materials Technologies*, *6*, 2001124.
381. Ren, Y., Liu, Z., Jin, G., Yang, M., Shao, Y., Li, W., Wu, Y., Liu, L. and Yan, F. (2021). Electric-field-induced gradient ionogels for highly sensitive, broad-range-response, and freeze/heat-resistant ionic fingers, *Advanced Materials*, *33*, 2008486.
382. Man, J., Chen, G. and Chen, J. (2022). Recent progress of biomimetic tactile sensing technology based on magnetic sensors, *Biosensors*, *12*, 1054.
383. Kim, Y. and Zhao, X. (2022). Magnetic soft materials and robots, *Chemical Reviews*, *122*, 5317–5364.
384. Wu, P., Yu, T., Chen, M., Kang, N. and Mansori, M.E. (2024). Electrically/magnetically dual-driven shape memory composites fabricated by multimaterial magnetic field-assisted 4D printing, *Advanced Functional Materials*, *34*, 2314854.
385. Alfadhel, A. and Kosel, J. (2015), Magnetic nanocomposite cilia tactile sensor, *Advanced Materials*, *27*, 7888–7892.
386. Zhou, Q., Ji, B., Wei, Y., Hu, B., Gao, Y., Xu, Q., Zhou, J. and Zhou, B. (2019). A bio-inspired cilia array as the dielectric layer for flexible capacitive pressure sensors with high sensitivity and a broad detection range, *Journal of Materials Chemistry A*, *7*, 27334–27346.
387. Zhou, Q., Ji, B., Hu, F., Luo, J. and Zhou, B. (2021). Magnetized micropillar-enabled wearable sensors for touchless and intelligent information communication, *Nano-Micro Letters*, *13*, 1–16.

388. Asghar, W., Li, F., Zhou, Y., Wu, Y., Yu, Z., Li, S., Tang, D., Han, X., Shang, J. and Liu, Y. (2020). Piezocapacitive flexible E-skin pressure sensors having magnetically grown microstructures, *Advanced Materials Technologies*, *5*, 1900934.
389. Ge, J., Wang, X., Drack, M., Volkov, O., Liang, M., Cañón Bermúdez, G.S., Illing, R., Wang, C., Zhou, S. and Fassbender, J. (2019). A bimodal soft electronic skin for tactile and touchless interaction in real time, *Nature Communications*, *10*, 4405.
390. Fang, D., Ding, S., Dai, Z., Zhong, J. and Zhou, B. (2023). Wearable patch with direction-aware sensitivity of in-plane force for self-powered and single communication channel based human-machine interaction, *Chemical Engineering Journal*, *468*, 143664.
391. Chauhan, A. and Singh, S. (2019). Management of delayed skin necrosis following hyaluronic acid filler injection using pulsed hyaluronidase, *Journal of Cutaneous and Aesthetic Surgery*, *12*, 183.
392. Wang, H., Zhu, Z., Jin, H., Wei, R., Bi, L. and Zhang, W. (2022). Magnetic soft robots: Design, actuation, and function, *Journal of Applied Science and Materials Chemistry and Physics*, *922*, 166219.

Chapter 4

Sensing Systems and Applications

Hongsen Niu[*], Peng Wang[†], Li Yang[‡,¶], and Shen Guozhen[§,**]

[*]School of Information Science and Engineering,
University of Jinan, Jinan 250022, China

[†]School of Mechanical Engineering, University of Jinan,
Jinan 250022, China

[‡]School of Integrated Circuits, Shandong University,
Jinan 250101, China

[§]School of Integrated Circuits and Electronics, Beijing Institute
of Technology, Beijing 100081, China

[¶]yang.li@sdu.edu.cn

[**]gzshen@bit.edu.cn

1. Introduction

Flexible sensor is a sensor technology with outstanding adaptability and flexibility, and its application scope covers numerous fields such as intelligent perception systems, human–machine interaction, high-resolution sensor arrays, the Internet of Things, virtual reality (VR)/augmented reality (AR), intelligent medical monitoring, and optoelectronic displays, among others.[1–9] With the rapid development of technology and the increasing demand for innovative technologies, flexible sensors, as an emerging sensor technology, are attracting more and more attention and finding widespread applications.

Intelligent perception systems refer to systems that integrate flexible sensors and machine learning (ML) technologies to create a system with thinking capabilities similar to the human brain. Such systems can explore and identify deeper levels of information, such as gestures, postures, speech, surface textures, object shapes, and materials. Intelligent perception systems are widely present in various artificial intelligence applications and can be categorized into the following six types based on their functions: intelligent hand gesture perception system, intelligent grasping object perception system, intelligent texture perception system, intelligent speech perception system, intelligent material perception system, and intelligent posture perception system. HMI refers to the process of information exchange and operation between humans and computers or smart devices. Traditional HMI methods mainly rely on input devices such as keyboards, mice, and touch screens, but these methods have fixed, single, and highly limiting characteristics. The emergence of flexible sensors has brought about new possibilities for HMI. By embedding flexible sensors in wearable devices, gloves, wristbands, and other carriers, real-time perception and recognition of actions such as gestures, touch, and pressure can be achieved, making HMI more natural, flexible, and intuitive. In the field of high-resolution sensor arrays, the flexibility and scalability of flexible sensors enable the construction of large-scale sensor networks. These sensors can form high-resolution sensor arrays used in image and video capture, object detection and tracking, gesture recognition, and other areas. The high resolution and sensitivity of flexible sensors make them widely applicable in medical imaging, security monitoring, and human behavior analysis, among other fields. As a vital component of Internet of Things (IoT) nodes, flexible sensors play a significant role in real-time monitoring and data acquisition (DAQ) of the environment and objects in the field of the IoT. By connecting with other sensors and IoT platforms, flexible sensors enable intelligent interaction and decision-making. The high flexibility and pliability of flexible sensors allow them to adapt to objects of different shapes and sizes, enabling the perception and monitoring of object movement, pressure, deformation, and other information. This provides IoT applications with more precise and comprehensive data collection capabilities, promoting the development of intelligent logistics management, environmental monitoring, smart homes, intelligent transportation, and other fields. In the field of VR/AR, the application of flexible sensors can enhance user immersion and interaction experience. Flexible sensors can be embedded in head-mounted displays, gloves, or body suits to real-time track user movements and postures, accurately mapping them into the virtual world. This real-time interactivity and enhanced realism bring more realistic and immersive experiences to VR games, training simulations, and design fields. In the field of intelligent medical monitoring, the application of flexible sensors has brought revolutionary changes to the healthcare industry. Flexible sensors can come into contact with the human body and monitor physiological parameters such as heart rate, respiration, and body temperature in real time. This non-invasive monitoring method allows patients to receive more comfortable and convenient medical care, while healthcare professionals can obtain real-time information about the patient's health status through remote monitoring systems, providing timely medical services. In the field of optoelectronic displays, the high flexibility and transparency of flexible sensors make them an ideal choice for the next generation of display technologies. Flexible sensors can be used as touch screens or gesture recognition devices to enable interactive control of displays. Furthermore, flexible sensors can also be used for bendable and foldable displays, bringing more portable and innovative display solutions for mobile devices and wearable devices.

In conclusion, the extensive application of flexible sensors covers multiple fields such as intelligent perception systems, human-machine interaction, high-resolution sensor arrays, the IoT, VR/AR, intelligent medical monitoring, and optoelectronic displays. With the continuous development and innovation of flexible sensor technology, it is believed that it will continue to drive progress and innovation in various fields, creating a more intelligent, convenient, and comfortable living environment for people. In the future, we can expect to see more applications of flexible sensors in various fields, bringing more surprises and conveniences to humanity.

2. Application of Flexible Sensor

2.1. *Intelligent perception systems*

An intelligent perception system based on ML algorithms is a system that utilizes ML techniques to process and analyze perceptual data. This system can obtain raw data from flexible sensors and use ML algorithms to process, analyze, and understand the data in order to extract useful information and knowledge from it.[10–12] Intelligent perception systems typically consist of the following main components: (i) Flexible sensors and DAQ: The DAQ stage is the starting point of the entire system, providing the raw data required for subsequent analysis. (ii) Feature extraction: After the perception data is collected, the system needs to extract meaningful features from the raw data. Feature extraction is the process of transforming the raw data into feature vectors or feature sets that can be processed by ML algorithms. (iii) ML algorithms: The extracted features serve as inputs to ML algorithms. These algorithms can be supervised learning, unsupervised learning, or reinforcement learning algorithms, depending on the specific task requirements. Commonly used algorithms include support vector machines (SVMs), decision trees, random forests, neural networks, etc. These algorithms infer patterns, and relationships, or perform classification and prediction from the perceptual data through training and learning. (iv) Model training and optimization: After selecting the ML algorithms, they need to be trained and optimized. In this stage, labeled or known-category data is used to train the model and optimize it based on predefined performance metrics. During the training process, the model gradually learns and adjusts its parameters to improve its generalization ability on unknown data.

2.2. Intelligent hand gesture perception system

Hand gestures, known as hand postures, have been used as a form of communication since ancient times and remain widely employed today. They represent a specific language system based on the positions and shapes of fingers, established by the human language center. Hand gestures possess a range of distinct meanings and offer rich expressiveness through long-term social practice. They have become one of the most potent means of human expression and hold a paramount position within the realm of nonverbal language.[13–19] The development of gesture perception systems for various applications has been inspired by the human gesture language system. Many of these systems employ a "master brain" approach, utilizing ML algorithms to intelligently decode different gestures. As an example, Tian et al.[20] presented a flexible hand gesture recognition glove (GRG) by integrating the ML algorithm, the proposed GRG system automatically recognizes 16 hand gestures underwater, significantly extending real-time and effective communication capabilities for divers (Figure 1).

Sign language, the most prevalent form of hand gestures, employs visual representations or finger configurations to simulate specific words, phrases, or meanings. It serves as the primary language for individuals with hearing impairments or those who are unable to speak (signers) to communicate and exchange ideas. However, individuals without prior knowledge of sign language (nonsigners) may struggle to receive and comprehend this form of conversation. Consequently, sign language interpretation plays a vital role in bridging the communication gap between

Sensing Systems and Applications 273

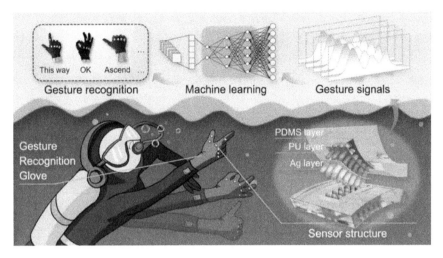

Figure 1. Flexible GRG. Reproduced with permission from Ref. 20 (Copyright 2024, American Chemical Society).

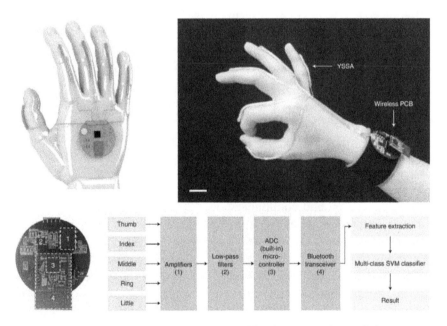

Figure 2. Schematic illustrations of the wearable sign-to-speech translation system. Reproduced with permission from Ref. 21 (Copyright 2020, Springer Nature).

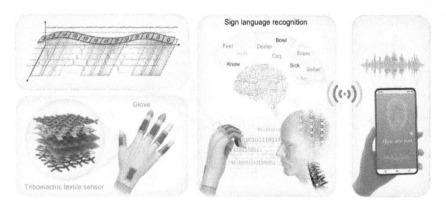

Figure 3. Sign language recognition and communication system. Reproduced with permission from Ref. 22 (Copyright 2021, Springer Nature).

signers and nonsigners. In pursuit of this goal, Chen et al.[21] developed a sign-to-speech translation system incorporating self-powered TENG gloves and wireless transmission blocks (Figure 2). With the assistance of ML, this system achieves high accuracy in recognizing 11 signs, including numbers, letters, and words, and displays the results unidirectionally on a mobile phone interface via Bluetooth. Nevertheless, while this system demonstrates an approach to implementing AI technology for the precise classification of a dozen gestures, it lacks an effective and practical method for real-time recognition of sign language sentences, which is crucial for practical communication between signers and nonsigners. Moreover, presenting sign language recognition results on mobile phones or personal computers can complicate the interaction between signers and non-signers.

Recently, Lee et al.[22] proposed an AI-supported system for sign language recognition and communication (Figure 3). This system incorporates sensing gloves, a deep learning block, and a VR interface. It employs a non-segmentation and segmentation-assisted deep learning model to accurately recognize 50 words and 20 sentences. Importantly, the system can identify previously unseen sentences created through the recombination of word elements with an average accuracy of 86.67%. The sign language recognition results are projected into a virtual space and translated into text and audio, enabling remote bidirectional communication between signers and nonsigners.

ML-based gesture recognition systems offer an alternative to traditional methods by utilizing a single-channel sensor attached to the wrist.

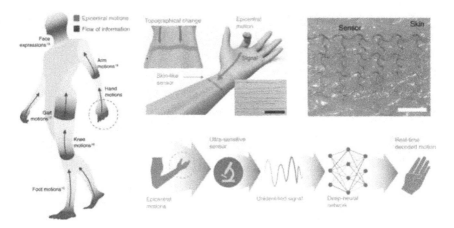

Figure 4. Design of the deep-learned skin-like sensor. Reproduced with permission from Ref. 23 (Copyright 2020, Springer Nature).

This sensor detects electromyography (EMG) signals, enabling the decoding of multiple finger movements instead of directly measuring joint motions. Through the inherent correlation between muscle movements and joint movements, ML algorithms establish meaningful connections. Ko et al.[23] took this concept further by combining a crack-based ultrasensitive sensor with rapid situation learning (RSL) to decode finger motion using transfer learning (Figure 4). Each finger's data collection takes approximately 8 s in their system, and a pre-trained network transmits parameters with a limited number of signals. Real-time demonstrations of results are presented using a virtual 3D hand, making this system suitable for applications like wearable haptics in VR.

To enhance the accuracy of gesture perception systems, Chen et al.[24] introduced a bioinspired somatosensory-visual (BSV) learning architecture (Figures 5(a) and 5(b)). This architecture integrates visual data with somatosensory data obtained from stretchable strain sensors resembling human skin. By employing a pruning strategy based on the Frobenius condition number, a sparse neural network is implemented, and the BSV learning architecture facilitates the learning process. Together, these components create a feedback system capable of automatic recognition for human-computer interaction. The BSV architecture was successfully employed to control a quadruped robot through a labyrinth, achieving low error rates of 1.7% under normal illumination (431 lux) and 3.3% under dark illumination (10 lux) (Figure 5(c)).

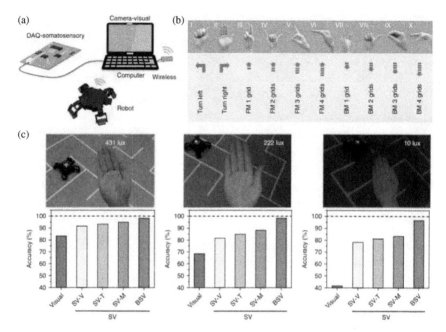

Figure 5. (a) Schematic showing the system consisting of a somatosensory-DAQ, a camera for capturing visual images, a computer, a wireless data transmission module and a quadruped robot. (b) Each of the 10 categories (I to X) of hand gestures was assigned a specific motor command to guide the movement of the quadruped robot. FM, forward move; BM, back move. (c) Performance accuracy of the robot using different recognition architectures under different illuminances (431, 222, and 10 lux). Reproduced with permission from Ref. 24 (Copyright 2022, Springer Nature).

2.3. Intelligent grasping object perception system

The human ability to perceive, weigh, and grip objects while deducing their shape and size relies on multiple sensory neuron subtypes within the skin, including the Ruffini ending, Pacinian corpuscle, Meissner corpuscle, and Merkel cell. Such perception necessitates long-term training, learning, and cognitive judgment by the human brain. Inspired by this phenomenon, researchers have developed grippers, robotic hands, and tactile gloves that mimic the grasping process. Furthermore, advancements in ML tools have endowed various object perception systems with the capability to recognize objects by imitating grasping actions.[25–33] As an example, a smart soft-clamp robot gripper system based on TENG sensors

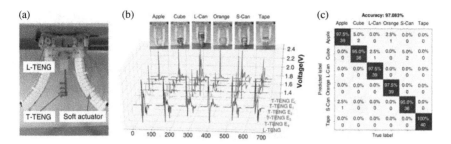

Figure 6. (a) Configuration of the gripper and sensors. (b) 3D plots of the output of the robotic sensor responding to different objects. (c) Confusion map for the ML outcome. Reproduced with permission from Ref. 34 (Copyright 2020, Springer Nature).

has been developed to retain crucial sensory information for classification. The symmetrical structure of the soft clamp, featuring a pair of T-TENG and L-TENG sensors arranged in separate soft actuators, reduces information complexity (Figure 6(a)).[34] Figure 6(b) displays the data collected by this gripper system for six different objects through repeated gripping and releasing motions. The dataset encompasses apples, cubes, long cans, oranges, short cans, and tape. After data collection, a multiclass SVM classifier, derived from the classical two-class SVM classifier through the one-against-all principle, classifies the gripped objects using hundreds of data points extracted from 600 data collection points. The gripper demonstrates high positive predictive value and true positive rate for object recognition (Figure 6(c)), achieving an overall recognition accuracy of 97.1%. To showcase the potential of the soft-robot sensory gripper system in future digital twin applications, a conceptual unmanned warehouse system based on a digital twin is proposed. This system enables automatic sorting and real-time monitoring in a camera-free environment, projecting real-time recognition results in real space into a virtual space.

Although the system accurately recognizes objects with varying shapes using single-mode sensors, it struggles with objects of identical shapes but different materials. Addressing this limitation, Zhu et al.[35] introduced a robot hand integrated with quadruple tactile sensors that ensures precise object recognition. They first determined the most frequently contacted positions of the robot hand when grasping different objects. Then, 10 quadruple tactile sensors were strategically placed on the five fingertips and palms of the humanoid robotic hand, as shown in Figure 7(a). The four-row signal graph represents the output signals of the

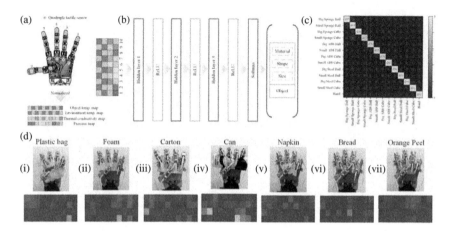

Figure 7. (a) Robot hand integrated with 10 quadruple tactile sensors. (b) MLP model structure for identifying object material, shape, and size. (c) Classification test confusion matrix of the test dataset. (d) Example signal maps when the robot hand grips seven types of garbage. Reproduced with permission from Ref. 35 (Copyright 2020, AAAS).

tactile sensors corresponding to object temperature, environment temperature, object thermal conductivity, and contact pressure. The robot hand was trained in object recognition by operating at room temperature and grasping a human hand and 12 standard objects of various sizes (small and big), shapes (cube and ball), and materials (steel, acrylonitrile butadiene styrene, and sponge). The corresponding output signals from the quadruple tactile sensors were recorded. Employing the PyTorch framework depicted in Figure 7(b), a multilayer perceptron (MLP) with three hidden layers was used to achieve object recognition based on size, shape, and material, exhibiting an overall classification accuracy of approximately 96% (Figure 7(c)). The smart hand was further applied to a garbage classification task (Figure 7(d)), achieving a classification accuracy of 94% in recognizing seven types of garbage.

In addition to grippers and robotic hands for object perception, Sundaram et al.[36] presented a scalable deep CNN tactile glove equipped with 548 uniformly distributed resistive sensors (Figure 8). This glove analyzes tactile patterns associated with object class and weight. When worn on the hand used for grasping objects, the glove's tactile mapping information is read by an external circuit, generating a tactile video with a frame rate of about 7.3 Hz. This method generates a large tactile video

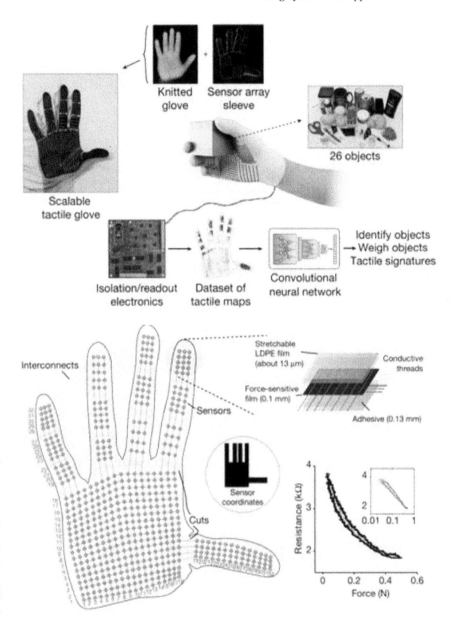

Figure 8. The STAG is a platform to learn from the human grasp. Reproduced with permission from Ref. 36 (Copyright 2019, Springer Nature).

dataset of 135,000 frames, with each frame capturing information from one hand. An improved residual neural network architecture is utilized to train this tactile video dataset. The system can predict object class and weight while sensing human grasping actions.

2.4. Intelligent speech perception system

Speech perception involves the study of speech and enables machines to automatically recognize and comprehend human spoken language through speech signal processing and pattern recognition. It is an advanced technology that transforms speech signals into corresponding text or commands through recognition and understanding processes.[37-43] Rogers et al.[44] utilized accelerometers integrated into the skin to successfully monitor seismocardiography signals. The sensors were created using ultralow-modulus elastomeric substrates (Silbione and Ecoflex) with respective moduli of 5 and 60 kPa. These sensors had a total weight of 213 mg and a thickness of 2 mm (refer to Figure 9).

To enhance the sensitivity of skin-attachable acoustic sensors, Nayeem et al.[45] proposed all-nanofiber-based acoustic sensors. These sensors consisted of polyurethane nanofibers as the substrate and PVDF nanofibers as the active layer. By incorporating nanoporous substrates that combined piezoelectricity and triboelectricity, these sensors achieved a remarkable sensitivity of 10050 mV Pa^{-1}. They were capable of detecting heart sounds for over 10 hours, with a signal-to-noise ratio (SNR) exceeding 38.2 dB (refer to Figure 10). Furthermore, skin-attachable acoustic sensors demonstrated the ability to detect body acoustic signals with high sensitivity, even in the presence of environmental noise. Le et al.[46] introduced a composite film made of graphene oxide and polydimethylsiloxane (PDMS) with microcracks and hierarchical surface textures. These sensors exhibited a high sensitivity, characterized by a gauge factor of 8699, and an exceptionally low detection limit (ε = 0.000064%) within the audible range of −20 to 20,000 Hz. When attached to the neck, these sensors were able to recognize various speech patterns, even in noisy environments where conventional microphones might fail to detect acoustic signals.

The application of deep learning in speech perception has significantly enhanced its performance and led to the widespread adoption of speech perception systems. For instance, Chen et al.[47] introduced a waterproof acoustic sensor (WAS) as a wearable translation interface for communication with machines. The WAS utilizes internal microparticles,

Sensing Systems and Applications 281

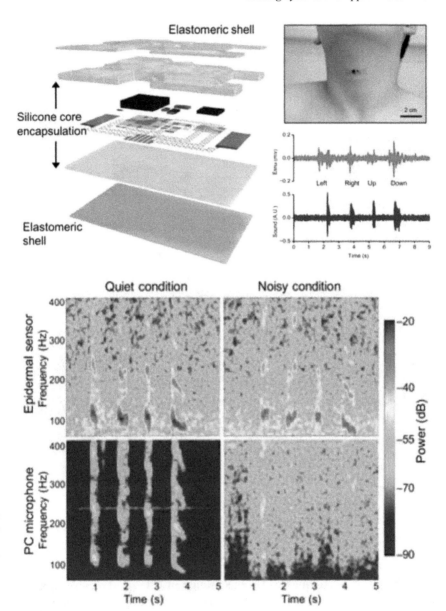

Figure 9. Accelerometer-based mechanoacoustic sensor. The sensor is capable of detection of voice or heart signal by attaching it to the neck or chest. Reproduced with permission from Ref. 44 (Copyright 2022, AAAS).

282   H. Niu et al.

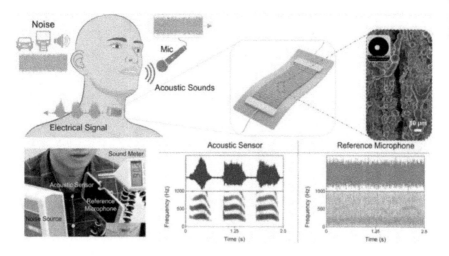

Figure 10. An acoustic sensor with microcracks and hierarchical surface textures. The sensor shows high sensitivity and ultralow detection limit. Reproduced with permission from Ref. 45 (Copyright 2020, ACCDON).

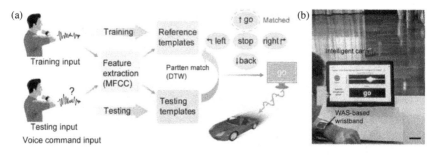

Figure 11. (a) Real-time wireless intelligent car control system. (b) Demonstration of the WAS-based intelligent car control system. Reproduced with permission from Ref. 47 (Copyright 2022, Wiley-VCH).

providing a broad frequency response range of 0.1–20 kHz, covering the entire human audible range. To validate the practicality of WAS in acoustic human-computer interaction, a real-time wireless intelligent car control system was developed. This system includes a sound-sensitive wristband based on WAS, an electric intelligent car, wireless modules, an amplifier, a DAQ card, and a computer. As shown in Figures 11(a) and 11(b), the wristband captures voice commands such as "go", "left", "right", "back", and "stop". The electrical signals from the wristband are amplified and

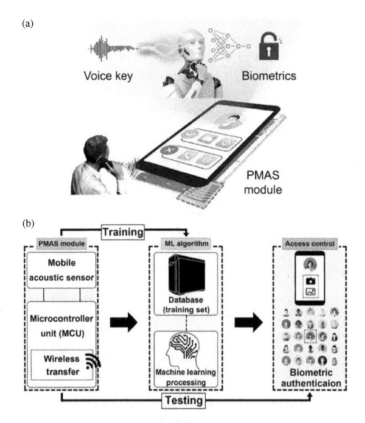

Figure 12. (a) Biometric authentication for mobile applications using the integrated acoustic module composed of the PMAS, ML processor, and wireless transmitter. (b) ML-based mobile biometric authentication using PMAS module. Reproduced with permission from Ref. 48 (Copyright 2021, AAAS).

captured by the DAQ. Using the designed AI algorithm, the speech commands are swiftly recognized, translated into text, and sent to the smart car via the wireless module to execute the corresponding actions.

Additionally, Lee et al.[48] created a highly sensitive and miniaturized piezoelectric mobile acoustic sensor (PMAS) for controlling multiresonant frequency bands. They demonstrated an ML-based biometric authentication technique using a smartphone-integrated sensor module consisting of PMAS, an algorithm processor, and a wireless transmitter (Figures 12(a) and 12(b)). With only 150 training data and 150 test data, the speaker identification error rate of the PMAS module was reduced by 56% compared to that of a commercial MEMS microphone.

Apart from speech perception, silent speech perception offers an alternative communication method for individuals with aphasia. Notably, silent speech interaction provides universal applicability across various environments compared to speech or visual interactions. Wang et al.[49] presented a lip-language interpretation system based on triboelectric sensors operating in contact separation mode, attached to the lip muscles. These sensors capture lip motion and transmit the measured electrical signals to the decoding system, where they are translated into a communication language, as depicted in Figure 13(a). Furthermore, they proposed a dilated recurrent neural network model based on prototype learning for lip language recognition, achieving a test accuracy of 94.5% for 20 classifications with 100 samples per classification (Figure 13(b)).

2.5. Intelligent texture perception system

Surface texture refers to a comprehensive characterization of the microscopic geometry and roughness of an object's surface, which deviates from an ideal smooth plane and exhibits local variations.[50–56] In the physical world, we perceive an object's texture directly through touch, enabling us to determine a range of material properties. When human skin comes into contact with the surface texture, it experiences tactile stimuli such as vibration and pressure that vary in time and space. This leads to deformations in the muscle tissue and activates tactile receptors, generating electrical impulses. These electrical signals are then transmitted through afferent nerve fibers to the central nervous system in the brain, where they are processed to form sensations like roughness, smoothness, density, and more. Initially, researchers mimicked the electrical signal-sensing mechanisms of human skin and differentiated textures by observing the differences in the acquired signals. In 2015, Ko et al.[57] developed piezoelectric e-skin with parallel ridges on the surface to amplify texture-induced vibrations (Figure 14(a)), enabling the detection of surfaces with varying roughness. Continuous scanning of surfaces such as sandpaper, paper, and glass produced different short-time Fourier transform features. Rougher surfaces, like sandpaper, exhibited high amplitudes across frequencies below 30 Hz, while smoother surfaces, like glass, lacked noticeable frequency features (Figure 14(b)).

However, with the rapid advancements in AI technology, artificial neural networks (ANNs) are now employed to train and learn from these

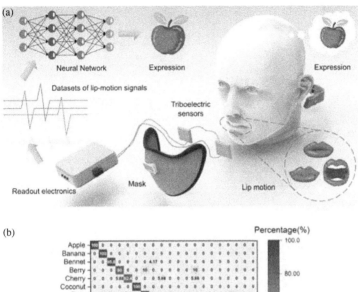

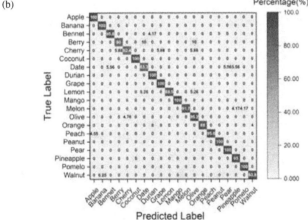

Figure 13. (a) Lip language decoding system and its components, including the triboelectric sensors, signal processing, and deep learning classifiers. (b) Confusion matrix for lip motion signals for 20 words. Reproduced with permission from Ref. 49 (Copyright 2022, Springer Nature).

data, facilitating the construction of texture perception systems for the autonomous identification of various textures. For instance, Choi et al.[58] introduced a self-powered flexible neural tactile sensor (NTS) composed of two laminated sensors (Figure 15(a)). They utilized deep learning techniques to analyze and classify the waveform patterns of vibration interactions between the NTS and complex textures. They tested 12 different fabrics to classify their texture patterns. The triboelectric output voltage

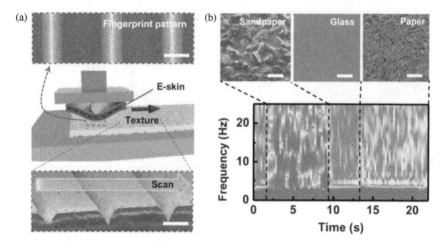

Figure 14. (a) Texture perception measurements. (b) Perception of textures with different levels of roughness. Reproduced with permission from Ref. 57 (Copyright 2022, Wiley-VCH).

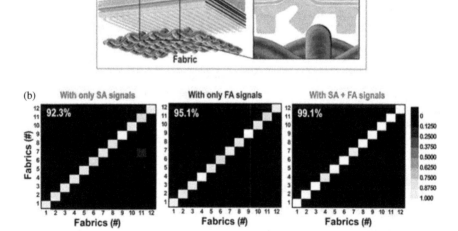

Figure 15. (a) Texture perception measurements. (b) Perception of textures with different levels of roughness. Reproduced with permission from Ref. 58 (Copyright 2019, American Chemical Society).

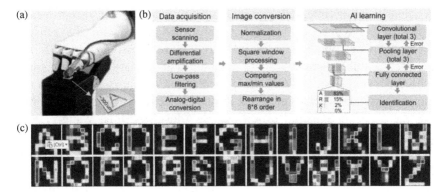

Figure 16. (a) A robotic finger with a microarray sensor touching the letter patterns. (b) Flowchart of AI identification, including DAQ, image conversion, and AI learning. (c) Collected letter images of A to Z through flow. Reproduced with permission. Reproduced with permission from Ref. 59 (Copyright 2020, AAAS).

waveforms analyzed through deep learning showed that among the 12 fabrics, the classification accuracy was 92.3% for slow adaptive (SA) sensors, 95.1% for fast adaptive (FA) sensors, and 99.1% for a combination of SA and FA sensors (Figure 15(b)).

Additionally, Choi et al.[59] developed tactile microarray sensors consisting of 64 sensors arranged in a 7 mm × 8 mm array, exhibiting high sensitivity and stability. They evaluated the sensor's ability to identify micro-sculpture patterns (modeling English letters A to Z) by integrating them onto a robotic fingertip and applying deep learning for data processing (Figure 16(a)). The test signals were acquired through a signal modification circuit (Figure 16(b)) and normalized to pixel values ranging from 0 to 255. Finally, the pixels were assembled into an 8 × 8 letter image based on their respective locations (Figure 16(c)).

2.6. Intelligent material perception system

Material perception involves effectively distinguishing and identifying different materials through the sense of touch. It is particularly challenging when dealing with materials that have smooth surfaces. The development of intelligent material perception systems has faced significant obstacles for a prolonged period.[60–62] This is mainly attributed to the absence of a sensing mechanism capable of extracting signal differences efficiently while interacting with various materials. In recent years, researchers have conducted extensive investigations into the sensing mechanism of triboelectric sensors. The findings reveal that different materials exhibit distinct voltage amplitudes and polarities upon contact with the triboelectric sensor. This phenomenon primarily arises from the coupling effect of contact electrification and electrostatic induction, forming the basis for material classification. Addressing this, Wang et al.[63] devised a material perception system utilizing a MATLAB lookup table algorithm, as depicted in Figure 17. In this system, five flat materials are affixed to a linear motor and brought into contact with the polytetrafluoroethylene-based sensor to generate voltage signals. These signals are then converted into a database containing maximum voltage values and polarities. Material species can be identified when the minimum absolute value of these materials surpasses the acceptable tolerance of 0.02 V. While this system successfully identifies the five materials using a simple algorithm, there is still room for improvement in terms of universality and automation for diverse materials.

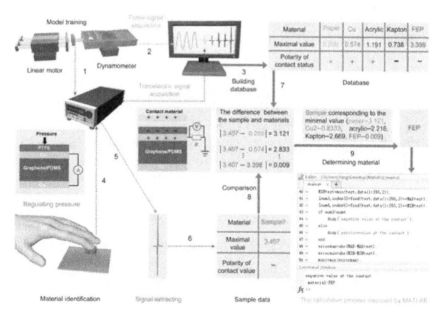

Figure 17. The two-factor material identification system consists of a training process (steps 1–3) and an identification process (steps 4–9). Reproduced with permission from Ref. 63 (Copyright 2020, AAAS).

Sensing Systems and Applications 289

Recently, motivated by AI, specifically ML algorithms, a series of intelligent material perception systems have been designed and developed for material species identification.[64–70] Li et al.[71] integrated the triboelectric effect with the supercapacitive iontronic effect to create a high-performance flexible and stretchable bionic electronic skin (e-skin) known as FSB e-skin. It features a stacked configuration of bionic vellus hair and epidermis-dermis-hypodermis double interlocked microcone structures. With the support of a linear motor system, a capacitance acquisition circuit, a field-programmable gate array high-speed DAQ circuit (comprising a signal amplification circuit, analog-digital conversion circuit, and USB serial communication), a six-layer MLP neural network, and a real-time display module, an intelligent material cognition system was constructed to identify 12 materials (Figure 18(a)). As illustrated in Figure 18, the 12 materials produce distinctive voltage signals under the same pressure, enabling material classification based on variations in voltage amplitude, duration time, and polarity (supercapacitive iontronic e-skin) as long as the applied voltage is determined by the capacitance signal (triboelectric e-skin). The classification results of the 12 materials

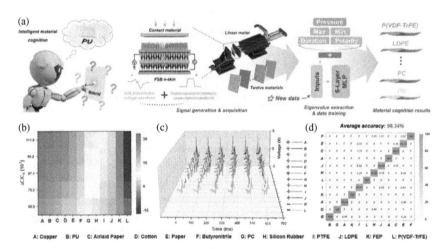

Figure 18. (a) Concept diagram of the designed intelligent material cognition and the neural network-assisted material cognition process flow. (b) Voltage amplitudes of villus hair-like structure triboelectric e-skin contacted with 12 materials at arbitrary same pressure. (c) Voltage waveforms generated by the 12 materials. (d) Confusion matrix displaying the corresponding classification accuracy (%) for the 12 materials. Reproduced with permission from Ref. 71 (Copyright 2022, Wiley-VCH).

are presented in the confusion matrix shown in Figure 18(d). Eight out of the 12 materials achieved a classification accuracy exceeding 99%, while the remaining four materials attained an accuracy of 85%. The average accuracy across all materials reached 98.34%.

2.7. Intelligent posture perception system

Posture refers to the body's alignment and positioning while engaged in muscular activities or maintaining stability through coordinated muscle actions. Posture perception involves determining the nature of a person's posture based on key body nodes, such as falling, sitting up, running, and jumping postures.[72–76] Among various methods of posture perception, gait perception has received significant attention due to its distinct and regular characteristics. Gait refers to the postural and behavioral features of the human body during walking, involving a series of continuous actions performed by the hip, knee, ankle, and toes to move in a specific direction. Normal gait exhibits stability, periodicity, rhythmicity, and individual variations. However, gait characteristics significantly change when individuals suffer from gait-related disorders and struggle to maintain a normal gait. While there have been extensive studies on developing insoles with energy collection and gait sensing capabilities, socks are considered the optimal choice for gait analysis due to their flexibility and comfort. Zhu et al.[77] presented a sock that is both self-powered and self-functional (referred to as S^2-sock). The sock utilizes a textile coated with poly(3,4-ethylenedioxythiophene):polystyrenesulfonate (PEDOT:PSS) and integrates a PZT piezoelectric sensor, as depicted in Figure 19(a). The researchers focused on the various phases involved in a complete foot-ground contact sequence during walking, which include heel contact, toe contact, heel leave, and toe leave (as illustrated in Figure 19(b)). By analyzing the data obtained from three participants during regular walking, the authors identified distinctive signal patterns corresponding to each individual, as well as contact forces and contact angles of the heel and forefoot. These findings demonstrate the potential of S^2-socks in monitoring Parkinson's disease. With the aid of ML processing, this sock can serve as an effective technique for accurate pattern recognition and monitoring of biomechanical activity. It is particularly valuable in detecting "freezing of gait", a measurable physiological signal.

Lee et al.[78] created gait-based triboelectric intelligent socks capable of capturing gait patterns from low-frequency body movements, wirelessly transmitting the sensor signals. These self-powered socks also serve as

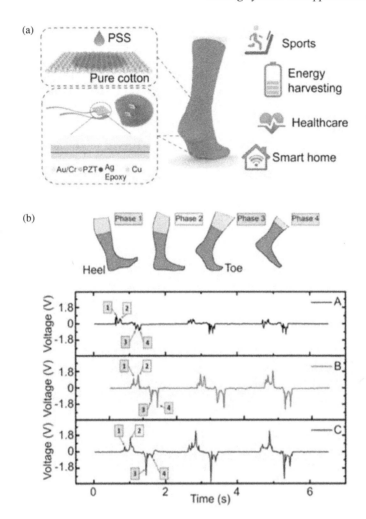

Figure 19. (a) Schematic of PEDOT:PSS-coated triboelectric S^2-sock integrated with lead zirconate titanate (PZT) force sensors for diversified applications, with the left side showing enlarged views of the triboelectric nanogenerator (TENG) textile and embedded PZT sensor. (b) Schematics of four phases of the typical contact cycle and corresponding signals: (1) heel contact, (2) forefoot/toe contact, (3) heel leave, and (4) forefoot/toe leave. Reproduced with permission from Ref. 77 (Copyright 2019, American Chemical Society).

wearable sensors to provide user information regarding identity, health status, and activities (Figure 20(a)). Moreover, using data samples from five participants, important gait features such as amplitude, frequency, and interval of gait cycle stages were extracted (Figure 20(b)). Assisted by

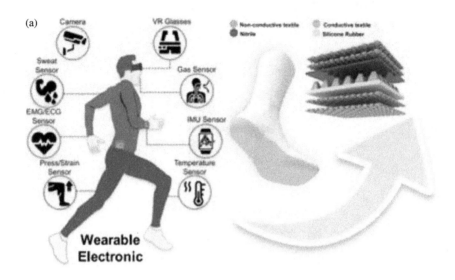

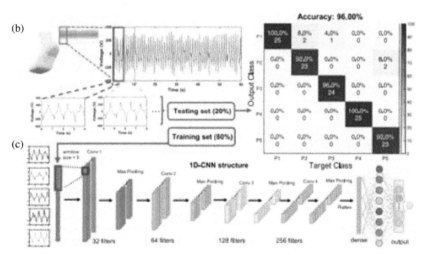

Figure 20. (a) Deep learning-enabled socks. (b) Process and parameters for constructing the 1D-CNN structure. (c) Confusion map of the prediction system with gait patterns of five participants. Reproduced with permission from Ref. 78 (Copyright 2020, Springer Nature).

a 1D convolutional neural network (CNN) in deep learning, the intelligent socks achieved gait recognition in five participants with an accuracy of 93.54% (Figure 20(c)).

Sensing Systems and Applications 293

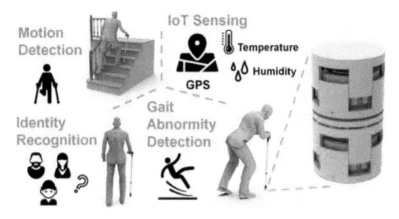

Figure 21. Caregiving walking sticks with multifunctional applications for users. Reproduced with permission from Ref. 79 (Copyright 2021, American Chemical Society).

Furthermore, Lee et al.[79] introduced a multifunctional walking stick powered by ultralow-frequency human motion, incorporating two self-powered triboelectric sensors to capture the motion features of the stick. Deep learning-based data analysis enabled high-precision augmented sensing functions, including identity recognition, disability evaluation, and motion status differentiation (Figure 21). Sitting posture perception is another significant research area that has led to the development of various corresponding perception systems.

Lee et al.[80] proposed a multifunctional AI toilet system as an integrated health monitoring system (Figure 22). With the aid of deep learning, this system accurately identified the biometric information of six users sitting on the toilet with over 90% accuracy.

Kim et al.[81] presented an 8×8 array of smart seat cushions for posture monitoring (Figure 23(a)). Using ML techniques, the random forest (Figure 23(b)) and ANN (Figure 23(c)) algorithms successfully distinguished six different sitting positions with respective accuracies of 96.7% and 97.2%. These systems provided real-time monitoring of the smart seat cushion, as depicted in Figure 23(d).

With the rapid advancement of AI technology and the proliferation of 5G communication within the IoT, meeting the current demands for enhanced living comfort and intelligence directly deployed on the human body or handheld wearable electronics has become challenging. There is a growing expectation for the widespread presence of multifunctional

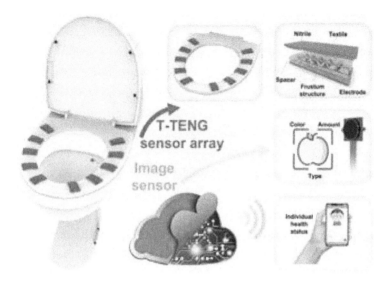

Figure 22. AI-enabled toilet for integrated health monitoring system. Reproduced with permission from Ref. 80 (Copyright 2021, Elsevier).

environment-aware devices in our living spaces, particularly in homes and offices. These distributed electronics can offer additional functionalities in a complementary and versatile manner, encompassing intelligent position monitoring, home automation, personalized healthcare, and authentication, ultimately facilitating the realization of truly "smart" buildings and homes. In line with this vision, Lee et al.[82] have developed an intelligent, cost-effective, and highly scalable floor monitoring system by integrating self-powered triboelectric smart mats with advanced deep-learning data analysis (Figure 24(a)). Leveraging prior data analysis within the implemented CNN model, this smart floor monitoring system enables real-time position sensing and identification. Furthermore, the position sensing information of each step is utilized to control the corresponding lighting, while the CNN model analyzes the complete walking signal to predict the validity of the person's room access and automatically regulate door entry (Figure 24(b)). Through extensive testing, this system achieved an average prediction accuracy of 96% for specific walking gaits within a 10-person CNN model with 1,000 data samples, thus demonstrating high accuracy in real-time scenarios.

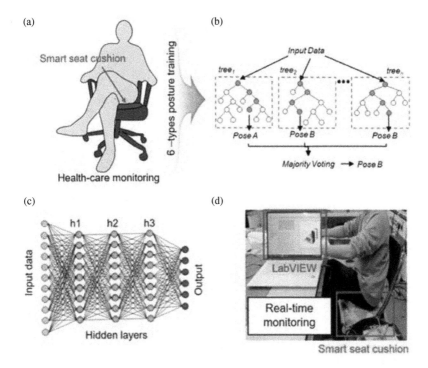

Figure 23. (a) Classification of sitting postures with the smart seat cushion. (b) Schematics of the random forest algorithm. (c) Schematic of the ANN algorithm. (d) Photograph of real-time monitoring of smart seat cushion. Reproduced with permission from Ref. 81 (Copyright 2021, American Chemical Society).

2.7.1. Human–machine interaction

HMI refers to the interaction and communication between humans and machines, including computers, robots, and other electronic devices. It encompasses a wide range of technologies and applications, from traditional input/output devices like keyboards and displays to advanced technologies such as speech recognition, gesture recognition, and brain–computer interfaces.[83–90] In our modern lifestyle, electronic products play an increasingly vital role, and human–machine interaction plays a crucial part in enhancing the user experience. Wearable HMIs have the potential to revolutionize user entertainment and improve the user experience, facilitated by the development of flexible pressure sensors that offer

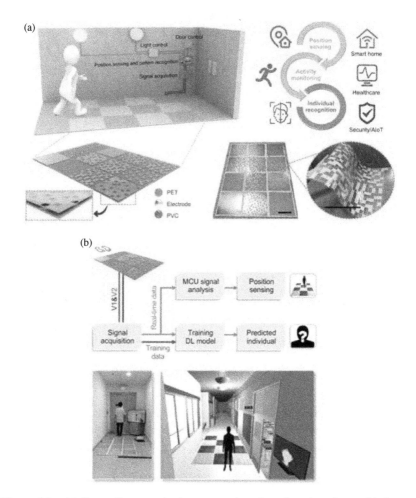

Figure 24. (a) Smart floor monitoring system based on deep learning-enabled smart mats. (b) Overall structure and data flow of the smart floor monitoring system for real-time position sensing and individual recognition in smart building/home applications, and demonstration of real-time position sensing and individual recognition. Reproduced with permission from Ref. 82 (Copyright 2020, Springer Nature).

deformability and stretchability. Fingers, being one of the most agile parts of the human body, are highly suitable as an interface for HMI. The combined operation of our 10 fingers allows for the definition of a significant amount of information and commands. This interaction mode is simple and intuitive, aligning perfectly with the logic of our brain. Pan et al.[91]

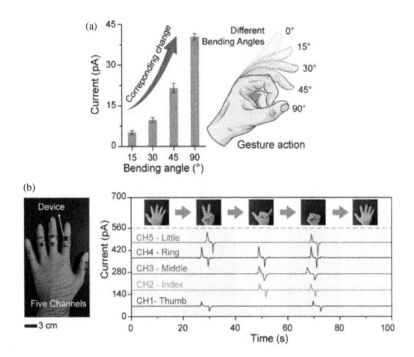

Figure 25. (a) Schematic of gesture action and the output property of the device. (b) Photograph of the devices bound to five fingers is constructed into a multi-channel system and the corresponding generated waveforms. Reproduced with permission from Ref. 91 (Copyright 2020, Elsevier).

employed a low-temperature hydrothermal growth method to synthesize variable aspect-ratio Sb-doped ZnO NW films. They utilized these films to construct a flexible self-powered strain sensor based on the piezoelectric effect. This sensor enables the perception and response to gesture movements. In Figure 25(a), electrical signals are obtained from different bending points of the index finger using a single device, and the output current shows a positive correlation with the degree of bending. To demonstrate gesture movement, a multi-channel DAQ system is used to simulate electrical signals driven by the motions of multiple fingers (Figure 25(b)).

In addition, Researchers have made significant advancements in utilizing flexible piezoelectric nanogenerators (PENGs) to recognize and synchronize human actions. By integrating flexible sensing, signal processing, and control execution, these studies have demonstrated the

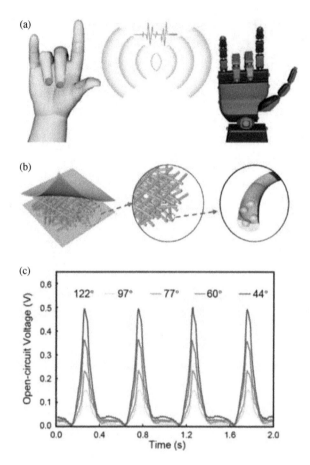

Figure 26. (a) Schematic diagram of self-powered piezoelectric sensor based on cowpea-structured PVDF/ZnO nanofibers towards the remote control of gestures. (b) Sketch of the designed device comprising a nanofiber film, and a single nanofiber. (c) The opencircuit voltage is generated by bending at different angles. Reproduced with permission from Ref. 92 (Copyright 2019, Elsevier).

transmission of gestures to robotic arms. For instance, Deng et al.[92] achieved remote gesture transmission to a robotic arm by combining PENG sensing, signal transmission, and executive control (Figure 26(a)). They utilized a functional layer composed of cowpea-structured PVDF/ZnO nanofibers (Figure 26(b)), which provided enhanced mechanical flexibility and electrical output. This design enabled the sensors to promptly and sensitively respond to bending angle motions (Figure 26(c)).

When the sensor was affixed to the finger knuckle, it converted finger-bending movements into electrical signals, allowing the robotic hand to mimic the corresponding gesture via a peripheral circuit module.

Likewise, Lim and colleagues[93] developed a closed-loop interactive human–machine interface system incorporating a specially designed sensor and actuator (Figure 27(a)). Their transparent and stretchable piezoelectric motion sensor consisted of poly(L-lactic acid), single-walled carbon nanotubes (CNTs), and silver nanowires (Figure 27(b)). This sensor monitored wrist movements and converted the collected data into electrical signals to control a robotic hand. As the human wrist bent, pressed, or relaxed, the robotic arm mirrored those movements. Figure 27(c) displays representative data collected from the motion sensor. Wearable PENGs prove particularly effective in detecting relatively large deformations associated with human movements, making them widely adopted for monitoring various body sites and exhibiting significant application potential.

One important application of flexible pressure sensors in human-machine interaction is intelligent recognition, such as image, voice, gesture, and face recognition. For instance, Syu et al.[94] developed a hybrid self-powered sensor by integrating a nanofiber-based piezoelectric sensor and a biomimetic triboelectric sensor with porous PVDF fibers. The synergy of this hybrid system effectively enhanced the energy harvesting capability, exhibiting significant voltage and current outputs. These sensors were integrated into garments like socks, gloves, and trousers to distinguish various human motions by recording average voltage signals. Gesture recognition was achieved through the implementation of a ML algorithm, resulting in an overall training accuracy of 82.3% (Figure 28).

Interactive control is another important application of human-machine interaction. The human–machine interface serves as a communication bridge between humans and machines, involving the acquisition of input signals from users and their conversion into instructions for specific machine actions. A wearable HMI system based on flexible pressure sensors significantly improves interactive accuracy and the user experience of the control process. For example, Zhong et al.[95] created a high-performance piezoresistive pressure sensor with a wide linear range and high sensitivity using a micro–nano hybrid conductive elastomer film with an arched micropatterned structure. These sensors were attached to a textile glove at the knuckle regions, establishing a human–machine interface (Figure 29). The smart glove system could record and process

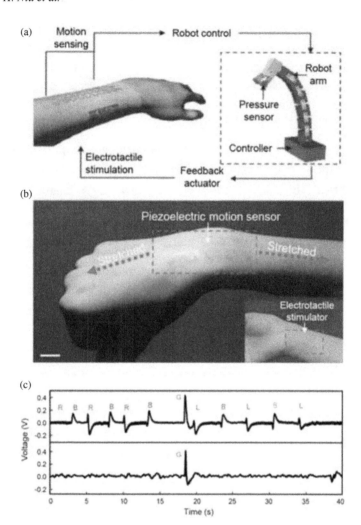

Figure 27. (a) Schematics of the closed-loop system of the interactive human–machine interface. (b) Images of a transparent piezoelectric motion sensor and the electroactive stimulator (inset). Scale bar, 20 mm. (c) Detected signals acquired by the transparent motion sensor (first row) and by the pressure sensor on the gripper (second row). Reproduced with permission from Ref. 93 (Copyright 2015, Wiley-VCH).

bending signals from each joint to control a robot hand in real-time, enabling gesture imitation and object grasping.

Furthermore, Wang et al.[96] designed a stretchable textile-based single-electrode TENG with high output performance. Integrated with

Sensing Systems and Applications 301

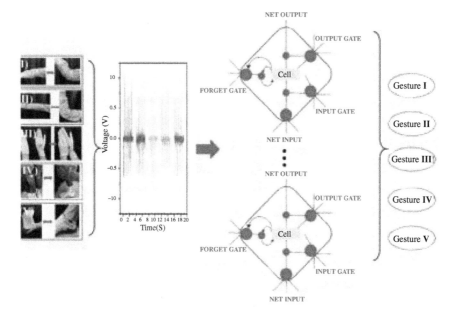

Figure 28. Application of a flexible self-powered pressure sensor in gesture recognition through a ML algorithm. Reproduced with permission from Ref. 94 (Copyright 2020, Elsevier).

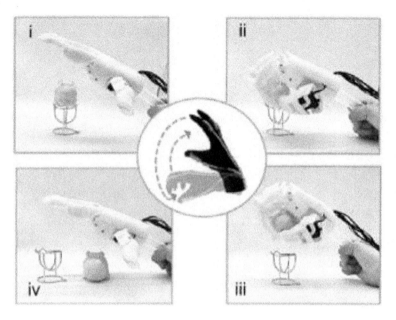

Figure 29. Application of the pressure sensor in real-time controlling of robot hand. Reproduced with permission from Ref. 95 (Copyright 2021, Elsevier).

microelectronic modules, a portable and wearable self-powered haptic controller was developed for HMI (Figure 30). This haptic controller, conveniently worn on an arm guard sleeve, allowed for the control of various devices such as lamps, electronic badges, slides, and humidifiers.

Moreover, Liu *et al.*[97] developed a novel core-sheath "pruney fiber" composed of a thermoplastic polyurethane fiber core and a wrinkled PPy conductive shell (Figure 31(a)). This piezoresistive fiber could be easily woven into an intelligent glove for HMI, enabling the control of car driving in computer games through different gestures. For example, specific finger contacts on the glove triggered corresponding electrical signals that rapidly increased, resulting in the car turning left, right, or braking (Figure 31(b)).

With the rapid evolution of AI and the IoT, wearable sensors have gained increasing importance as a means of HMI. In the realm of glove-based HMIs, hybrid nanogenerators exhibit remarkable adaptability. These nanogenerators are particularly valuable for achieving precise control through immersive experiences and comprehensive sensations, thanks to their well-designed haptic feedback capabilities.[98] In one study,

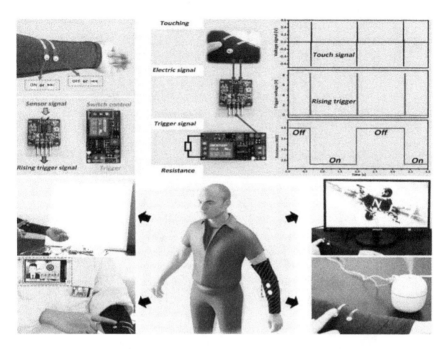

Figure 30. Application of the TENG-based pressure sensor in wearable haptic controllers such as controlling light switch states, electronic badges, slides, and humidifiers. Reproduced with permission from Ref. 96 (Copyright 2021, Elsevier).

researchers developed a cost-effective smart glove for intuitive HMI applications. This glove incorporated an elastomer-based triboelectric tactile sensor and a PZT piezoelectric haptic mechanical stimulator. The index finger was equipped with six sensors to measure upward and downward bending, while additional sensors were added for left–right bending detection in the proximal phalanx (Figure 32(a)). Through ML-assisted adaptive performance tracking (APT), finger movements could be accurately detected and recognized, meeting the requirements of various operations. The authors of the study tested the glove's applicability in VR surgical training programs and AR-based human-humanoid interactions. They allocated control of the entire arm and hand movement, as well as the operation mode switching, to the left glove, while the right glove facilitated object recognition and surgical operations (Figure 32(b)). Beyond VR training programs, smart gloves offer the potential for intuitive interactions in virtual space communication, particularly when

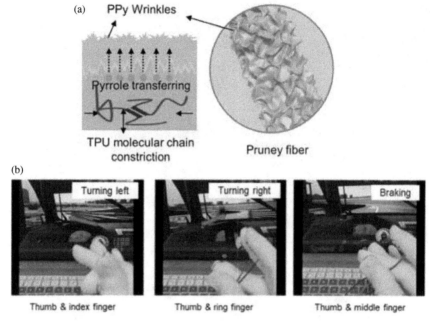

Figure 31. (a) Schematic of the formation of the pruney fiber. (b) Digital photos of the virtual game show how different gestures are applied to controlling a car. Reproduced with permission from Ref. 97 (Copyright 2022, Elsevier).

combined with AR technologies. Consequently, HMIs find applications in VR training, entertainment, social networking, and robotic control.

As robotic manipulators play a crucial role in diverse industries, novel designs for robotic wrists have emerged to enable specialized tasks. A recently proposed triboelectric sensor system, featuring patterned electrodes and gear-structured length sensors, enhances the intelligence of soft manipulators.[99] Supported by ML technology, these manipulators exhibit capabilities such as automatic sorting and real-time monitoring in non-camera environments. This concept is embodied in a digital twin-based unmanned storage system, which projects real space into a virtual one for real-time visualization. The process of establishing and utilizing this digital twin-based unmanned warehouse system is depicted in Figure 33(a). By employing the digital twin's virtual projection, objects randomly positioned in the virtual space can be grasped by a soft gripper (Figure 33(b)). The trained SVM models, utilizing input signals collected from triboelectric sensors, aid in object recognition of the gripped items with the assistance of ML technology. With the improved intelligence of the soft

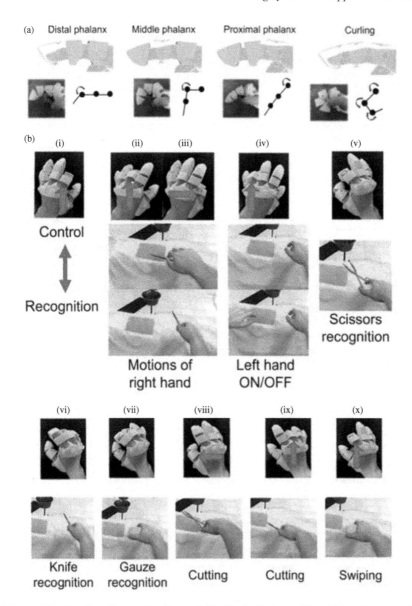

Figure 32. (a) Bending mechanism of (i) distal phalanx, (ii) middle phalanx, (iii) proximal phalanx, and (iv) all three phalanges (curling). (b) Photographs and screenshots of the finger motions for realizing (I) mode switching, (II,III) motions of right-hand, (IV) display of left-hand, (V) recognition of scissors, (VI) recognition of a knife, (VII) recognition of gauze, (VIII) operation of scissors (cutting), (IX) operation of knife (cutting), and (X) operation of gauze (swiping). Reproduced with permission from Ref. 98 (Copyright 2020, AAAS).

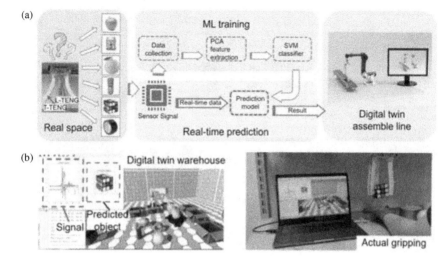

Figure 33. (a) Process flow from sensory information collection to ML training and real-time prediction in digital twin systems. L-TENG = length triboelectric nanogenerator; T-TENG = tactile triboelectric nanogenerator; PCA = principal component analysis; SVM = support vector machine. (b) System interface integrated with object recognition and its digital twin warehouse application. Reproduced with permission from Ref. 99 (Copyright 2020, Springer Nature).

gripper, the digital twin model simulates robotic manipulation and real-time object recognition within a duplicated virtual environment. This technology can further be applied to assembly lines, contributing to production control management in the next generation of smart floor management in unmanned warehouses.

2.7.2. High-resolution sensor arrays

A high-resolution flexible sensor array is a collection composed of multiple flexible sensors, characterized by high accuracy and sensitivity. These sensors can be embedded in the form of soft, bendable, and deformable structures on curved or bent surfaces, such as skin, clothing, robot surfaces, and so on.[100–107] The sensor array typically consists of numerous tiny sensor units, which can perceive the intensity and location of multiple point stimuli simultaneously through tactile mapping, providing more comprehensive tactile feedback. Among them, the integration of multi-touch sensing in the tactile sensor network is one of the main challenges in developing

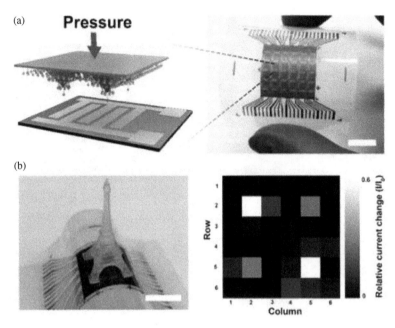

Figure 34. (a) Photo of the flexible sensor arrays by integrating the v-AuNW/PDMS microarrays with the printed interdigitated AgNP electrodes. (b) Relative current change distribution of the sensor array by the plastic Eiffel Tower model. Reproduced with permission from Ref. 108 (Copyright 2019, American Chemical Society).

large-scale tactile sensor arrays. Therefore, array integration processes have become a research hotspot, and array structural designs can be divided into planar dot-style and planar cross-style structures. The planar dot structure refers to integrating several tactile sensors on a plane according to certain rules, where the sensors are independent of each other, and each sensor has its dedicated surface electrode and circuit. Cheng et al.[108] developed a 6×6 pixel tactile sensor array utilizing an Au NWs/PDMS microarray film that was affixed to printed electrode patterns. Each sensor unit operated independently, as depicted in Figure 34(a). Through experiments, it was demonstrated that the sensor array successfully detected the position and pressure of a plastic Eiffel Tower model placed on it, with no interference between sensors (Figure 34(b)). These findings highlight the array's capability for precise and crosstalk-free measurements.

Guo et al.[109] presented a compact tactile sensor featuring a remarkable capacitance density, as depicted in Figure 35. This sensor exhibits

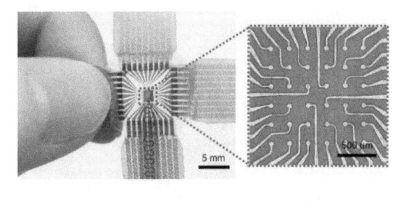

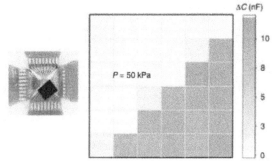

Figure 35. Spatial resolution of graded intrafillable architecture (GIA)-based microsensor arrays, and the pressure distribution mapping of the microsensor array in the shapes of a triangle (50 kPa). Reproduced with permission from Ref. 109 (Copyright 2020, Springer Nature).

outstanding stability and maintains reliable sensing capabilities, producing high signal intensity while keeping noise levels negligible. It covers an area of at least 50 μm × 50 μm. As a result, this innovative iontronic sensor enables high-resolution pressure mapping within approximately 100 μm. The pressure sensing array consists of 6 × 6 pixels, occupying a 1.6 mm × 1.6 mm area. The spacing between pixels measures 150 μm, and each sensing pixel has a diameter of 60 μm. The sensor accurately records and maps the position information within the image.

However, in this array structure, each tactile sensor requires an independent signal acquisition, transmission, and reception circuit, which poses a major challenge to the system design and implementation of high-precision hardware circuits. For the planar intersection structure, the array

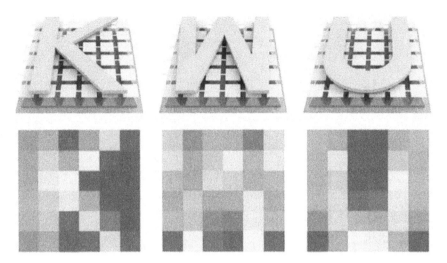

Figure 36. Schematic of the 3D printed letters on the flexible perception array and corresponding spatial pressure mapping. Reproduced with permission from Ref. 110 (Copyright 2023, Elsevier).

consists of two overlapping strip electrodes, one horizontal and one vertical, and a sensing layer sandwiched between them, and each sensor unit is located at the rectangular intersection of the two strip electrodes. This array structure is favored by many researchers because of its simple circuit and easy implementation. Niu et al.[110] developed a versatile perception array comprising 36 sensing units. This array was constructed by vertically intersecting two strip electrodes, each containing 6 arrays. The mapping information was obtained through cross-measurements of the top and bottom electrode arrays. To showcase the spatial mapping capability of this flexible perception array, 3D-printed letter blocks were positioned on the array. Notably, the array effectively perceived the shapes of the letter blocks, as depicted in Figure 36.

Bao et al.[111] introduced a capacitive tactile sensing array that utilized a lithography template. This template consisted of a hill-shaped bottom electrode, an intermediate dielectric layer, and a pyramid-shaped top electrode, as illustrated in Figure 37. The array design proposed by the researchers emulated a high-density tactile sensor, reminiscent of a mechanoreceptor. Each hill on the bottom electrode encompassed 25 sensors, with one positioned at the top of the hill, four on the slopes, four on the "corners", and sixteen surrounding the hill. To ensure precise sensor

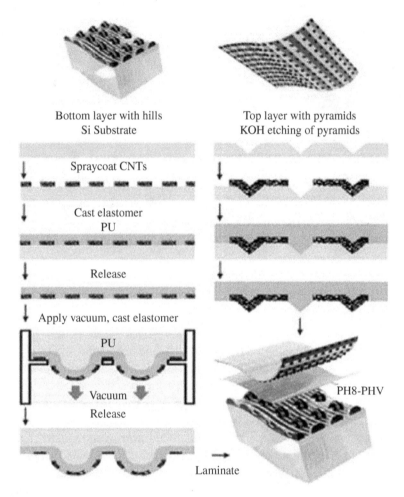

Figure 37. E-skin fabrication and assembly. The device consists of three layers, assembled by lamination: (i) a bottom 1-mm-thick PU layer with an array of hills (hill diameter, 1 mm; height, 200 mm), (ii) an intermediate 10-mm-thick PHB-PHV dielectric layer used as a spacer between the top and bottom electrodes, and (iii) a top 60-mm-thick PU layer with an array of pyramids. The electrodes were made of spray-coated and photolithograph-patterned conducting CNTs embedded into the PU matrix (electrode width, 300 mm; separation distance between two electrodes, 50 mm). The construct was reinforced with tape at the sides, and no sliding of the layers was observed when shear force was applied. For our current sensor size, we observed that the use of tape was sufficient to stabilize the system for laboratory experiments. If we were to scale the sensor array, the proper adhesion between layers would need to be implemented to ensure mechanical stability. Reproduced with permission from Ref. 111 (Copyright 2018, AAAS).

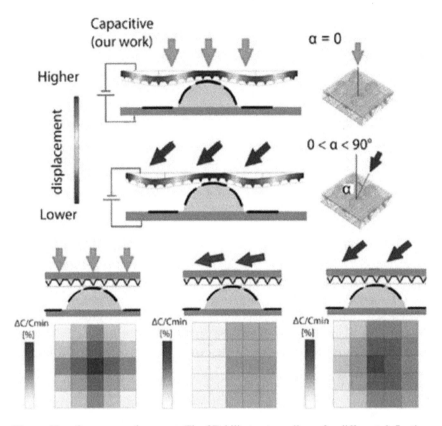

Figure 38. Our proposed concept. The 3D hill structure allows for different deflection capabilities on the top and around the hills, thus differentiating capacitive responses to a pressure event from different directions. Black lines are side views of electrodes. Schematics showing the advantages associated with our proposed design. With our design, it is possible to measure and discriminate in real-time normal and shear forces and forces applied in various directions. The tilt force is a combination of normal and shear force. In the case of a robotic application, the data of a fraction of the 25 pixels could provide sufficient information (for instance, nine of them—one on the top of the hill, four on the sides, and four in the corners), but the 25 pixels show the significance of our concept. Reproduced with permission from Ref. 111 (Copyright 2018, AAAS).

placement, careful alignment was employed. This unique arrangement of the tactile sensing array facilitated the accurate detection of various applied force directions, including normal force, shear force, and tilt force (Figure 38).

With the continuous evolution of micro-nano processing technology and the continuous improvement of intelligent requirements, array miniaturization will gradually become a new trend. However, current tactile sensor arrays based on crossover structures do not have a competitive advantage, which is mainly attributed to the fact that the miniaturization of the array will infinitely amplify the crosstalk between adjacent sensing units, thereby limiting the sensing density. Wu et al.[112] presented a sensing mechanism based on the Fowler-Nordheim tunneling effect in a high-density sensing array composed of orderly packed, monodispersed spiky carbon nanospheres. the sensor array film can be easily attached to different objects, providing detailed strain distribution akin to Finite Element Analysis (FEA) simulations.

Guo et al.[113] presented a novel approach in their research, wherein they embedded an array of intronic sensing elements within a flexible rubbery matrix. This innovative design allowed them to achieve exceptional characteristics such as high sensitivity (>174 kPa^{-1} from 0.15 Pa to 400 kPa), remarkable mechanical stability, compact pixel arrangement (28 × 28 pixels within a 10 cm by 10 cm area), and virtually no cross-talk interference. The skin-like structure employed PDMS as a stretchable matrix, with each microstructured sensing element precisely positioned within an individual cavity (totaling 28 × 28 cavities). The embedded configuration involved isolated chambers that housed microstructured ionic gels

Figure 39. Soft sensing films for strain field mapping. Reproduced with permission from Ref. 112 (Copyright 2024, Springer Nature).

Sensing Systems and Applications 313

(IMIGs), resulting in a modified iontronic interface. This modification significantly enhanced the capacitance-to-pressure sensitivity while ensuring signal integrity by eliminating cross-talk. The researchers successfully demonstrated the skin's ability to accurately map pressure distribution during both static and dynamic loading scenarios, thanks to the suppressed intercell cross-talk. In the case of static pressure detection, they placed a resin model with a specific pattern on the sensor array and recorded static capacitance values for all pixels. The recorded capacitance signal pattern exhibited strong correspondence with the pattern of the printed model (Figure 40(a)), with minimal noise observed. Moreover, the intronic skin proved capable of dynamic pressure mapping, as shown in Figure 40(b). Volunteers interacted with the skin by touching it with a single finger and later with two fingers, and a real-time visual interface displayed the dynamic pressure distribution during this process. The researchers also subjected the skin to the impact of a hammer, recording the output signals, yet no damage was observed on the skin's surface.

Regarding the flexible tactile sensing arrays discussed earlier, they demonstrated exceptional sensing capabilities and high resolution. However, researchers aspire to develop a sensing array with a greater number of pixels. A recent advancement in this area was proposed by Park et al.[114] who introduced an innovative and impactful interactive skin display (Figure 41). This display incorporates the epidermal stimuli electrode (ISDEE) as its central component, enabling the simultaneous detection of multiple epidermal stimuli on a single device. As a result, diverse body information can be visualized. The ISDEE consists of a straightforward two-layer architecture, comprising a micropyramid elastic polymer composite material embedded with light-emitting ZnS: Cu phosphors. Two electrodes are positioned at a specific parallel distance. The ISDEE exploits the conductance-dependent field-induced electroluminescence phenomenon, instantly projecting a highly detailed fingerprint when a transparent surface is gently touched by a fingertip equipped with the ISDEE.

To the current arrays of piezocapacitive and piezoresistive tactile sensors, while they demonstrate exceptional sensing capabilities and high resolution, most of them are limited to the millimeter scale. However, to fulfill the demand for tactile sensor arrays with a spatial resolution surpassing 50 μm, Pan et al.[115] introduced a flexible pressure sensor array based on p-GaN film/n-ZnO nanowire heterojunction light-emitting diodes (LEDs) (Figure 42(a)). This innovative design achieves a remarkable spatial resolution of 2.6 μm and a rapid response time of 180 ms by

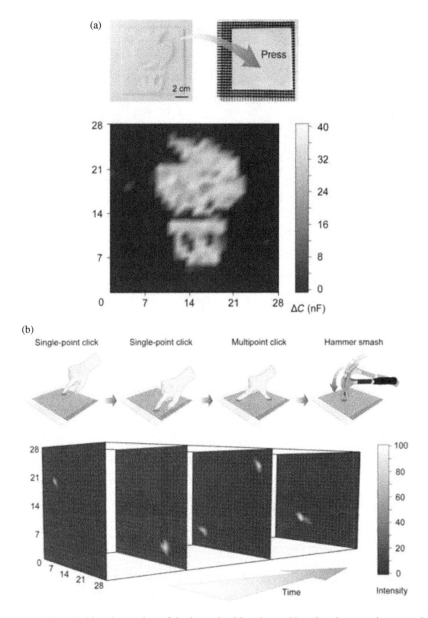

Figure 40. (a) Signal mapping of the intronic skin when a 3D-printed pattern is covered by applying 5 kPa static pressure. (b) Schematic of dynamic stimuli including single-point touch, multipoint touch, and hammer smashing; Dynamic signal mapping of the dynamic process using the signal acquisition system. Reproduced with permission from Ref. 113 (Copyright 2024, AAAS).

Sensing Systems and Applications 315

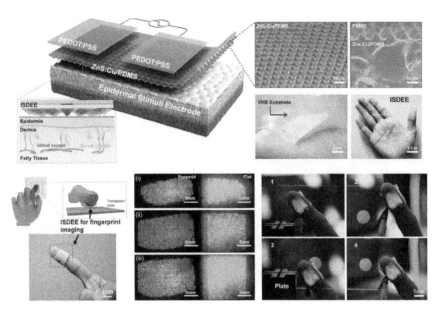

Figure 41. Schematic of device architecture of ISDEE, and imaging of high-resolution fingerprint using an ISDEE. Reproduced with permission from Ref. 114 (Copyright 2019, Wiley-VCH).

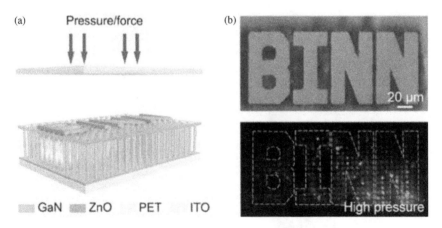

Figure 42. (a) Schematic of pressure distribution mapping performance. (b) Optical image of the convex-character sapphire seal of "BINN" used in the pressure measurement, and electroluminescence images of the device at zero, low and high pressure, respectively. Reproduced with permission from Ref. 115 (Copyright 2019, Elsevier).

simultaneously monitoring the illumination intensities of all LED pixels. Consequently, it generates a pressure distribution map (Figure 42(b)). When the sensor experiences compressive strain, the piezo-phototronic effect induces the generation of positive piezoelectric polarization charges at the local interface of the p-n junction. This process leads to alterations in the band structure and facilitates the recombination of electrons and holes, thus enhancing the luminous intensity. Consequently, due to their exceptional flexibility and ultrahigh spatial resolution, these devices hold great potential for application as large-scale tactile sensor arrays in various intelligent systems.

In addition to flexible arrays based on tactile sensing, temperature-based multifunctional flexible arrays have also been greatly developed and have broad prospects in fields such as flexible robots and intelligent prosthetics. Detecting both thermal and mechanical signals simultaneously poses a challenge due to the coupling problem, but it is indeed possible. Researchers have developed five mainstream multisensory detection modes that incorporate both mechanical and thermal sensing. The conventional approach involves placing temperature and pressure sensors on different parts of a prosthesis, as illustrated in Figures 43(a) and 43(b).

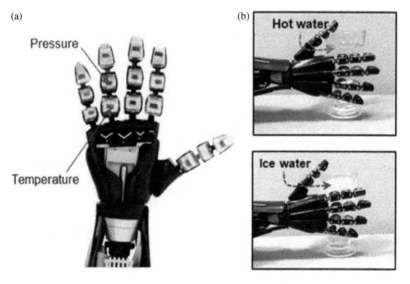

Figure 43. (a) A prosthetic hand integrated with pressure and temperature sensors on different regions. (b) The robotic hand touching and identifies hot and cold cups. Reproduced with permission from Ref. 116 (Copyright 2017, AAAS).

Sensing Systems and Applications 317

However, integrating these two sensors into a single substrate offers higher spatial resolution. Unfortunately, these approaches do not inherently enable the simultaneous measurement of two signals at one point, resulting in the inefficient utilization of valuable space, particularly in critical areas like fingertips.[116]

To address this issue, the development of novel dual-parameter sensors has emerged. These sensors are capable of transducing different stimuli into separate signals, thereby minimizing signal interference and allowing for the concurrent detection of temperature and pressure at a single point without the need for decoupling analysis. Figure 44 demonstrates one such solution, leveraging independent thermoelectric and piezoresistive effects to transduce temperature and pressure stimuli into distinct electrical signals.[117]

Another potential solution involves measuring pressure and temperature within a single parameter, facilitating more efficient data collection. Figure 45(a) showcases the first stretchable e-skin that decouples temperature and strain within this single parameter. The e-skin comprises a simple electrode-electrolyte-electrode structure, where temperature and strain are

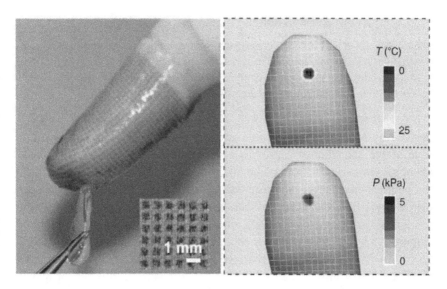

Figure 44. An e-finger assembled with flexible dual-parameter temperature–pressure sensors touching an ice cube and corresponding photograph temperature and pressure mappings. Reproduced with permission from Ref. 117 (Copyright 2015, Springer Nature).

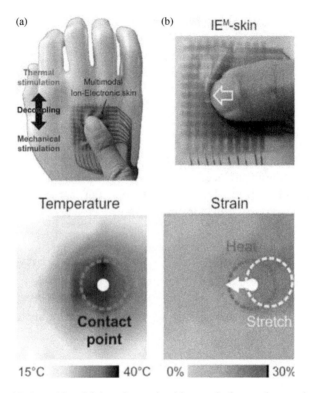

Figure 45. (a) A multimodal ion-electronic skin attached to a dummy hand and its temperature/strain sensor responses under a weak unidirectional shear. (b) Photo of finger press and corresponding schematic of temperature and strain variation. Reproduced with permission from Ref. 118 (Copyright 2020, AAAS).

derived by analyzing the ion relaxation dynamics. The charge relaxation time of the capacitance serves as a strain-insensitive feature to measure absolute temperature, while the normalized capacitance functions as a temperature-insensitive extrinsic feature to measure strain. This e-skin can provide real-time thermal information, force directions, and strain graphics during various tactile motions, such as shear, pinch, spread, and torsion (Figure 45(b)). Additionally, a promising multimodal detection method based on ML is the "cross-reactive" sensor matrix. Unlike traditional "lock and key" sensors that respond to specific stimuli, this sensor matrix exhibits the ability to respond to a wide range of stimuli. By employing ML techniques to directly analyze the coupled multimodal data, these devices can achieve a certain level of decoupling. This approach significantly simplifies the complexity of the sensor mechanism and structure. Overall, these dual-monitoring devices greatly enhance the acquisition of external information by prosthetic haptics, expanding the capabilities of tactile sensing in the realms of temperature, pressure, and strain.[118]

2.7.3. Internet of Things

The IoT refers to physical devices, sensors, software, and other technologies that connect and communicate through the Internet, enabling them to interact and share data to achieve intelligent interconnection. The IoT based on flexible sensors combines flexible sensors with the IoT technology and realizes real-time monitoring and data collection of objects by integrating flexible sensors on the surface of various objects or embedding them in objects. These flexible sensors can sense the deformation, pressure, temperature, humidity and other information of objects, and transmit the collected data to the cloud or other terminal devices through wireless communication.[9,119-124] With the support of the IoT, these data can be analyzed and processed in real-time to provide real-time monitoring, early warning, control and decision support. The IoT based on flexible sensors has a wide range of applications, including smart healthcare, sports monitoring, and smart homes. It has the advantages of flexibility, scalability and low cost, and can realize high-precision monitoring and control of objects, bringing more convenience and intelligent experience to people's lives and work.

Wearable health monitoring integrates flexible sensors, computers, communication technologies, and more to enable personalized and remote health monitoring. It encompasses various aspects such as pulse monitoring, respiratory monitoring, and muscle monitoring. In the field of medicine, the arterial pulse refers to the rhythmic sensation caused by the blood flowing out of the heart's aorta during cardiac contraction. It provides valuable information about human health, including physiological activity intensity and psychological well-being. For instance, changes in pulse amplitude and frequency can be observed before and after exercise, while anxiety disorders can lead to increased heart rate and irregular pulse, and depression can result in a slower heart rate due to frequent feelings of sadness. In recent years, researchers have made significant advancements in developing tactile sensors that can effectively monitor the human pulse. These sensors are constructed using advanced solution-synthesis methods, creating a hierarchical structure that grants them high sensitivity and a low detection limit. Consequently, the collected pulse waveform exhibits high fidelity and distinct features. Despite these advancements, the realization of remote and personalized health monitoring has been hindered by the absence of wireless transmission and real-time display modules. To address this limitation and enable real-time monitoring for doctors and family members, Chen et al.[125] developed a honeycomb-structure-inspired wearable pulse sensor is reported which not only performs ambulant cardiovascular monitoring but also realizes biometric authentication utilizing the acquired individual pulse wave profiles. In practical application, the sensor can record pulse signals continuously and accurately from individuals aged between 27 and 57 years, especially including a 29-year-old pregnant woman (Figure 46).

Similar to monitoring pulse and respiratory rates, muscle monitoring offers valuable insights into the health status and pathological characteristics of the human body from diverse perspectives. For example, it enables precise mapping of gait variations among individuals. Real-time monitoring and analysis of gait have found extensive application in various medical monitoring scenarios, such as dynamic monitoring of Parkinson's disease and stroke, as well as the early diagnosis and rehabilitation evaluation of myasthenia. Lee et al.[126] designed a wireless

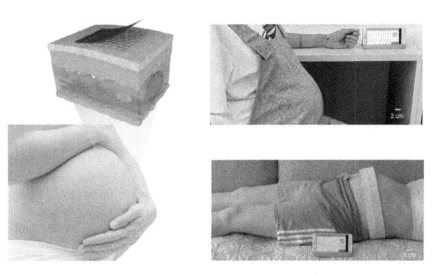

Figure 46. Continuous pulse wave monitoring in a pregnant individual. Reproduced with permission from Ref. 125 (Copyright 2024, Wiley-VCH).

smartphone controlled by non-contact mechanisms based on bacterial cellulose (BC)-BC/graphene (Gr) helical fibers and integrated it into the lab gown, allowing users to respond to or make emergency calls without physical contact during experiments. Figure 47(a) illustrates 12 groups of non-contact sensors composed of BC-BC/Gr helical fibers integrated into a wearable keyboard (I), the Bluetooth-controlled integrated module used for processing and transmitting signals (II), and the display interface of the smartphone APP (III). Additionally, due to their unique flexibility, the non-contact sensors can be sewn onto the sleeve of volunteer test clothing. The volunteer approached the number "3-6-9-9" with a finger, then approached "del" to delete the last number "9", and finally approached "dial", successfully making a call (Figure 47(b)). This design significantly reduces pollution and corrosion of buttons from chemicals or experimental materials, lowers potential harm to humans, and minimizes unnecessary contact and the spread of bacteria and viruses.

By utilizing a multitude of self-powered sensors within the IoT framework, our homes can also achieve various monitoring functions with minimal power consumption. These functions involve the integration of temperature sensors, gas sensors, motion sensors, humidity sensors, and more. In Figure 48(a), noncontact TENG sensors demonstrate contactless tracking of the motion of elderly and visually impaired individuals within their homes. This type of sensor comprises a negatively charged PDMS film and a flexible aluminum sheet that generates electron flow when a positively charged object approaches. Consequently, this noncontact triboelectric sensor can detect people's movements in a room based on different output characteristics. Moreover, it can serve as an anticollision

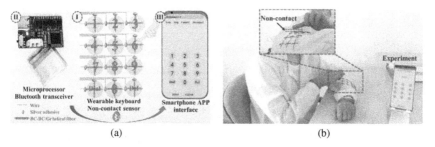

(a) (b)

Figure 47. (a) Schematic of a wireless smartphone controlled by non-contact mechanisms. (b) Volunteers use non-contact mode to control wireless smartphones to make a call. Reproduced with permission from Ref. 126 (Copyright 2024, Elsevier).

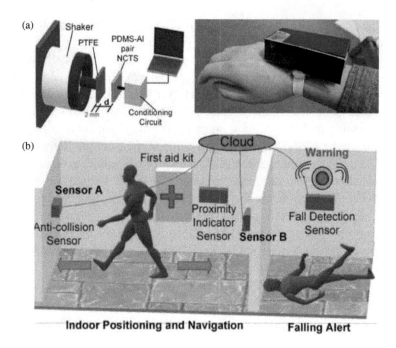

Figure 48. (a) Portable Alarm hardware for assisting blind people. (b) Applications using one or more sensors include blind people navigation assistance, indoor positioning, and falling sensor detection. Reproduced with permission from Ref. 127 (Copyright 2021, Elsevier).

system, fall detector, and indoor position platform for blind and visually impaired individuals (Figure 48(b)).[127]

When integrated with AI technology, these sensors offer enhanced convenience for multiple Artificial Intelligence of Things (AIoT) smart home applications, effectively acting as caregivers for users' healthcare monitoring. In a study by Zhang et al.[128] a wearable TENG-based sensor is presented for lower-limb and waist rehabilitation, as depicted in Figures 49(a) and 49(b). For gait analysis, a TENG-based insole equipped with two triboelectric sensors accurately detects walking patterns, achieving a high accuracy rate of 98.4% across five different individuals through the utilization of ML technology. To address waist rehabilitation, a belt embedded with four triboelectric sensors enables real-time robotic manipulation and immersive-enhanced virtual gaming by recognizing waist motion. These two functions culminate in the realization of a lower-limb

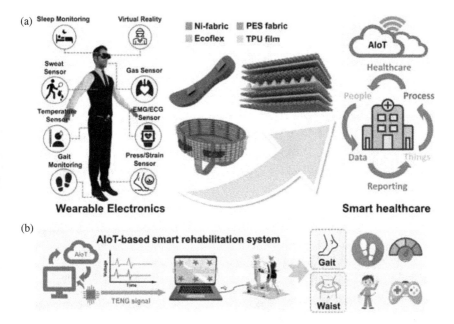

Figure 49. (a) Wearable electronics including proposed TENG sensors for smart healthcare. (b) Schematics of the AIoT-based smart rehabilitation system with gait detection and waist motion capture. Reproduced with permission from Ref. 128 (Copyright 2022, Wiley-VCH).

rehabilitation robot capable of user recognition and motion monitoring. The lower-limb sensory system exhibits excellent performance in robot and gaming-based rehabilitation applications, demonstrating considerable potential within IoT-based healthcare applications.

2.7.4. Virtual reality/augmented reality

Combining VR and AR technologies with flexible sensors can provide a more immersive interactive experience. Flexible sensors are sensors that can bend, stretch, and twist, allowing them to perceive the deformation and movement of objects. Their flexibility and adaptability make them highly suitable for application in the VR and AR fields (Figure 50). VR is a computer-generated 3D environment that allows users to immerse themselves and interact with it. Flexible sensors can be integrated into VR headsets to detect the motion and posture of the user's head. By perceiving

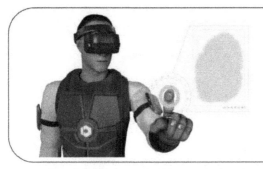

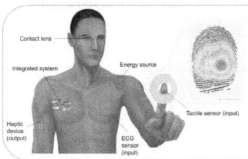

Figure 50. Schematic illustrations of conventional and skin electronic VR/AR devices and the comparison. Reproduced with permission from Ref. 136 (Copyright 2021, Wiley-VCH).

the user's head movements, the system can adjust the rendering and projection of the virtual environment in real-time, making the user feel more realistic and immersed.[129–135] For example, when the user turns their head, the system can correspondingly change the perspective and field of view of the virtual environment, enhancing the user's perception and experience of the surroundings. AR is a technology that overlays virtual information onto the real world, allowing users to simultaneously perceive and interact with both the real environment and virtual information. Flexible sensors can be integrated into AR glasses or gloves to sense the motion of the user's hands and the position of their fingers. By capturing the user's hand movements, the system can interact with virtual objects or interfaces based on the user's hand actions, enabling a more intuitive and natural way of control. Users can manipulate virtual objects, perform virtual operations, or interact with virtual information using gestures and finger movements. The advantages of flexible sensors lie in their lightweight, flexible, and wearable characteristics. They can adapt to the shape and movements of the human body, providing a more natural and comfortable interaction method. Additionally, flexible sensors can be applied to other parts of the body such as the arms, legs, and other joints, further enhancing the interactive performance of VR and AR. In summary, VR and AR technologies based on flexible sensors can provide a more immersive and intuitive interactive experience by perceiving the user's head posture and hand movements.[136] They have broad application prospects and can play an important role in fields such as gaming, education, healthcare, and industry, bringing users a new immersive and practical interactive experience.

By incorporating flexible sensors, known as VR, into the human experience, a scenario is created wherein individuals can perceive information from other dimensions of time and space. While visual and auditory senses are typically used to engage with virtual worlds, akin to seeing and hearing things on television, the tactile sense plays a paramount role in establishing profound, deep, and emotionally connective experiences between people. Undoubtedly, enhancing the VR experience can be achieved by developing tactile senses and haptic interfaces. One groundbreaking method to attain remote, contactless "touch" is through the utilization of vibration-like sensations generated by mechanical stimulation, allowing individuals to feel their surroundings through Spatio-temporal patterns of force applied to the skin. Several years ago, Novich et al.[137] designed a wearable vibrotactile vest that presented an effective approach to encoding vibrotactile data in both space and time, enabling information transfer beyond visualization. Nevertheless, these wearable vibrotactile vests are often cumbersome due to wire connections, motors, and batteries. Recently, researchers successfully demonstrated the first-ever skin-integrated wireless haptic interfaces. A skin-integrated wireless haptic interface was developed for VR/AR (as depicted in Figure 51). This interface enables users to experience touch simply by wearing a bandage-like, thin, soft, and adhesive device.[138] The development of such a complex system necessitates expertise from various disciplines such as materials science, electrical engineering, mechanical engineering, and biomedical engineering.

The sense of touch, which is perceived by the body, is replicated through the simulation of millimeter-scale vibrations by haptic actuators. Leveraging advanced mechanical designs, these haptic actuators require less than 2 mW of power to produce notable sensory vibrations. Chip-scale integrated circuits and antennas enable wireless powering and control of the actuators. By integrating the actuators and hundreds of

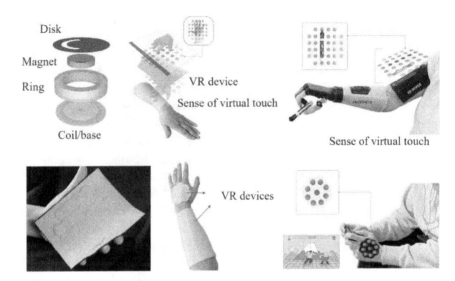

Figure 51. Skin-integrated wireless haptic interfaces for virtual and AR. Epidermal VR device with a haptic actuator array that can sense virtual touch. Amputees can distinguish the shape characteristics of objects with the aid of a prosthetic hand and this VR device. Players can experience real tactile senses in the game. Reproduced with permission from Ref. 138 (Copyright 2019, Springer Nature).

functional components into a thin, soft cloth substrate, skin-integrated wireless haptic interfaces with a thickness of only 3 mm were achieved. These interfaces are breathable, reusable, and retain functionality even during a full range of bending and twisting motions. Additionally, the haptic actuators are driven by wirelessly harvested energy through a flexible antenna, allowing users to move freely. The exploration of skin-integrated sensory interfaces significantly enriches our life experiences. In the clinical field, this interface can find applications in surgical training, virtual scene development, prosthetic control, and rehabilitation. For instance, amputees can experience touch at the fingertips of their prosthetics through sensory inputs on the upper arm, with the sensation transmitted to their brains via the upper arm. In our daily lives, this innovation holds substantial potential in communication, social media interaction, gaming, and multimedia entertainment. For example, it may become feasible to sense a hug from friends and family members during a video call. Virtual touch can be accomplished using pressures and patterns applied through a touchscreen interface. In summary, skin-integrated haptic interfaces have

Sensing Systems and Applications   327

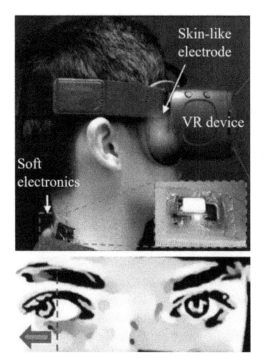

Figure 52. Wearable electronics for real-time detection of eye vergence in VR. Reproduced with permission from Ref. 139 (Copyright 2020, AAAS).

the capacity to serve as a significant component of VR, enhancing social interactions, clinical medicine, and a multitude of other applications. In Figure 52, a wireless and flexible wearable device for real-time detection of eye vergence in VR195 is depicted. This innovative technology combines skin-conformal sensors with the VR system to accurately track eye movements with high sensitivity. By integrating electronic components onto the skin, this VR case leverages skin-integrated electronics and vision to enhance the user experience.[139]

A similar concept is applied in Figure 53, where an e-skin compass system is developed based on the omnipresent geomagnetic field.[140] This compass, fabricated on a thin polymeric film, can be easily integrated onto the skin. As a person rotates from the magnetic north (N) to the magnetic south (S) via the west, the output voltage of the e-skin compass changes accordingly. This property makes the e-skin compass a valuable tool for creating interactive devices in VR and AR. For instance, it can be utilized

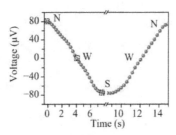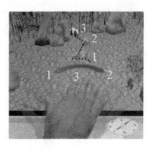

Figure 53. E-skin compass and geomagnetic interaction with VR environment. Reproduced with permission from Ref. 140 (Copyright 2018, Springer Nature).

for touchless control of virtual units in a game engine. Denys Makarov's research group explored the touchless tactile sense and control enabled by the magnetic field's touchless property.

Figure 54 showcases a bimodal soft e-skin that allows tactile and touchless interaction in real-time. The system combines a giant magnetoresistance sensor on a thin film, a round silicon-based polymer cavity, and a flexible permanent magnet with pyramid-like tips. By placing the sensor on the fingertip and interacting with a glass plate containing a permanent magnet, the user can select virtual buttons without physical contact.[141]

Moreover, the touchless manipulation of objects based on interaction with magnetic fields is realized as shown in Figure 55. By using magnetosensitive e-skins with directional perception on the hand, the user can control the light intensity of a virtual bulb by turning their hand relative to the direction of the magnetic field lines produced by a permanent magnet.[142] These technologies racking, sports, and gaming interaction.

In the future, the development of tactile senses in VR will require further advancements compared to visual and auditory senses. One potential approach is to explore thermal stimulus as a suitable substitute for electromagnetism-induced mechanical vibration, considering the skin's sensitivity to temperature. Additionally, interconverting different perceptions is an intriguing concept to explore. For example, integrating a sound sensor and haptic interface into the body of a deaf person could enable the conversion of auditory signals into tactile signals during conversations. Furthermore, the auditory sense system could translate detected sounds into visual texts, enhancing effective communication for the deaf person. This integration of multiple senses opens up new possibilities for understanding and interaction. However, developing an integrated system that

Sensing Systems and Applications 329

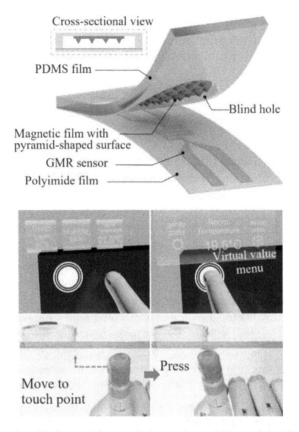

Figure 54. Soft e-skin for touchless tactile interaction and 3D touch in AR. Reproduced with permission from Ref. 141 (Copyright 2019, Springer Nature).

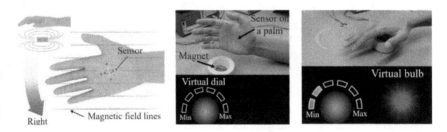

Figure 55. Magnetosensitive e-skins with directional perception for AR. A sensor is applied to the palm for light dimming via a virtual bulb. Reproduced with permission from Ref. 142 (Copyright 2018, AAAS).

intelligently selects the use of different senses in specific situations presents a challenging yet promising task. Such a system would significantly contribute to the advancement of skin-integrated electronics in the context of AR/VR.

2.7.5. *Intelligent medical monitoring*

With the continuous advancement of technology and the development of artificial intelligence, intelligent medical detection applications are receiving increasing attention. These applications utilize technologies such as artificial intelligence, big data, and sensors to bring revolutionary changes to the medical field.[143–147] They not only improve the accuracy and efficiency of medical detection but also provide more convenient and personalized medical services for doctors and patients. Intelligent medical detection applications cover a wide range of areas, including disease diagnosis, prevention and monitoring, and health management. By utilizing artificial intelligence algorithms to analyze and interpret medical data, intelligent medical detection applications can quickly and accurately diagnose various diseases, helping doctors make better treatment decisions. At the same time, they can also assist patients in self-monitoring at home, promptly detecting and addressing potential health issues, and improving their quality of life. One of the core technologies of intelligent medical detection applications is sensor technology. Sensors can collect patients' physiological data, environmental data, and other information, transmitting them to intelligent devices or cloud servers for processing and analysis. For example, wearable devices can monitor physiological indicators such as heart rate, blood pressure, and sleep quality, while smart homes can detect indoor temperature, humidity, and other environmental indicators. These data can be compared with historical data, and valuable information can be extracted using algorithms such as ML and data mining, providing support for medical diagnosis and prevention. Furthermore, intelligent medical detection applications also leverage big data analysis techniques to mine and analyze massive amounts of medical data. By comparing and statistically analyzing data from patient populations, intelligent medical detection applications can identify underlying causes and risk factors, predict the development trends of diseases, and provide a basis for personalized medical services.[119,148–152] Compared to traditional medical detection methods, intelligent medical detection applications can process large amounts of medical data more quickly and analyze and interpret the data through intelligent algorithms, thereby reducing the workload of doctors and improving the accuracy and efficiency of medical detection.

Additionally, intelligent medical detection applications have good scalability and adaptability, allowing for customized development and application based on different medical needs and data characteristics. However, intelligent medical detection applications also face some challenges and issues. One of them is the protection of data privacy and security. Medical data is sensitive personal information, and ensuring the secure storage and transmission of data to protect patients' privacy rights is an important issue. Additionally, intelligent medical detection applications need to integrate with the traditional healthcare system to ensure trust and acceptance from doctors and patients. Despite facing these challenges, the development prospects of intelligent medical detection applications remain promising. With the continuous advancement of artificial intelligence and big data technologies, we have reasons to believe that intelligent medical detection applications will bring more innovation and breakthroughs to the healthcare industry, improving people's health conditions and quality of life. In the realm of personalized healthcare, the monitoring of vital signs holds utmost importance. It not only enables accurate assessment of an individual's physiological state but also serves as a reference for diagnosing associated diseases. Flexible sensor, an active sensor, has been extensively developed to continuously and non-invasively monitor human vital signs, including pulse frequency, respiration rate, blood flow, and more, in real-time. Among its various applications, pulse measurement stands out as a common practice. Continuously monitoring arterial pulsation aids in the early detection of cardiovascular diseases and provides an assessment of overall health. One notable development in this field is the PZT film-based PENG, designed by Park et al.,[153] which allows real-time pulse wave monitoring. The PZT film, obtained through the LLO technique by exfoliating it from a rigid Sapphire substrate, is then transferred to an ultrathin PET substrate (Figure 56). When this ultrathin PENG is attached to the human wrist, it can detect pulse information and transmit it to a smartphone for diagnosis. Numerous reports have emerged concerning pulse frequency measurement using different piezoelectric materials.

Recently, Dai et al.[154] introduced a 2D layered van der Waals indium selenide device that exhibits a significant piezoelectric response and long-term mechanical durability. This device can be utilized to monitor pulse

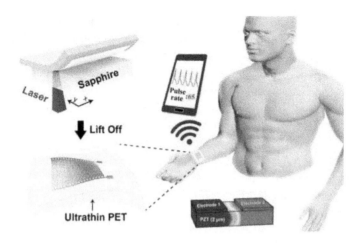

Figure 56. Schematic illustration of the fabrication process for a self-powered pressure sensor based on PZT film. Reproduced with permission from Ref. 153 (Copyright 2017, Wiley-VCH).

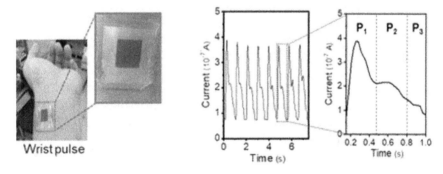

Figure 57. The pressure sensor is attached to the human wrist for monitoring wrist pulse. Reproduced with permission from Ref. 155 (Copyright 2024, American Chemical Society).

frequency and respiration rate in real-time by attaching it to the wrist and chest. Cho and his collaborators[155] developed a low-cost, facile, and scalable approach to fabricating a highly strain-tolerant and linearly sensitive soft piezoresistive pressure sensor. The pressure sensor was attached to a human wrist, it successfully monitored pulse waves and clearly distinguished the waveforms with P_1, P_2, and P_3 signals–those that can reflect the health condition of radial arteries (Figure 57).

Sensing Systems and Applications 333

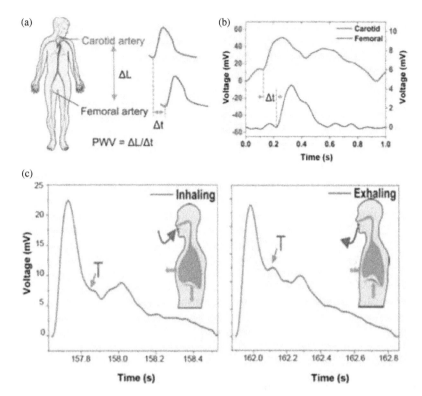

Figure 58. (a) Schematic illustration of the calculation of PWV. (b) Carotid-femoral pulse waveform used for PWV calculation. (c) Brachial artery pulse waveforms during inhaling and exhaling periods in a deep breathing process. Reproduced with permission from Ref. 156 (Copyright 2019, Wiley-VCH).

To extract more information from the pulse waveform, Chen et al.[156] developed a flexible PENG based on a single-crystalline group III-nitride thin film. By attaching these PENGs to different pulse sites, pulse waveforms can be recorded simultaneously, enabling the calculation of pulse wave velocity (PWV) (Figure 58(a)). Based on the test results, it is possible to extrapolate a time delay of approximately 98 ms between the carotid and femoral arteries based on the signal. This time delay corresponds to a PWV of 4.5 m/s and an estimated path length of 44 cm (Figure 58(b)). Additionally, this sensor, owing to the high sensitivity of the III-nitride thin film, can monitor breathing rate and identify subtle differences in pulse waveforms during inhalation and exhalation (Figure 58(c)).

Chu et al.[157] constructed a high-precision PENG with a sandwich-structured piezoelectric, capable of detecting weak vibration patterns of the human radial artery even when nearby. A sandwich structure composed of two layers of fluorinated ethylene propylene (FEP) film and a middle layer of Ecoflex film, with circular holes, along with two opposite electrodes, forms the PENG (Figure 59(a)). Through a corona charging process, the PENG exhibits an ultra-high sensitivity with an equivalent piezoelectric coefficient d33 of approximately 4100 pC N^{-1}. Consequently, a three-channel pulse sensor array can simulate the three fingers used by a Traditional Chinese Medicine doctor to record pulse waves at corresponding mapping points on the radial artery, namely the Cun, Guan, and Chi positions (Figure 59(b)). By analyzing the recorded waveforms, specific frequency, amplitude,s, providing a valuable tool for medical diagnosis (Figure 59(c)). In summary, the PENG, with its exceptional sensitivity and rapid responsiveness, offers significant advantages in monitoring vital signs associated with dynamic weak physical deformations, providing valuable data insights for personalized medical diagnosis.

Sweat, a physiological signal containing valuable physiological information, has emerged as a promising target for noninvasive and continuous monitoring of the human body. Utilizing the mechanical energy generated during motion, researchers have introduced a novel approach for integrated self-powered electronic skin (e-skin) by leveraging sweat as a sensing platform. In a study by Song et al. (Figure 60), a flexible circuit board processing technology was employed to seamlessly integrate freestyle TENGs and flexible circuit modules. This platform effectively harnesses mechanical energy from human motion, converting it into electrical energy to enable the stable operation of electrochemical sensing units. It facilitates wireless transmission of dynamic biomarker indicators (e.g., pH, sodium ions) found in sweat.[158]

However, the use of a freestyle TENG for energy collection results in a bulkier, less flexible, and less comfortable platform. To overcome this limitation, the researchers also tapped into the energy present in sweat itself, employing biofuel cells as a power supply device. They integrated biological and physical sensors onto ultra-thin and transparent polyimide (PI) substrates, thereby constructing a new generation of self-powered e-skin. Key metabolic biomarkers like glucose, urea, NH$_4$, and pH were measured, and the personalized information was wirelessly transmitted to a user interface via Bluetooth Low Energy. In Figure 61(a), the biofuel cell was enhanced through the integration of 0D to 3D nanomaterials and

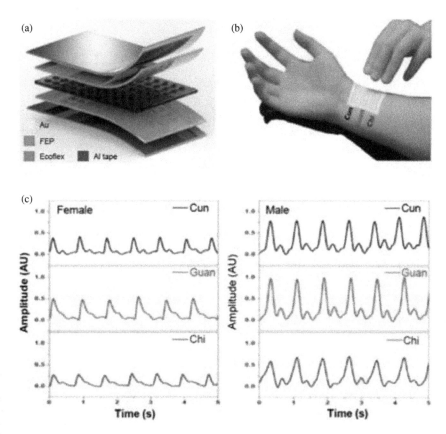

Figure 59. (a) The schematic diagram of the pulse sensing device using the FEP/Ecoflex/FEP sandwich-structured piezoelectric film. (b) Mimicking the Traditional Chinese Medicine pulse palpation acquisition scheme of a real doctor by using three fingers with the pulse sensing system at the Cun, Guan, and Chi positions. (c) Measured typical pulse waveforms at the Cun, Guan, and Chi positions for a female and male volunteer, respectively. Reproduced with permission from Ref. 157 (Copyright 2018, Wiley-VCH).

modifications involving lactic acid oxidase and platinum cobalt alloy nanoparticles. This allowed the biofuel cell to directly extract a power density of 3.5 mW cm^{-2} from sweat, providing continuous energy to the e-skin for up to 60 h.[159] Benefiting from the abundant energy supply, the researchers further optimized the energy utilization efficiency. They achieved this by implementing control mechanisms within the Bluetooth module, enabling wireless data transmission through sleep/activation

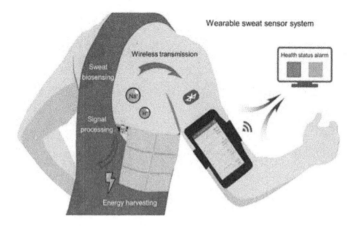

Figure 60. Diagram of the sensor system for real-time health monitoring. Reproduced with permission from Ref. 158 (Copyright 2020, AAAS).

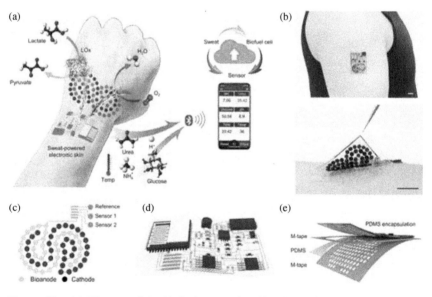

Figure 61. (a) Diagram of the biofuel-powered e-skin. (b) Optical photographs of the e-skin on a healthy adult's arm. Schematic plots of (c) the flexible sensor and (d) the soft e-skin interface. (e) System-level encapsulation for biofluid sampling. Reproduced with permission from Ref. 159 (Copyright 2020, AAAS).

mode switching. This effectively reduced the energy demand of the e-skin and extended its continuous operational duration. As depicted in Figure 61(b), the device is exceptionally thin, ensuring a barely perceptible presence when worn. It consists of two primary components: (i) a flexible electrochemical patch housing biosensors (Figure 61(c)) for energy harvesting and sweat analysis, and (ii) ultra-thin PI plates containing rigid electronics (Figure 61(d)) responsible for power management, signal processing, and wireless communication. To prevent sweat from reaching the electronic components, the entire setup is encapsulated in PDMS (Figure 61(e)). By exploiting the high concentration of lactic acid present in sweat, the e-skin relies on a biofuel cell to generate sufficient and sustained power for its operation.

In addition to the ex vivo pressure signals generated by physiological activities, various internal pressures can be detected by implanted flexible pressure sensors. These include ureteral peristalsis pulse, gastric peristalsis, intracranial pressure (ICP), and intraocular pressure (IOP). For instance, Zhao and colleagues[160] designed a piezoresistive pressure sensor using PVA as a crosslinker to connect MXene sheets into a layered network structure through strong hydrogen bonding. This sensor demonstrated stable performance lasting over six months in harsh environments, including aqueous, strongly acidic, and alkaline conditions. Furthermore, both *in vitro* and *in vivo* tests confirmed the pressure sensor's excellent biocompatibility, which is crucial for biomedical applications within living organisms. To evaluate its functionality, the sensor was implanted into a BALB/c mouse for in vivo testing. Initially, the sensor was attached to the epicardial tip of the heart to capture a robust electrocardiograph (ECG) signal with high sensitivity and SD (Figure 62).

Subsequently, the sensor was positioned on the serous membrane of the outermost gastric wall to detect gastric peristalsis. In surgical procedures such as gynecologic, colorectal, and pelvic surgeries, identifying the ureter visually can be challenging due to its inconspicuous anatomical location. To overcome this issue, Wang and colleagues[161] proposed tubular porous pressure sensors with excellent performance to identify the ureter intraoperatively. These sensors were fabricated using uniform silicon dioxide microspheres as sacrificial templates that could be dissolved with hydrofluoric acid, leaving behind uniform pores. The uniform porous structure contributed to high sensitivity (448.2 kPa^{-1}), reproducibility, and low sensor-to-sensor variation (3.29%). By incorporating these sensors

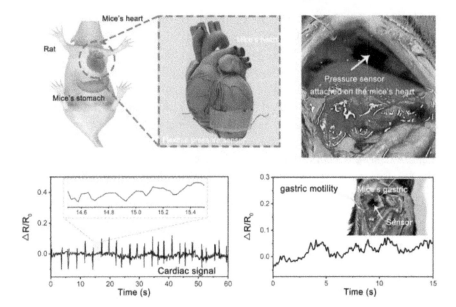

Figure 62. Application of flexible pressure sensor in study heart and gastric dynamics of mice. Reproduced with permission from Ref. 160 (Copyright 2021, Elsevier).

with forceps, it was possible to monitor ureteral peristalsis pulses (6 times/min) and carotid artery pulses (60 times/min) in a female Bama minipig in situ intraoperatively (Figure 63). Consequently, the ureter could be identified in real-time by monitoring the frequency of pressure pulses.

Accurate measurement of in vivo pressures plays a crucial role in diagnosing life-threatening medical conditions. ICP, for instance, is a key parameter for monitoring patients with traumatic brain injuries, as an increase in ICP can impede blood flow and lead to ischemia. Lu and colleagues[162] recently introduced a bioresorbable, wireless pressure sensor based on passive inductor-capacitor resonance circuits with optimal sensitivity (200 kHz/mmHg) and resolution (1 mmHg). In a rat model, the ICP measurement was conducted by attaching the sensor over a burr hole drilled through the skull, connecting the cranial cavity to the sensor (Figure 64). This sensor effectively captured changes in ICP after applying pressure to the rat's flank, and the in vivo measurement remained stable for up to 96 h (4 days).

Additionally, the sensor gradually dissolved completely in biofluids due to its bioresorbable constituent materials, eliminating the need for

Sensing Systems and Applications 339

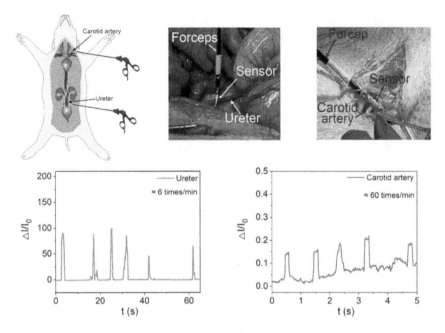

Figure 63. Application of flexible piezoresistive pressure sensor in *in situ* monitoring the carotid artery and ureter of a female Bama minipig. Reproduced with permission from Ref. 161 (Copyright 2021, American Chemical Society).

extraction surgeries. Another important indicator in vivo health monitoring is IOP, which, when elevated due to glaucoma, can lead to blindness. Therefore, precise and continuous monitoring of IOP is crucial for early diagnosis and treatment of this condition. Kim and colleagues[163] developed a transparent contact lens sensor by incorporating two inductive spirals made of graphene-Ag NW hybrid electrodes on both sides of a silicone elastomer film. These sensors enabled simultaneous monitoring of glucose levels in tears and IOP by analyzing changes in electrical signals. Elevated IOP increases the corneal radius of curvature, which, in turn, enhances both the inductance through the biaxial lateral expansion of the spiral coils and the capacitance through the thinning of the dielectric. Consequently, the elevated IOP reduces the resonance frequency of the sensor. The sensor was tested in vitro by transferring it onto a contact lens worn by a bovine eyeball (Figure 65). It exhibited a nearly linear frequency response for relatively small pressures and demonstrated good reproducibility with negligible hysteresis.

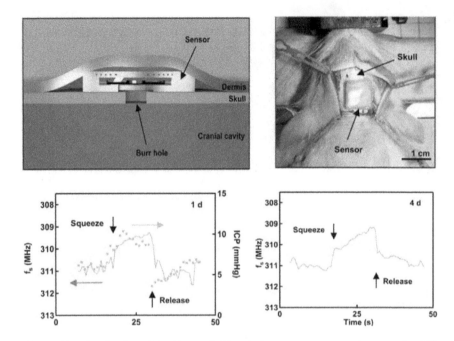

Figure 64. Application of the bioresorbable wireless pressure sensor in measuring ICP of rats. Reproduced with permission from Ref. 162 (Copyright 2021, Elsevier).

In addition, self-powered sensors can also be used for clinical diagnosis and treatment of diseases. Dagdeviren and colleagues[164] introduced a thin and stretchable network consisting of PZT nanoribbons to measure the viscoelasticity in the near-surface regions of the epidermis. In their study, depicted in Figure 66(a), a serpentine metal electrode connected to the PZT nanoribbon was supported by a thin elastomer. This device could be easily attached to the skin or other biological tissues using passive van der Waals forces. By placing the device on a forearm lesion, as shown in Figure 66(b), they were able to obtain modulus mapping (Figure 66(c)), providing a non-invasive and rapid method for dermatologic investigation.

Yu and colleagues[165] further improved this ultra-thin piezoelectric microsystem by engineering it into a needle-shaped form (Figure 67(a)), allowing for real-time quantitative monitoring of tissue modulus when injected or fixed onto a conventional biopsy needle (Figure 67(b)). As illustrated in Figure 67(c), this miniaturized modulus-sensing device could be percutaneously inserted into various biological tissues, including

Sensing Systems and Applications   341

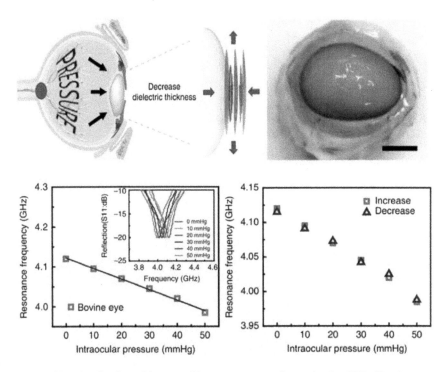

Figure 65. Application of the wearable pressure sensor in monitoring IOP of bovine eye. Reproduced with permission from Ref. 163 (Copyright 2017, Springer Nature).

the spleen, liver, kidneys, lungs, and cutaneous fat, providing information on their corresponding modulus and demonstrating the device's ability to utilize the piezoelectric effect.

Taking inspiration from the kirigami concept, Sun and colleagues[166] introduced a stretchable strain monitoring system based on a piezoelectric PVDF film for self-powered cardiac monitoring. When this device was placed on a pig's heart (Figure 68(a)), the sensor could monitor heart activity by producing varying voltage outputs, as depicted in the lower part of Figure 68(b). In another study, Dagdeviren and colleagues[167] developed a flexible piezoelectric device using a PZT ribbon to detect the mechanical deformation of the stomach. The designed sensor, shown in Figure 68(c), was attached conformally to the stomach and responded to its peristaltic action by generating varying voltage outputs, which were linked to digestive activity. These approaches demonstrated the

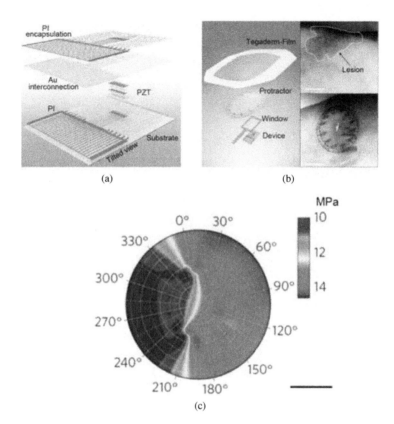

Figure 66. (a) Exploded-view schematic illustration of the modulus sensor based on nanoribbons of PZT. (b) Schematic illustration of a system and photograph of a forearm without and with a mounted device. (c) Representative data from experiments. Scale bar, 2 cm. Reproduced with permission from Ref. 164 (Copyright 2015, Springer Nature).

significant potential of piezoelectric devices in diagnosing and treating gastrointestinal motility disorders.

However, there is an urgent need for wearable and implantable electronic products with the required functionality, operational safety, and long-term stability for diverse applications in the healthcare field. In this situation, electricity becomes a key issue for sustainable electronic systems. Conventional batteries are inadequate for meeting the growing demands of energy storage in wearable or implantable devices. As a result, various strategies have been proposed to overcome the limitations of bulky batteries, and one such approach involves harnessing energy

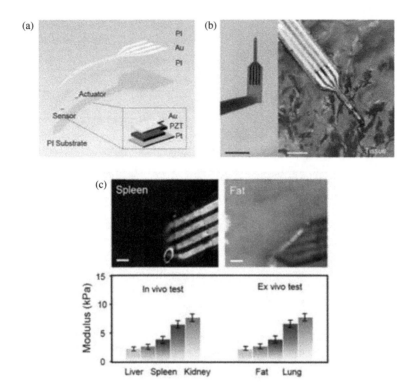

Figure 67. (a) Schematic illustrations of a device based on ultrathin PZT actuators and sensors. (b) Optical images of said device (left, Scale bar, 5 mm) and the device placed on a biological tissue (right, Scale bar, 2 mm). (c) Optical images and results of *in vivo* modulus measurements on a live rat and *ex vivo* results of the same organs after explanation. Scale bars, 1 mm. Reproduced with permission from Ref. 165 (Copyright 2018, Springer Nature).

from the sun. Solar energy, being the cleanest and most abundant renewable energy source, has found widespread adoption in self-powered wearable systems. A self-powered sensor system has been developed, comprising a solar cell on a plastic substrate. This setup converts incoming light into electricity, serving as the power supply for the entire sensing system (refer to Figure 69).[168] To ensure continuous operation of the sensing system without being reliant on surrounding conditions such as insufficient light intensity, planar MnO_2-based supercapacitors are integrated into the system as intermediate energy storage units. Combining these supercapacitors with a SnO_2 gas sensor has led to the development of a

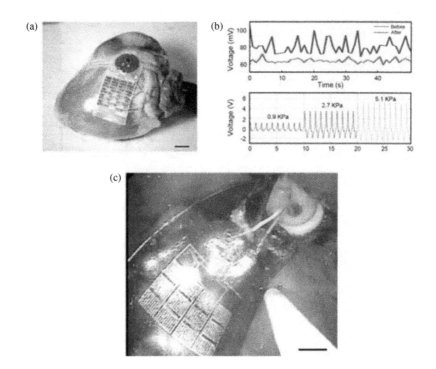

Figure 68. (a) Potential application of implantable sensors on tissue (pig heart) surface. Scale bar, 1 cm. (b) The voltage output of the sensor on the pig heart under different pressures with the pulse and heartbeat-like inputs on the air-driven platform (bottom). Voltage output graphs of before and after milk ingestion from in vivo evaluation in a Yorkshire swine model (top). (c) Photograph of the PZT-based sensor with percutaneous endoscopic gastrostomy tube inside the stomach. Scale bar, 1 cm. Reproduced with permission from Ref. 167 (Copyright 2017, Springer Nature).

self-powered ethanol/acetone sensor with remarkable sensitivity. This advancement showcases the system's immense potential for a wide range of biomedical monitoring applications.

Moreover, mechanical energy is another readily available power source present in our surroundings, making it promising for self-sustainable wearable and implantable electronics. A high-output magneto-mechano-TENG has recently been reported, as capable of generating electricity from alternating magnetic fields. This nanogenerator effectively powers an indoor wireless positioning system (depicted in Figure 70).[169] The generated electrical energy is connected to a storage unit

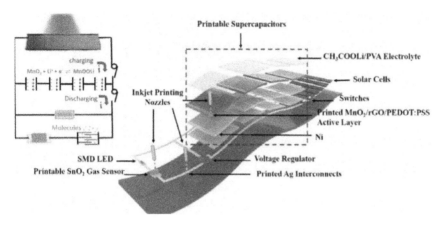

Figure 69. Self-powered gas monitoring system with embedded solar cells as the energy source. Reproduced with permission from Ref. 168 (Copyright 2019, Wiley-VCH).

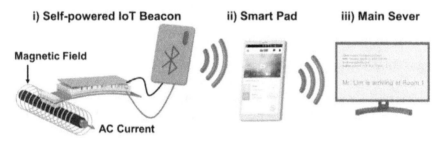

Figure 70. Self-powered indoor IoT positioning system integrated with energy harvesting and storage units. Reproduced with permission from Ref. 169 (Copyright 2019, Royal Society of Chemistry).

through a power management circuit, facilitating the uninterrupted operation of an IoT Bluetooth beacon for wireless signal transmission.

Furthermore, converting human body heat into usable energy offers a reliable approach for self-powered wearable systems. In Figure 71, a flexible thermoelectric generator (TEG) has been developed, featuring a polymer-based heat sink assembled on its top surface to enhance the output power density from 8 to 38 $\mu W\ cm^{-2}$.[170] Additionally, an ECG sensing circuit is fabricated on a flexible PCB substrate and powered by the wearable TEG, utilizing body heat as the power source. Although body heat possesses a lower energy density compared to solar or mechanical energy,

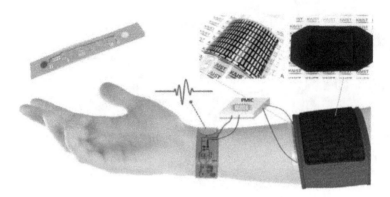

Figure 71. Self-powered wearable ECG system powered by a wearable TEG. Reproduced with permission from Ref. 170 (Copyright 2018, American Chemical Society).

it remains consistently available with minimal interference from the ambient environment.

For implantable electronic devices, the power sources may face limitations in terms of their applications and usage scenarios. However, by separating the power source from the implanted functional elements, a self-powered implantable rehabilitation system becomes feasible by harnessing the substantial output of wearable energy harvesters. A demonstration of such a self-powered system involves a stack-layered TENG and a multiple-channel epimysial electrode for direct muscle stimulation (see Figure 72).[171] The generated current output is directly connected to the muscle tissue through the inserted multi-channel electrodes. The results indicate that this approach yields a more stable force output compared to conventional square wave stimulation or enveloped high-frequency stimulation. Moreover, the unique current waveforms of the TENG effectively reduce force fluctuations caused by synchronous motoneuron recruitment at the simulation electrodes. These findings suggest a promising future for utilizing TENG in direct muscle stimulation, enabling self-powered rehabilitation and treatment of muscle function loss.

To eliminate the need for wired connections from external sources to the body's interior, the inclusion or delivery of electrical power becomes a significant challenge for implantable medical devices. One potential solution involves harvesting biomechanical energy from cardiac motion, respiratory movement, and blood flow. However, this approach has limitations in terms of low output power density and limited implantation sites.

Sensing Systems and Applications 347

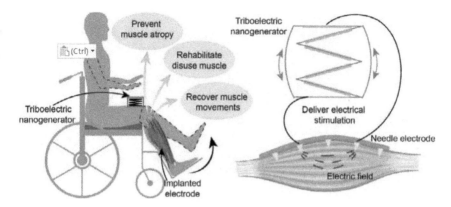

Figure 72. Self-powered muscle stimulation system with a stacked-layer TENG as the power source. Reproduced with permission from Ref. 171 (Copyright 2019, Wiley-VCH).

Alternatively, the most promising technology involves delivering mechanical energy from outside the body to the implanted devices. Recently, a capacitive triboelectric technology that utilizes ultrasound for energy delivery has been reported (see Figure 73).[172] The implantable TENG is designed to be inserted beneath the skin and consists of a perfluoroalkoxy (PFA) membrane capable of vibrating under ultrasound pressure. The TENG device integrates a rectifier, transformer, voltage regulator, and battery, forming an energy conversion and storage unit. This prototype can generate milliwatts of output power to charge capacitors and Li-ion batteries, showcasing its potential to provide continuous energy to implanted medical devices.

Another strategy to eliminate the need for rechargeable batteries involves using radio-frequency coupling to wirelessly operate the implanted sensing elements. For instance, a biodegradable and flexible arterial-pulse sensor has been proposed for wireless blood flow monitoring (see Figure 74).[173] The device comprises a capacitive sensor and a bilayer coil for radio-frequency data transmission. Variations in vessel diameter, as blood flows through, result in changes in capacitance, causing a shift in the resonant frequency of the inductor-capacitor-resistor circuit. This shift can be wirelessly monitored through inductive coupling with the external reading coil, eliminating the need for batteries inside the body. Additionally, the sensor's biodegradable nature circumvents the issue of secondary procedures for implant removal. Soon, this battery-free

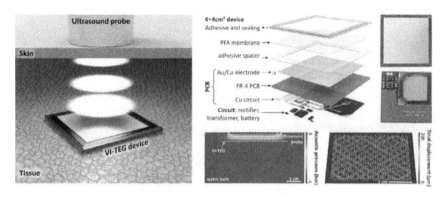

Figure 73. Implantable TENG for ultrasound energy harvesting through skin and liquids. Reproduced with permission from Ref. 172 (Copyright 2019, AAAS).

sensing technique holds promise for providing biomedical functionalities to implantable electronics, greatly benefiting personalized healthcare monitoring and rehabilitation.

2.7.6. Optoelectronic displays

The optoelectronic display, an essential component of electronic systems, plays a crucial role in projecting information through texts, images, and videos for intuitive visualization, aiding human cognition.[85,174–179] Currently valued at over $100 billion, the display market is expected to exceed $200 billion by 2025. Traditional displays predominantly employ solid-state LED and liquid crystal display (LCD) technologies, finding widespread use in consumer electronics such as televisions, laptops, mobile phones, and tablets. Over the past decade, wearable electronics have seen significant advancements, driven by the demand for personalized healthcare monitoring systems and immersive wearable VR/AR entertainment experiences.[180–185] Consequently, display technology has also made strides to accommodate the conformal nature of human skin, introducing mechanical flexibility and stretchability. In this regard, wearable displays can achieve such characteristics through three methods, akin to those employed in wearable electronics. These methods involve reducing display film thickness, substituting rigid electronic interconnects with stretchable counterparts while keeping tiny active devices rigid, or employing stretchable

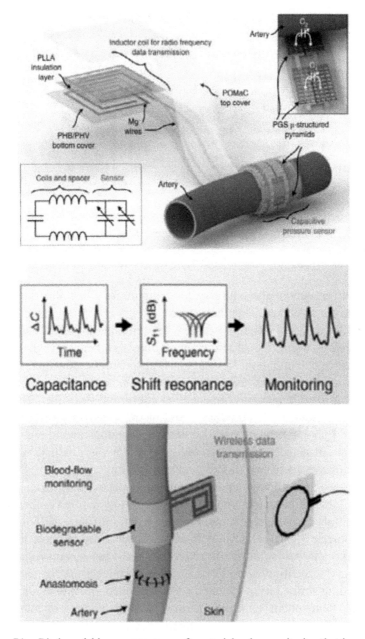

Figure 74. Biodegradable pressure sensor for arterial pulse monitoring that is operated wirelessly. Reproduced with permission from Ref. 173 (Copyright 2019, Springer Nature).

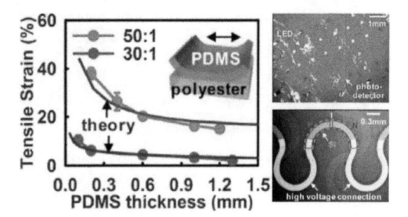

Figure 75. Flexible LED enabled by reducing the film thickness and using meander structure. Reproduced with permission from Ref. 186 (Copyright 2011, AAAS).

materials throughout the entire system. Figure 75 illustrates the critical tensile strain at which delamination occurs in a structure composed of a PDMS film on a polyester substrate.[186] By reducing the PDMS thickness to less than 300 μm, the structure can endure significantly higher tensile strain before delamination. Leveraging film thickness reduction, an array of LEDs, photodetectors, and silicon solar cell power sources can be successfully affixed to human skin, enabling conformal flexibility. The wearable system showcased in the same work incorporates not only solar cells, LEDs, and photodetectors, but also electrophysiological and temperature sensors, strain sensors, transistors, radio frequency (RF) inductors, capacitors, oscillators, rectifying diodes, and wireless coils.

The second method entails replacing rigid electrical connections with flexible ones while maintaining the rigidity of tiny active components to preserve their high performance. Figure 76 demonstrates the bonding of an elastomeric microlens array with a stretchable array of photodiodes.[186] Within the photodiode array, the active photodiodes remain rigid, while filamentary serpentine wires serve as the interconnect matrix, capable of withstanding strain during stretching. Once bonded, the integrated camera takes on a hemisphere shape, mimicking the compound eye of anthropods. Each microlens aligns with a specifically designed photodiode, allowing the combination of elastomeric compound optical elements with deformable arrays of thin, rigid silicon photodetectors to form integrated sheets with flexibility and stretchability. Through this method, micro-LED and

Sensing Systems and Applications 351

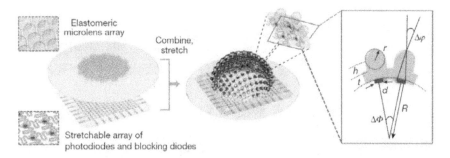

Figure 76. Stretchable photodetector array enabled by connecting the rigid active devices using stretchable interconnects. Reproduced with permission from Ref. 186 (Copyright 2013, Springer Nature).

photodetector arrays achieve excellent stretchability, enabling integration with diverse substrates. Furthermore, with suitable encapsulation, waterproofing capability has been demonstrated.

The third method involves employing stretchable materials throughout the entire system. As early as 2009, a rubber-like active-matrix organic LED (OLED) display with 16 × 16 pixels was developed. It integrated OLED for lighting, organic transistors as drivers, and printed elastic conductors as interconnects. Due to its composition solely comprising stretchable components, this system exhibits remarkable mechanical robustness, remaining functional even when folded or crumpled. In addition to OLED, there is another type of active display known as polymeric light-emitting devices (PLEDs). Unlike OLED, which relies on small molecules, PLED utilizes polymers and offers the advantages of flexibility and large-area display. In 2013, Liang et al. presented a stretchable elastomeric PLED that demonstrated several desirable properties. This PLED is semi-transparent and exhibits excellent surface electrical conductivity and smoothness. It also boasts a straightforward all-solution-based fabrication process. Moreover, the PLED is foldable, capable of withstanding a maximum linear strain of 120%, and can endure 1000 cycles of stretching at a 30% strain repeatedly. Despite the impressive performance and broad applicability of LED-based display systems, their ultimate strain capabilities are limited to less than 120%. This poses a challenge for their application in soft robotics and on human joints. On the other hand, elastomers can achieve ultimate strains as high as 400% to 700%. Therefore, elastomer-based electroluminescent (EL) materials are employed in conjunction with OLED and PLED

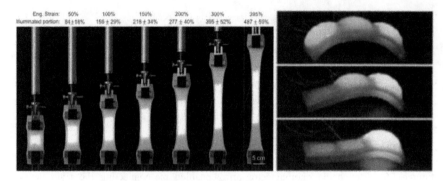

Figure 77. Stretchable transparent elastomeric polymer LED. Reproduced with permission from Ref. 187 (Copyright 2016, AAAS).

when high strain resistance is required. Larson et al. reported the development of a highly stretchable EL skin for optical signaling and tactile sensing. The active EL layer in this skin consists of a ZnS phosphor-doped dielectric elastomer layer called Ecoflex.[187] The two electrodes utilized are PAM-LiCl hydrogels, and the entire device is encapsulated by two Ecoflex layers. Even when subjected to a high strain of close to 400%, the device continues to exhibit high luminescence (Figure 77). By selectively doping the EL phosphor layer, the emission of light at different wavelengths can be achieved. A panel comprising three pixels with different emission colors, controllable independently, has been successfully demonstrated. In applications such as skin treatment, light detection ranging, and high-intensity displays that demand high temporal and spatial coherence of light, lasers are an indispensable light source.

In 2024, Pan et al.[188] developed a contactless user-interactive sensing display (CUISD) based on dynamic alternating current electroluminescence (ACEL) that responds to humidity. Subsecond humidity-induced luminescence is achieved by integrating a highly responsive hydrogel into the ACEL layer. The patterned silver nanofiber electrode and luminescence layer, produced through electrospinning and microfabrication, result in a stretchable, large-scale, high-resolution, multicolor, and dynamic CUISD (Figure 78).

In addition to developing devices at the individual level, wearable displays are also incorporated into wearable electronic systems to provide visual functions on a system-wide scale. Traditional E-skin, particularly those focused on pressure sensors, typically utilize electronic readout

Sensing Systems and Applications 353

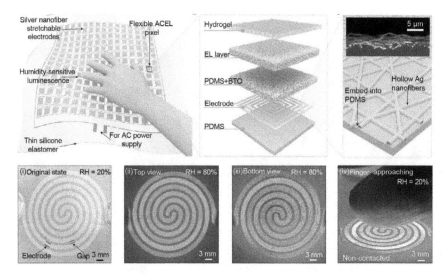

Figure 78. Design and schematic of the CUISD. Reproduced with permission from Ref. 188 (Copyright 2018, Springer Nature).

schemes that lack intuitiveness. C. Wang et al.[189] introduced an interactive E-skin that not only relays applied pressure information but also delivers immediate visual feedback through an integrated wearable display system. The complete system consists of a 16×16 pixel configuration with a total size of 3×3.5 cm². Each pixel comprises a thin-film transistor (TFT) based on a CNT network, an OLED, and a pressure-sensitive rubber (PSR). The PSR establishes electrical contact with the OLED cathode, enabling the emission intensity of the underlying OLED to vary based on changes in the PSR's conductivity due to applied pressure. The emission color of each pixel is determined by employing different emissive layer materials. To visualize spatial pressure information intuitively without the need for complex DAQ circuits and electronic boards, PDMS slabs in the shapes of the letters C, A, and L are utilized to apply specific pressure to the system. The OLED responds optically, providing a direct optical response that maps the spatial pressure information for human interpretation. To minimize the discomfort experienced by the skin, an ultrathin, ultra-flexible, and high-performance integrated PLED/organic photodetector (OPD) system has been developed to serve both display and sensing functions.[190] As depicted in Figure 79, the optoelectronic thin film system, including the substrate and encapsulation layer, measures only 3 μm in

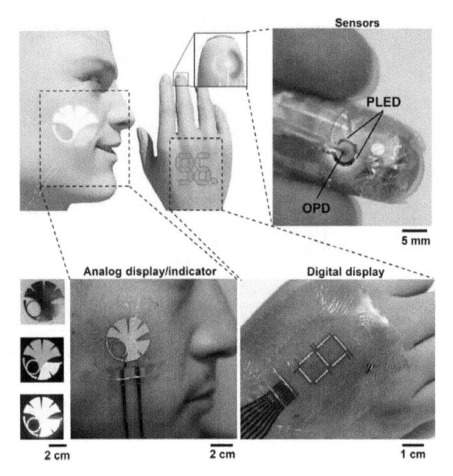

Figure 79. Ultraflexible organic photonic skin with sensor and analog/digital display. Reproduced with permission from Ref. 190 (Copyright 2016, AAAS).

thickness — roughly one order of magnitude thinner than the epidermal layer. In this integrated system, light emitted from the PLED permeates the skin and interacts with the blood, subsequently being reflected back and detected by the OPD. Consequently, the system can inconspicuously measure blood oxygen concentration when worn on a finger. Additionally, another PLED assists in directly visualizing the sensing data on the body.

The same research team went on to advance a smart display based on quantum dots (QDs) that could visualize various information, including

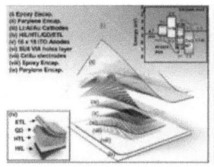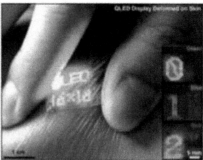
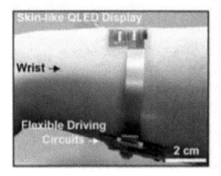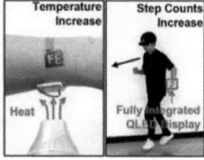

Figure 80. A QD-based smart display. Reproduced with permission from Ref. 190 (Copyright 2017, Wiley-VCH).

temperature and acceleration data obtained from wearable sensors.[190] The LED used in this display consisted of an indium tin oxide ITO anode, a Li-Al/Au cathode, and a layered charge transport structure comprising a hole injection layer, a hole transport layer, and the QDs. With a thickness of approximately 5.5 μm, the resulting LED display could conform to human skin. To boost the brightness and minimize nonradiative recombination, QDs with relatively thick shells were employed, achieving an impressive brightness of up to 44719 cd m^{-2} at an operational voltage of 9 V. The researchers integrated touch sensors, temperature sensors, and an accelerometer into a flexible wrist-worn device (Figure 80) to create a comprehensive smart LED system. They successfully demonstrated real-time visualization of temperature and step counts while the user was at rest or engaged in physical activity, such as running.

References

1. Rus, D. and Tolley, M.T. (2015). Design, fabrication and control of soft robots, *Nature*, *521*, 467–475.
2. He, Y., Cheng, Y., Yang, C. and Guo, C.F. (2024). Creep-free polyelectrolyte elastomer for drift-free iontronic sensing, *Nature Materials*, *23*, 1–8.
3. Rumley, E.H., Preninger, D., Shomron, A.S., Rothemund, P., Hartmann, F., Baumgartner, M., Kellaris, N., Stojanovic, A., Yoder, Z., Karrer, B., *et al.* (2023). Biodegradable electrohydraulic actuators for sustainable soft robots, *Science Advances*, *9*, eadf5551.
4. Luo, Y., Liu, C., Lee, Y.J., DelPreto, J., Wu, K., Foshey, M., Rus, D., Palacios, T., Li, Y. and Torralba, A. (2024). Adaptive tactile interaction transfer via digitally embroidered smart gloves, *Nature Communications*, *15*, 868.
5. Kim, K.K., Kim, M., Pyun, K., Kim, J., Min, J., Koh, S., Root, S.E., Kim, J., Nguyen, B.-N.T. and Nishio, Y. (2023). A substrate-less nanomesh receptor with meta-learning for rapid hand task recognition, *Nature Electronics*, *6*, 64–75.
6. Lee, J.P., Jang, H., Jang, Y., Song, H., Lee, S., Lee, P.S. and Kim, J. (2024). Encoding of multi-modal emotional information via personalized skin-integrated wireless facial interface, *Nature Communications*, *15*, 530.
7. Wang, F., Hu, F., Dai, M., Zhu, S., Sun, F., Duan, R., Wang, C., Han, J., Deng, W. and Chen, W. (2023). A two-dimensional mid-infrared optoelectronic retina enabling simultaneous perception and encoding, *Nature Communications*, *14*, 1938.
8. Zhao, W., Ni, H., Ding, C., Liu, L., Fu, Q., Lin, F., Tian, F., Yang, P., Liu, S. and He, W. (2023). 2D Titanium carbide printed flexible ultrawideband monopole antenna for wireless communications, *Nature Communications*, *14*, 278.
9. Yang, Y., Guo, X., Zhu, M., Sun, Z., Zhang, Z., He, T. and Lee, C. (2023). Triboelectric nanogenerator enabled wearable sensors and electronics for sustainable internet of things integrated green earth, *Advanced Energy Materials*, *13*, 2203040.
10. Silver, D., Schrittwieser, J., Simonyan, K., Antonoglou, I., Huang, A., Guez, A., Hubert, T., Baker, L., Lai, M. and Bolton, A. (2017). Mastering the game of go without human knowledge, *Nature*, *550*, 354–359.
11. Esteva, A., Kuprel, B., Novoa, R.A., Ko, J., Swetter, S.M., Blau, H.M. and Thrun, S. (2017). Dermatologist-level classification of skin cancer with deep neural networks, *Nature*, *542*, 115–118.
12. Zhu, M., He, T. and Lee, C. (2020). Technologies toward next generation human machine interfaces: From machine learning enhanced tactile sensing to neuromorphic sensory systems, *Advanced Physics Research*, *7*, 031305.

13. Moin, A., Zhou, A., Rahimi, A., Menon, A., Benatti, S., Alexandrov, G., Tamakloe, S., Ting, J., Yamamoto, N. and Khan, Y. (2021). A wearable biosensing system with in-sensor adaptive machine learning for hand gesture recognition, *Nature Electronics*, *4*, 54–63.
14. Tan, P., Han, X., Zou, Y., Qu, X., Xue, J., Li, T., Wang, Y., Luo, R., Cui, X. and Xi, Y. (2022). Self-powered gesture recognition wristband enabled by machine learning for full keyboard and multicommand input, *Advanced Materials*, *34*, 2200793.
15. Araromi, O.A., Graule, M.A., Dorsey, K.L., Castellanos, S., Foster, J.R., Hsu, W.-H., Passy, A.E., Vlassak, J.J., Weaver, J.C. and Walsh, C.J. (2020). Ultra-sensitive and resilient compliant strain gauges for soft machines, *Nature*, *587*, 219–224.
16. Liu, Y., Yiu, C., Song, Z., Huang, Y., Yao, K., Wong, T., Zhou, J., Zhao, L., Huang, X. and Nejad, S.K. (2022). Electronic skin as wireless human-machine interfaces for robotic VR, *Science Advances*, *8*, eabl6700.
17. Tchantchane, R., Zhou, H., Zhang, S. and Alici, G. (2023). A review of hand gesture recognition systems based on noninvasive wearable sensors, *Advanced Intelligent Systems*, *5*, 2300207.
18. Gu, W., Yan, S., Xiong, J., Li, Y., Zhang, Q., Li, K., Hou, C. and Wang, H. (2023). Wireless smart gloves with ultra-stable and all-recyclable liquid metal-based sensing fibers for hand gesture recognition, *Chemical Engineering Journal*, *460*, 141777.
19. Wang, T., Zhao, Y. and Wang, Q. (2023). A flexible iontronic capacitive sensing array for hand gesture recognition using deep convolutional neural networks, *Scientific Reports*, *10*, 443–453.
20. Liu, J., Wang, L., Xu, R., Zhang, X., Zhao, J., Liu, H., Chen, F., Qu, L. and Tian, M. (2024). Underwater gesture recognition meta-gloves for marine immersive communication, *ACS Nano*, *18*, 10818–10828.
21. Zhou, Z., Chen, K., Li, X., Zhang, S., Wu, Y., Zhou, Y., Meng, K., Sun, C., He, Q. and Fan, W. (2020). Sign-to-speech translation using machine-learning-assisted stretchable sensor arrays, *Nature Electronics*, *3*, 571–578.
22. Wen, F., Zhang, Z., He, T. and Lee, C. (2021). AI enabled sign language recognition and VR space bidirectional communication using triboelectric smart glove, *Nature Communications*, *12*, 5378.
23. Kim, K.K., Ha, I., Kim, M., Choi, J., Won, P., Jo, S. and Ko, S.H. (2020). A deep-learned skin sensor decoding the epicentral human motions, *Nature Communications*, *11*, 2149.
24. Wang, M., Yan, Z., Wang, T., Cai, P., Gao, S., Zeng, Y., Wan, C., Wang, H., Pan, L. and Yu, J. (2020). Gesture recognition using a bioinspired learning architecture that integrates visual data with somatosensory data from stretchable sensors, *Nature Electronics*, *3*, 563–570.
25. Gould, J. (2018). Superpowered skin, *Nature*, *563*, S84–S85.

26. Zimmerman, A., Bai, L. and Ginty, D.D. (2014). The gentle touch receptors of mammalian skin, *Science*, *346*, 950–954.
27. Lumpkin, E.A. and Caterina, M.J. (2007). Mechanisms of sensory transduction in the skin, *Nature*, *445*, 858–865.
28. Wang, M., Luo, Y., Wang, T., Wan, C., Pan, L., Pan, S., He, K., Neo, A. and Chen, X. (2021). Artificial skin perception, *Advanced Materials*, *33*, 2003014.
29. Qiu, Y., Wang, Z., Zhu, P., Su, B., Wei, C., Tian, Y., Zhang, Z., Chai, H., Liu, A. and Liang, L. (2023). A multisensory-feedback tactile glove with dense coverage of sensing arrays for object recognition, *Chemical Engineering Journal*, *455*, 140890.
30. Jin, J., Wang, S., Zhang, Z., Mei, D. and Wang, Y. (2023). Progress on flexible tactile sensors in robotic applications on objects properties recognition, manipulation and human-machine interactions, *Smart Sensors*, *3*, 8–34.
31. Deng, M., Fan, F. and Wei, X. (2023). Learning-based object recognition via a eutectogel electronic skin enabled soft robotic gripper, *IEEE Robotics and Automation Letters*, *8*, 7424–7431.
32. Shahabi, E., Visentin, F., Mondini, A. and Mazzolai, B. (2023). Octopus-inspired suction cups with embedded strain sensors for object recognition, *Advanced Intelligent Systems*, *5*, 2200201.
33. Wang, S., Zhang, Z., Yang, B., Zhang, X., Shang, H., Jiang, L., Liu, H., Zhang, J. and Hu, P. (2023). High sensitivity tactile sensors with ultrabroad linear range based on gradient hybrid structure for gesture recognition and precise grasping, *Chemical Engineering Journal*, *457*, 141136.
34. Jin, T., Sun, Z., Li, L., Zhang, Q., Zhu, M., Zhang, Z., Yuan, G., Chen, T., Tian, Y., Hou, X., et al. (2020). Triboelectric nanogenerator sensors for soft robotics aiming at digital twin applications, *Nature Communications*, *11*, 5381.
35. Li, G., Liu, S., Wang, L. and Zhu, R. (2020). Skin-inspired quadruple tactile sensors integrated on a robot hand enable object recognition, *Science Advances*, *5*, eabc8134.
36. Sundaram, S., Kellnhofer, P., Li, Y., Zhu, J.-Y., Torralba, A. and Matusik, W. (2019). Learning the signatures of the human grasp using a scalable tactile glove, *Nature*, *569*, 698–702.
37. Han, J.H., Bae, K.M., Hong, S.K., Park, H., Kwak, J.-H., Wang, H.S., Joe, D.J., Park, J.H., Jung, Y.H. and Hur, S. (2018). Machine learning-based self-powered acoustic sensor for speaker recognition, *Nature Electronics*, *53*, 658–665.
38. Park, J., Kang, D.-H., Chae, H., Ghosh, S.K., Jeong, C., Park, Y., Cho, S., Lee, Y., Kim, J. and Ko, Y. (2022). Frequency-selective acoustic and haptic smart skin for dual-mode dynamic/static human-machine interface, *Science Advances*, *8*, eabj9220.

39. Jiang, Y., Zhang, Y., Ning, C., Ji, Q., Peng, X., Dong, K. and Wang, Z.L. (2022). Ultrathin eardrum-inspired self-powered acoustic sensor for vocal synchronization recognition with the assistance of machine learning, *Small*, *18*, 2106960.
40. Liu, Y., Li, H., Liang, X., Deng, H., Zhang, X., Heidari, H., Ghannam, R. and Zhang, X. (2023). Speech recognition using intelligent piezoresistive sensor based on polystyrene sphere microstructures, *Advanced Intelligent Systems*, *5*, 2200427.
41. Zhou, J., Chen, T., He, Z., Sheng, L. and Lu, X. (2023). Stretchable, ultralow detection limit and anti-interference hydrogel strain sensor for intelligent throat speech recognition using Resnet50 neural network, *Journal of Materials Chemistry C*, *11*, 13476–13487.
42. Yang, Q., Jin, W., Zhang, Q., Wei, Y., Guo, Z., Li, X., Yang, Y., Luo, Q., Tian, H. and Ren, T.-L. (2023). Mixed-modality speech recognition and interaction using a wearable artificial throat, *Nature Machine Intelligence*, *5*, 169–180.
43. Xu, S., Yu, J.-X., Guo, H., Tian, S., Long, Y., Yang, J. and Zhang, L. (2023). Force-induced ion generation in zwitterionic hydrogels for a sensitive silent-speech sensor, *Nature Communications*, *14*, 219.
44. Liu, Y., Norton, J., Qazi, R. and Zhang, Z. (2016). Epidermal mechanoacoustic sensing electronics for cardiovascular diagnostics and human-machine interfaces, *Science Advances*, *2*, e1601185.
45. Nayeem, M.O.G., Lee, S., Jin, H., Matsuhisa, N. and Someya, T. (2020). All-nanofiber-based, ultrasensitive, gas-permeable mechanoacoustic sensors for continuous long-term heart monitoring, *Proceedings of the National Academy of Sciences*, *117*, 201920911.
46. Le, Q.V., Suh, J., Choi, J.J., Park, G.T., Lee, J.W., Shim, G. and Oh, Y.K. (2024). In Situ Nanoadjuvant-Assembled Tumor Vaccine for Preventing Long-Term Recurrence. *ACS Nano*, *7*, 7442–74627.
47. Ziefle, M., Röcker, C., Rennies, J., Goetze, S. and Appell, J.E. (2011). Personalized acoustic interfaces for human-computer interaction. *Human-Centered Design of E-Health Technologies*, *22*, 1–200.
48. Peng, X., Dong, K., Ye, C., Jiang, Y., Zhai, S., Cheng, R., Liu, D., Gao, X., Wang, J. and Wang, Z.L. (2020). A breathable, biodegradable, antibacterial, and self-powered electronic skin based on all-nanofiber triboelectric nanogenerators, *Science Advances*, *6*, eaba9624.
49. Hou, C., Xu, Z., Qiu, W., Wu, R., Wang, Y., Xu, Q., Liu, X.Y. and Guo, W. (2019). A Biodegradable and stretchable protein-based sensor as artificial electronic skin for human motion detection, *Small*, *15*, 1805084.
50. Ghosh, S.K., Park, J., Na, S., Kim, M.P. and Ko, H. (2021). A fully biodegradable ferroelectric skin sensor from edible porcine skin gelatine, *Advanced Science*, *8*, 2005010.

51. Shin, Y.E., Park, Y.J., Ghosh, S.K., Lee, Y., Park, J. and Ko, H. (2022). Ultrasensitive multimodal tactile sensors with skin-inspired microstructures through localized ferroelectric polarization, *Advanced Science*, *9*, 2105423.
52. Hou, S., Huang, Q., Zhang, H., Chen, Q., Wu, C., Wu, M., Meng, C., Yao, K., Yu, X. and Roy, V.A. (2024). Biometric-tuned e-skin sensor with real fingerprints provides insights on tactile perception: Rosa Parks had better surface vibrational sensation than Richard Nixon, *Advanced Science*, *11*, 2400234.
53. Noor, A., Sun, M., Zhang, X., Li, S., Dong, F., Wang, Z., Si, J., Zou, Y. and Xu, M. (2024). Recent advances in Triboelectric tactile sensors for robot hand, *Materials Today Physics*, *46*, 101496.
54. Bai, N., Xue, Y., Chen, S., Shi, L., Shi, J., Zhang, Y., Hou, X., Cheng, Y., Huang, K. and Wang, W. (2023). A robotic sensory system with high spatiotemporal resolution for texture recognition, *Nature Communications*, *14*, 7121.
55. Xing, P., An, S., Wu, Y., Li, G., Liu, S., Wang, J., Cheng, Y., Zhang, Y. and Pu, X. (2023). A triboelectric tactile sensor with flower-shaped holes for texture recognition, *Nano Energy*, *116*, 108758.
56. Qiao, H., Sun, S. and Wu, P. (2023). Non-equilibrium-growing aesthetic ionic skin for fingertip-like strain-undisturbed tactile sensation and texture recognition, *Advanced Materials*, *35*, 2300593.
57. Park, J., Kim, M., Lee, Y., Lee, H.S. and Ko, H. (2015). Fingertip skin-inspired microstructured ferroelectric skins discriminate static/dynamic pressure and temperature stimuli, *Science Advances*, *1*, e1500661.
58. Chun, S., Son, W., Kim, H., Lim, S.K., Pang, C. and Choi, C. (2019). Self-powered pressure- and vibration-sensitive tactile sensors for learning technique-based neural finger skin, *Nano Letters*, *5*, 3305–3312.
59. Pang, K., Song, X., Xu, Z., Liu, X. and Gao, C. (2020). Hydroplastic foaming of graphene aerogels and artificially intelligent tactile sensors, *Science Advances*, *6*, eabd4045.
60. Zhang, H., Li, H. and Li, Y. (2024). Biomimetic electronic skin for robots aiming at superior dynamic-static perception and material cognition based on triboelectric-piezoresistive effects, *Nano Letters*, *24*, 4002–4011.
61. Sun, Q., Ren, G., He, S., Tang, B., Li, Y., Wei, Y., Shi, X., Tan, S., Yan, R. and Wang, K. (2024). Charge dispersion strategy for high-performance and rain-proof triboelectric nanogenerator, *Advanced Materials*, *36*, 2307918.
62. Zhao, H., Zhang, Y., Han, L., Qian, W., Wang, J., Wu, H., Li, J., Dai, Y., Zhang, Z. and Bowen, C.R. (2024). Intelligent recognition using ultralight multifunctional nano-layered carbon aerogel sensors with human-like tactile perception, *Nano-Micro Letters*, *16*, 11.

63. Wu, T., Ding, M., Shi, C., Qiao, Y., Wang, P., Qiao, R., Wang, X. and Zhong, J. (2020). Resorbable polymer electrospun nanofibers: History, shapes and application for tissue engineering, *Chinese Chemical Letters, 31*, 617–625.
64. Wei, X., Li, H., Yue, W., Gao, S., Chen, Z., Li, Y. and Shen, G. (2022). A high-accuracy, real-time, intelligent material perception system with a machine-learning-motivated pressure-sensitive electronic skin, *Microsystems & Nanoengineering, 5*, 1481–1501.
65. Li, Y., Zhang, Y., Yi, J., Peng, X., Cheng, R., Ning, C., Sheng, F., Wang, S., Dong, K. and Wang, Z.L. (2022). Large-scale fabrication of core-shell triboelectric braided fibers and power textiles for energy harvesting and plantar pressure monitoring, *EcoMat*, e12191.
66. Wei, X., Wang, B., Wu, Z. and Wang, Z.L. (2022). An open-environment tactile sensing system: Toward simple and efficient material identification, *Advanced Materials, 34*, 2203073.
67. Zhu, M., Sun, Z. and Lee, C. (2022). Soft modular glove with multimodal sensing and augmented haptic feedback enabled by materials' multifunctionalities, *ACS Nano, 16*, 14097–14110.
68. Shu, S., Wang, Z., Chen, P., Zhong, J., Tang, W. and Wang, Z.L. (2023). Machine-learning assisted electronic skins capable of proprioception and exteroception in soft robotics, *Advanced Materials, 35*, 2211385.
69. Cao, L., Ye, C., Zhang, H., Yang, S., Shan, Y., Lv, Z., Ren, J. and Ling, S. (2023). An artificial motion and tactile receptor constructed by hyperelastic double physically cross-linked silk fibroin ionoelastomer, *Advanced Functional Materials, 33*, 2301404.
70. Li, L., Zhao, S., Ran, W., Li, Z., Yan, Y., Zhong, B., Lou, Z., Wang, L. and Shen, G. (2022). Dual sensing signal decoupling based on tellurium anisotropy for VR interaction and neuro-reflex system application, *Nature Communications, 13*, 5975.
71. Niu, H., Li, H., Gao, S., Li, Y., Wei, X., Chen, Y., Yue, W., Zhou, W. and Shen, G. (2022). Perception-to-cognition tactile sensing based on artificial-intelligence-motivated human full-skin bionic electronic skin, *Advanced Materials, 34*, 2202622.
72. Liu, J., Zhao, W., Li, J., Li, C., Xu, S., Sun, Y., Ma, Z., Zhao, H. and Ren, L. (2024). Multimodal and flexible hydrogel-based sensors for respiratory monitoring and posture recognition, *Biosensors and Bioelectronics, 243*, 115773.
73. Zhang, H., Chen, X., Liu, Y., Yang, C., Liu, W., Qi, M. and Zhang, D. (2024). PDMS film-based flexible pressure sensor array with surface protruding structure for human motion detection and wrist posture recognition, *ACS Applied Materials & Interfaces, 16*, 2554–2563.

74. Yu, J., Xian, S., Zhang, Z., Hou, X., He, J., Mu, J., Geng, W., Qiao, X., Zhang, L. and Chou, X. (2023). Synergistic piezoelectricity enhanced BaTiO3/polyacrylonitrile elastomer-based highly sensitive pressure sensor for intelligent sensing and posture recognition applications, *Nano Research*, *16*, 5490–5502.
75. Sun, P., Cai, N., Zhong, X., Zhao, X., Zhang, L. and Jiang, S. (2021). Facile monitoring for human motion on fireground by using MiEs-TENG sensor, *Nano Energy*, *89*, 106492.
76. Turner, A. and Hayes, S. (2019). The classification of minor gait alterations using wearable sensors and deep learning, *IEEE Transactions on Biomedical Engineering*, *66*, 3136–3145.
77. Zhu, M., Shi, Q., He, T., Yi, Z. and Yiming, M. (2019). Self-powered and self-functional cotton sock using piezoelectric and triboelectric hybrid mechanism for healthcare and sports monitoring, *ACS Nano*, *2*, 1940–1952.
78. Zhang, Z., He, T., Zhu, M., Sun, Z., Shi, Q., Zhu, J., Dong, B., Yuce, M.R. and Lee, C. (2020). Deep learning-enabled triboelectric smart socks for IoT-based gait analysis and VR applications, *NPJ Flexible Electronics*, *4*, 29.
79. Guo, X., He, T., Zhang, Z., Luo, A. and Lee, C. (2021). Artificial intelligence-enabled caregiving walking stick powered by ultra-low-frequency human motion, *ACS Nano*, *12*, 19054–19069.
80. Zhang, Z., Shi, Q., He, T., Guo, X., Dong, B., Lee, J. and Lee, C. (2021). Artificial intelligence of toilet (AI-Toilet) for an integrated health monitoring system (IHMS) using smart triboelectric pressure sensors and image sensor, *Nano Energy*, *90*, 106517.
81. Lee, H.J., Yang, J.C., Choi, J., Kim, J., Lee, G.S., Sasikala, S.P., Lee, G.-H., Park, S.-H.K., Lee, H.M., Sim, J.Y., et al. (2021). Hetero-dimensional 2D Ti_3C_2Tx MXene and 1D graphene nanoribbon hybrids for machine learning-assisted pressure sensors, *ACS Nano*, *15*, 10347–10356.
82. Shi, Q., Zhang, Z., He, T., Sun, Z., Wang, B., Feng, Y., Shan, X., Salam, B. and Lee, C. Deep learning enabled smart mats as a scalable floor monitoring system, *Nature Communications*, *11*, 4609.
83. Cheng, Y., Zhan, Y., Guan, F., Shi, J., Wang, J., Sun, Y., Zubair, M., Yu, C. and Guo, C.F. (2024). Displacement-pressure biparametrically regulated softness sensory system for intraocular pressure monitoring, *National Science Review*, *11*, nwae050.
84. Zarei, M., Lee, G., Lee, S.G. and Cho, K. (2023). Advances in biodegradable electronic skin: Material progress and recent applications in sensing, robotics, and human-machine interfaces, *Advanced Materials*, *35*, 2203193.
85. Wang, S., Wang, X., Wang, Q., Ma, S., Xiao, J., Liu, H., Pan, J., Zhang, Z. and Zhang, L. (2023). Flexible optoelectronic multimodal proximity/pressure/temperature sensors with low signal interference, *Advanced Materials*, *35*, 2304701.

86. Zhang, T., Ding, Y., Hu, C., Zhang, M., Zhu, W., Bowen, C.R., Han, Y. and Yang, Y. (2023). Self-powered stretchable sensor arrays exhibiting magnetoelasticity for real-time human-machine interaction, *Advanced Materials*, *35*, 2203786.
87. Shi, Y., Guan, Y., Liu, M., Kang, X., Tian, Y., Deng, W., Yu, P., Ning, C., Zhou, L. and Fu, R. (2024). Tough, antifreezing, and piezoelectric organohydrogel as a flexible wearable sensor for human-machine interaction, *ACS Nano*, *18*, 3720–3732.
88. Wang, Q., Li, Y., Xu, Q., Yu, H., Zhang, D., Zhou, Q., Dhakal, R., Li, Y. and Yao, Z. (2023). Finger-coding intelligent human-machine interaction system based on all-fabric ionic capacitive pressure sensors, *Nano Energy*, *116*, 108783.
89. Dai, N., Guan, X., Lu, C., Zhang, K., Xu, S., Lei, I.M., Li, G., Zhong, Q., Fang, P. and Zhong, J. (2023). A Flexible self-powered noncontact sensor with an ultrawide sensing range for human-machine interactions in harsh environments, *ACS Nano*, *17*, 24814–24825.
90. Shi, Y., Guan, Y., Liu, M., Kang, X., Tian, Y., Deng, W., Yu, P., Ning, C., Zhou, L., Fu, R., *et al.* (2024). Tough, antifreezing, and piezoelectric organohydrogel as a flexible wearable sensor for human-machine interaction, *ACS Nano*, *18*, 3720–3732.
91. Huo, Z., Wang, X., Zhang, Y., Wan, B., Wu, W., Xi, J., Yang, Z., Hu, G., Li, X. and Pan, C. (2020). High-performance Sb-doped p-ZnO NW films for self-powered piezoelectric strain sensors, *Nano Energy*, *73*, 104744.
92. Deng, W., Yang, T., Jin, L., Yan, C., Huang, H., Chu, X., Wang, Z., Xiong, D., Tian, G. and Gao, Y. (2019). Cowpea-structured PVDF/ZnO nanofibers based flexible self-powered piezoelectric bending motion sensor towards remote control of gestures, *Nano Energy*, *55*, 516–525.
93. Lim, S., Son, D., Kim, J., Lee, Y.B., Song, J.-K., Choi, S., Lee, D.J., Kim, J.H., Lee, M., Hyeon, T., *et al.* (2015). Transparent and stretchable interactive human machine interface based on patterned graphene heterostructures, *Advanced Functional Materials*, *25*, 375–383.
94. Syu, M.H., Lo, Y.J., Fuh, W.C. and Kuen, Y. (2020). Biomimetic and porous nanofiber-based hybrid sensor for multifunctional pressure sensing and human gesture identification via deep learning method, *Nano Energy*, *76*, 105029.
95. Zhong, M., Liu, L., Zhou, X., Zhang, M., Wang, Y., Yang, L. and Wei, D. (2021). Wide linear range and highly sensitive flexible pressure sensor based on multistage sensing process for health monitoring and human-machine interfaces, *Chemical Engineering Journal*, *412*, 128649.
96. Wang, J., He, J., Ma, L., Yao, Y., Zhu, X., Peng, L., Liu, X., Li, K. and Qu, M. (2021). A humidity-resistant, stretchable and wearable textile-based triboelectric nanogenerator for mechanical energy harvesting and

multifunctional self-powered haptic sensing, *Chemical Engineering Journal, 423*, 130200.
97. Chen, S., Li, J. and Liu, H. (2022). Pruney fingers-inspired highly stretchable and sensitive piezoresistive fibers with isotropic wrinkles and robust interfaces, *Chemical Engineering Journal, 430*, P433.
98. Zhu, M., Sun, Z., Zhang, Z., Shi, Q., He, T., Liu, H., Chen, T. and Lee, C. (2020). Haptic-feedback smart glove as a creative human-machine interface (HMI) for virtual/augmented reality applications, *Advanced Science, 6*, eaaz8693.
99. Jin, T., Sun, Z., Li, L., Zhang, Q. and Lee, C. (2020). Triboelectric nanogenerator sensors for soft robotics aiming at digital twin applications, *Nature Communications, 11*, 5381.
100. Gao, F.-L., Liu, J., Li, X.-P., Ma, Q., Zhang, T., Yu, Z.-Z., Shang, J., Li, R.-W. and Li, X. (2023). Ti3C2Tx MXene-Based Multifunctional Tactile Sensors for Precisely Detecting and Distinguishing Temperature and Pressure Stimuli, *ACS Nano, 17*, 16036–16047.
101. Zhang, Y., Lu, Q., He, J., Huo, Z., Zhou, R., Han, X., Jia, M., Pan, C., Wang, Z.L. and Zhai, J. (2023). Localizing strain via micro-cage structure for stretchable pressure sensor arrays with ultralow spatial crosstalk, *Nature Communications, 14*, 1252.
102. Hua, Q., Sun, J., Liu, H., Bao, R., Yu, R., Zhai, J., Pan, C. and Wang, Z.L. (2018). Skin-inspired highly stretchable and conformable matrix networks for multifunctional sensing, *Nature Communications, 9*, 244.
103. Zhi, X., Ma, S., Xia, Y., Yang, B., Zhang, S., Liu, K., Li, M., Li, S., Wang, P. and Wang, X. (2024). Hybrid tactile sensor array for pressure sensing and tactile pattern recognition, *Nano Energy, 125*, 109532.
104. Zhang, J., Yan, K., Huang, J., Sun, X., Li, J., Cheng, Y., Sun, Y., Shi, Y. and Pan, L. (2024). Mechanically robust, flexible, fast responding temperature sensor and high-resolution array with ionically conductive double cross-linked hydrogel, *Advanced Functional Materials, 34*, 2314433.
105. Bian, Y., Zhu, M., Wang, C., Liu, K., Shi, W., Zhu, Z., Qin, M., Zhang, F., Zhao, Z. and Wang, H. (2024). A detachable interface for stable low-voltage stretchable transistor arrays and high-resolution X-ray imaging, *Nature Communications, 15*, 2624.
106. Xiang, S., Chen, G., Wen, Q., Li, H., Luo, X., Zhong, J., Shen, S., Di Carlo, A., Fan, X. and Chen, J. (2024). Fully addressable textile sensor array for self-powered haptic interfacing, *Matter, 7*, 82–94.
107. Gao, W., Liu, W., Huang, S., Wang, L., Jian, Y., Li, Y., Wu, W. and Shang, L. (2024). Portable hydrogel-based tri-channel fluorescence sensor array for visual detection of multiple explosives, *Nano Research, 17*, 6483–6492.
108. Zhu, B., Ling, Y., Yap, L.W., Yang, M., Lin, F., Gong, S., Wang, Y., An, T., Zhao, Y. and Cheng, W. (2019). Hierarchically structured vertical gold

nanowire array-based wearable pressure sensors for wireless health monitoring, *ACS Applied Materials & Interfaces*, *11*, 29014–29021.
109. Bai, N., Wang, L., Wang, Q., Deng, J., Wang, Y., Lu, P., Huang, J., Li, G., Zhang, Y. and Yang, J. (2020). Graded intrafillable architecture-based iontronic pressure sensor with ultra-broad-range high sensitivity, *Nature Communications*, *11*, 209.
110. Niu, H., Wei, X., Li, H., Yin, F., Wang, W., Seong, R.-S., Shin, Y.K., Yao, Z., Li, Y. and Kim, E.-S. (2024). Micropyramid array bimodal electronic skin for intelligent material and surface shape perception based on capacitive sensing, *Advanced Materials*, *11*, 2305528.
111. Boutry, C.M., Negre, M., Jorda, M., Vardoulis, O. and Bao, Z. (2018). A hierarchically patterned, bioinspired e-skin able to detect the direction of applied pressure for robotics, *Science Robotics*, *3*, eaau6914.
112. Mei, S., Yi, H., Zhao, J., Xu, Y., Shi, L., Qin, Y., Jiang, Y., Guo, J., Li, Z. and Wu, L. (2024). High-density, highly sensitive sensor array of spiky carbon nanospheres for strain field mapping, *Nature Communications*, *15*, 3752.
113. Shi, J., Dai, Y., Cheng, Y., Li, G., Liu, Y., Wang, J., Zhang, R., Bai, N., Cai, M., Zhang, Y., et al. (2023). Embedment of sensing elements for robust, highly sensitive, and cross-talk-free iontronic skins for robotics applications, *Science Advances*, *9*, eadf8831.
114. Park, C. (2019). Interactive skin display with epidermal stimuli electrode. In: *International Workshop on Active-Matrix Flatpanel Displays and Devices*, *6*, 1802351.
115. Peng, Y., Que, M., Lee, H.E., Bao, R., Wang, X., Lu, J., Yuan, Z., Li, X., Tao, J. and Sun, J. (2019). Achieving high-resolution pressure mapping via flexible GaN/ZnO nanowire LEDs array by piezo-phototronic effect, *Nano Energy*, *58*, 633–640.
116. Kim, H.J., Sim, K., Thukral, A. and Yu, C. (2017). Rubbery electronics and sensors from intrinsically stretchable elastomeric composites of semiconductors and conductors, *Science Advances*, *3*, e1701114.
117. Zhang, F., Zang, Y., Huang, D., Di, C.-A. and Zhu, D. (2015). Flexible and self-powered temperature–pressure dual-parameter sensors using microstructure-frame-supported organic thermoelectric materials, *Nature Communications*, *6*, 8356.
118. You, I., Mackanic, D.G., Matsuhisa, N., Kang, J. and Jeong, U. (2020). Artificial multimodal receptors based on ion relaxation dynamics, *Science*, *370*, 961–965.
119. Mao, J., Zhou, P., Wang, X., Yao, H., Liang, L., Zhao, Y., Zhang, J., Ban, D. and Zheng, H. (2023). A health monitoring system based on flexible triboelectric sensors for intelligence medical internet of things and its applications in virtual reality, *Nano Energy*, *118*, 108984.

120. Kachouei, M.A., Kaushik, A. and Ali, M.A. (2023). Internet of Things-enabled food and plant sensors to empower sustainability, *Advanced Intelligent Systems*, *5*, 2300321.
121. Luo, Y., Abidian, M.R., Ahn, J.-H., Akinwande, D., Andrews, A.M., Antonietti, M., Bao, Z., Berggren, M., Berkey, C.A. and Bettinger, C.J. (2023). Technology roadmap for flexible sensors, *ACS Nano*, *17*, 5211–5295.
122. Lu, P., Liao, X., Guo, X., Cai, C., Liu, Y., Chi, M., Du, G., Wei, Z., Meng, X. and Nie, S. (2024). Gel-Based triboelectric nanogenerators for flexible sensing: principles, properties, and applications, *Nano-Micro Letters*, *16*, 1–47.
123. Chen, X., Hu, H., Zhou, J., Li, Y., Wan, L., Cheng, Z., Chen, J., Xu, J. and Zhou, R. (2024). Indoor photovoltaic materials and devices for self-powered Internet of Things applications, *Materials Today Energy*, *44*, 101621.
124. Kong, L., Li, W., Zhang, T., Ma, H., Cao, Y., Wang, K., Zhou, Y., Shamim, A., Zheng, L. and Wang, X. (2024). Wireless technologies in flexible and wearable sensing: from materials design, system integration to applications, *Advanced Materials*, *36*, 2400333.
125. Meng, K., Liu, Z., Xiao, X., Manshaii, F., Li, P., Yin, J., Wang, H., Mei, H., Sun, Y., He, X., *et al.* Bioinspired Wearable Pulse Sensors for Ambulant Cardiovascular monitoring and biometric authentication, *Advanced Materials*, *34*, 2403163.
126. Liang, Q., Zhang, D., He, T., Zhang, Z., Wu, Y., Zhang, G., Xie, R., Chen, S., Wang, H. and Lee, C. (2023). A multifunctional helical fiber operated in non-contact/contact dual-mode sensing aiming for HMI/VR applications, *Nano Energy*, *117*, 108903.
127. Dong, B., Kim, Z.A., Li, T., Chen, L.D., Mou, R.Y. and Tao, A. (2021). Contactless tracking of humans using non-contact triboelectric sensing technology: Enabling new assistive applications for the elderly and the visually impaired, *Nano Energy*, *9*, 106486.
128. Zhang, Q., Jin, T., Cai, J., Xu, L., He, T., Wang, T., Tian, Y., Li, L., Peng, Y., Lee, C. (2022). Wearable triboelectric sensors enabled gait analysis and waist motion capture for IoT-based smart healthcare applications. *Advanced Science*, *9*, 2103694.
129. Shi, Y. and Shen, G. (2024). Haptic sensing and feedback techniques toward virtual reality, *Research*, *7*, 0333.
130. Yin, F., Niu, H., Kim, E.S., Shin, Y.K., Li, Y. and Kim, N.Y. (2023). Advanced polymer materials-based electronic skins for tactile and non-contact sensing applications, *iScience*, *5*, e12424.
131. Sun, T., Yao, C., Liu, Z., Huang, S., Huang, X., Zheng, S., Liu, J., Shi, P., Zhang, T. and Chen, H. (2024). Machine learning-coupled vertical graphene triboelectric pressure sensors array as artificial tactile receptor for finger action recognition, *Nano Energy*, *123*, 109395.

132. Dai, N., Lei, I.M., Li, Z., Li, Y., Fang, P. and Zhong, J. (2023). Recent advances in wearable electromechanical sensors — Moving towards machine learning-assisted wearable sensing systems, *Nano Energy*, *105*, 108041.
133. Fang, H., Wang, L., Fu, Z., Xu, L., Guo, W., Huang, J., Wang, Z.L. and Wu, H. (2023). Anatomically designed triboelectric wristbands with adaptive accelerated learning for human-machine interfaces, *Advanced Science*, *10*, 2205960.
134. Shang, J., Tang, L., Guo, K., Yang, S., Cheng, J., Dou, J., Yang, R., Zhang, M. and Jiang, X. (2023). Electronic exoneuron based on liquid metal for the quantitative sensing of the augmented somatosensory system, *Microsystems & Nanoengineering*, *9*, 112.
135. Bai, Y., Yin, L., Hou, C., Zhou, Y., Zhang, F., Xu, Z., Li, K. and Huang, Y. (2023). Response regulation for epidermal fabric strain sensors via mechanical strategy, *Advanced Functional Materials*, *33*, 2214119.
136. Kim, J.J., Wang, Y., Wang, H., Lee, S., Yokota, T. and Someya, T. (2021). Skin electronics: Next-generation device platform for virtual and augmented reality, *Advanced Functional Materials*, *31*, 2009602.
137. Novich, S.D. and Eagleman, D.M. (2015). Using space and time to encode vibrotactile information: Toward an estimate of the skin's achievable throughput, *Experimental Brain Research*, *233*, 2777–2788.
138. Yu, X., Xie, Z., Yu, Y., Lee, J. and Rogers, J.A. (2019). Skin-integrated wireless haptic interfaces for virtual and augmented reality, *Nature*, *575*, 473–479.
139. Mishra, S., Kim, Y.S., Intarasirisawat, J., Kwon, Y.T., Lee, Y., Mahmood, M., Lim, H.R., Herbert, R., Yu, K.J. and Ang, C.S. (2020). Soft, wireless periocular wearable electronics for real-time detection of eye vergence in a virtual reality toward mobile eye therapies, *Science Advances*, *6*, eaay1729.
140. Cañón Bermúdez, G.S., Fuchs, H., Bischoff, L., Fassbender, J. and Makarov, D. (2018). Electronic-skin compasses for geomagnetic field-driven artificial magnetoreception and interactive electronics, *Nature Electronics*, *1*, 589–595.
141. Ge, J., Wang, X., Drack, M., Volkov, O., Liang, M., Bermúdez, G.S.C., Illing, R., Wang, C., Zhou, S. and Group, J.F. (2019). A bimodal soft electronic skin for tactile and touchless interaction in real time, *Nature Physics*, *10*, 4405.
142. Cañón Bermúdez, G.S., Karnaushenko, D.D., Karnaushenko, D., Lebanov, A., Bischoff, L., Kaltenbrunner, M., Fassbender, J., Schmidt, O.G. and Makarov, D. (2018). Magnetosensitive e-skins with directional perception for augmented reality, *Science Advances*, *4*, eaao2623.
143. Lai, Q.T., Zhao, X.H., Sun, Q.J., Tang, Z., Tang, X.G. and Roy, V.A. (2023). Emerging MXene-Based flexible tactile sensors for health monitoring and haptic perception, *Small*, *19*, 2300283.

144. Xu, D., Ouyang, Z., Dong, Y., Yu, H.-Y., Zheng, S., Li, S. and Tam, K.C. (2023). Robust, breathable and flexible smart textiles as multifunctional sensor and heater for personal health management, *Advanced Fiber Materials*, *5*, 282–295.
145. Xue, P., Chen, Y., Xu, Y., Valenzuela, C., Zhang, X., Bisoyi, H.K., Yang, X., Wang, L., Xu, X. and Li, Q. (2022). Bioinspired MXene-based soft actuators exhibiting angle-independent structural color, *Nano-Micro Letters*, *15*, 1.
146. Liu, T., Gou, G.-Y., Gao, F., Yao, P., Wu, H., Guo, Y., Yin, M., Yang, J., Wen, T., Zhao, M., et al. (2023). Multichannel flexible pulse perception array for intelligent disease diagnosis system, *ACS Nano*, *17*, 5673–5685.
147. Tran, V.V., Phung, V.-D. and Lee, D. (2024). Recent advances and innovations in the design and fabrication of wearable flexible biosensors and human health monitoring systems based on conjugated polymers, *Bio-Design and Manufacturing*, *7*, 1–41.
148. Yu, A., Zhu, M., Chen, C., Li, Y., Cui, H., Liu, S. and Zhao, Q. (2024). Implantable flexible sensors for health monitoring, *Advanced Healthcare Materials*, *13*, 2302460.
149. Huang, X., Liu, Y., Park, W., Li, J., Ma, J., Yiu, C.K., Zhang, Q., Li, J., Wu, P. and Zhou, J. (2023). Intelligent soft sweat sensors for the simultaneous healthcare monitoring and safety warning, *Advanced Healthcare Materials*, *12*, 2202846.
150. Sun, T., Feng, B., Huo, J., Xiao, Y., Wang, W., Peng, J., Li, Z., Du, C., Wang, W. and Zou, G. (2024). Artificial intelligence meets flexible sensors: Emerging smart flexible sensing systems driven by machine learning and artificial synapses, *Nano-Micro Letters*, *16*, 14.
151. Sun, Q.J., Lai, Q.T., Tang, Z., Tang, X.G., Zhao, X.H. and Roy, V.A. (2023). Advanced functional composite materials toward E-skin for health monitoring and artificial intelligence, *Advanced Materials Technologies*, *8*, 2201088.
152. Zhang, Y., Xu, Z., Yuan, Y., Liu, C., Zhang, M., Zhang, L. and Wan, P. (2023). Flexible antiswelling photothermal-therapy MXene hydrogel-based epidermal sensor for intelligent human-machine interfacing, *Advanced Functional Materials*, *33*, 2300299.
153. Park, D.Y., Joe, D.J., Kim, D.H., Park, H., Han, J.H., Jeong, C.K., Park, H., Park, J.G., Joung, B. and Lee, K.J. (2017). Self-powered real-time arterial pulse monitoring using ultrathin epidermal piezoelectric sensors, *Advanced Materials*, *29*, 1702308.
154. Dai, M., Wang, Z., Wang, F., Qiu, Y. and Hu, P.A. (2019). Two-dimensional van der Waals materials with aligned in-plane polarization and large piezoelectric effect for self-powered piezoelectric sensors, *Nano Letters*, *19*, 5410–5416.

155. Cho, M.B.K., Cho, K. and Lee, S.G. (2024). Mechanically robust and linearly sensitive soft piezoresistive pressure sensor for a wearable human-robot interaction system, *ACS Nano*, *18*, 3151–3160.
156. Chen, J., Liu, H., Wang, W., Nabulsi, N., Zhao, W., Kim, J.Y., Kwon, M.-K. and Ryou, J.-H. (2019). High durable, biocompatible, and flexible piezoelectric pulse sensor using single-crystalline III-N thin film, *Advanced Materials*, *29*, 1903162.
157. Chu, Y., Zhong, J., Liu, H., Ma, Y., Liu, N., Song, Y., Liang, J., Shao, Z., Sun, Y. and Dong, Y. (2018). Human pulse diagnosis for medical assessments using a wearable piezoelectret sensing system, *Advanced Materials*, *28*, 1803413.
158. Song, Y., Min, J., Yu, Y., Wang, H. and Gao, W. (2020). Wireless battery-free wearable sweat sensor powered by human motion, *Science Advances*, *6*, eaay9842.
159. Yu, Y., Nassar, J., Xu, C., Min, J. and Gao, W. (2020). Biofuel-powered soft electronic skin with multiplexed and wireless sensing for human-machine interfaces, *Science Robotics*, *5*, eaaz7946.
160. Zhao, L., Wang, L., Zheng, Y., Zhao, S. and Han, W. (2021). Highly-stable polymer-crosslinked 2D MXene-based flexible biocompatible electronic skins for in vivo biomonitoring, *Nano Energy*, *84*, 105921.
161. Wang, S., Chen, G., Yao, B., Chee, A.J.Y. and Yu, A.C.H. (2021). *In situ* and intraoperative detection of the ureter injury using a highly sensitive piezoresistive sensor with a tunable porous structure, *ACS Applied Materials & Interfaces*, *13*, 21669–21679.
162. Lu, P., Wang, L., Zhu, P., Huang, J. and Guo, C.F. (2021). Iontronic pressure sensor with high sensitivity and linear response over a wide pressure range based on soft micropillared electrodes, *Science Bulletin*, *66*, 1091–1100.
163. Kim, J., Kim, M., Lee, M.S., Kim, K., Ji, S., Kim, Y.T., Park, J., Na, K., Bae, K.H. and Kim, H.K. Wearable smart sensor systems integrated on soft contact lenses for wireless ocular diagnostics, *Nature Communications*, *8*, 14997.
164. Dagdeviren, C., Shi, Y., Joe, P., Ghaffari, R., Balooch, G., Usgaonkar, K., Gur, O., Tran, P.L., Crosby, J.R. and Meyer, M. (2015). Conformal piezoelectric systems for clinical and experimental characterization of soft tissue biomechanics, *Nature Materials*, *14*, 728–736.
165. Yu, X., Wang, H., Ning, X., Sun, R., Albadawi, H., Salomao, M., Silva, A.C., Yu, Y., Tian, L. and Koh, A. (2018). Needle-shaped ultrathin piezoelectric microsystem for guided tissue targeting via mechanical sensing, *Nature Biomedical Engineering*, *2*, 165–172.
166. Sun, R., Carreira, S.C., Chen, Y., Xiang, C. and Rossiter, J. (2019). Stretchable piezoelectric sensing systems for self-powered and wireless health monitoring, *Advanced Materials Technologies*, *4*, 1900100.

167. Dagdeviren, C., Traverso, C.G. and Langer, R.S. Flexible piezoelectric devices for gastrointestinal motility sensing, *Nature Biomedical Engineering*, *1*, 807–817.
168. Lin, Y., Chen, J., Tavakoli, M.M., Gao, Y., Zhu, Y., Zhang, D., Kam, M., He, Z. and Fan, Z. (2019). Printable fabrication of a fully integrated and self-powered sensor system on plastic substrates, *Advanced Materials*, *31*, 1804285.
169. Lim, K.W., Peddigari, M., Park, C.H., Lee, H.Y., Min, Y., Kim, J.W., Ahn, C.W., Choi, J.J., Hahn, B.D. and Choi, J.H. (2019). A high output magneto-mechano-triboelectric generator enabled by accelerated water-soluble nano-bullets for powering a wireless indoor positioning system, *Energy & Environmental Science*, *12*, 666–674.
170. Kim, C.S., Yang, H.M., Lee, J., Lee, G.S., Choi, H., Kim, Y., Lim, S.H., Cho, S.H. and Cho, B.J. (2018). Self-powered wearable electrocardiography using a wearable thermoelectric power generator, *ACS Energy Letters*, acsenergylett.7b01237. *3*, 501–507.
171. Wang, J., Wang, H., He, T., He, B., Thakor, N.V. and Lee, C. (2019). Investigation of low-current direct stimulation for rehabilitation treatment related to muscle function loss using self-powered TENG system, *Advanced Science*, *6*, 1900149.
172. Ronan, H., Hong-Joon, Y., Hanjun, R., Moo-Kang, K., Eue-Keun, C., Dong-Sun, K. and Sang-Woo, K. (2020). Transcutaneous ultrasound energy harvesting using capacitive triboelectric technology, *Science*, *365*, 491–494.
173. Boutry, C.M., Beker, L., Kaizawa, Y., Vassos, C., Tran, H., Hinckley, A.C., Pfattner, R., Niu, S., Li, J. and Claverie, J. (2019). Biodegradable and flexible arterial-pulse sensor for the wireless monitoring of blood flow, *Nature Biomedical Engineering*, *3*, 47–57.
174. Chang, S., Koo, J.H., Yoo, J., Kim, M.S., Choi, M.K., Kim, D.-H. and Song, Y.M. (2024). Flexible and stretchable light-emitting diodes and photodetectors for human-centric optoelectronics, *Chemical Reviews*, *124*, 768–859.
175. Yu, S., Park, T.H., Jiang, W., Lee, S.W., Kim, E.H., Lee, S., Park, J.E. and Park, C. (2023). Soft human-machine interface sensing displays: Materials and devices, *Advanced Materials*, *35*, 2204964.
176. Sun, L., Wang, J., Matsui, H., Lee, S., Wang, W., Guo, S., Chen, H., Fang, K., Ito, Y. and Inoue, D. (2024). All-solution-processed ultraflexible wearable sensor enabled with universal trilayer structure for organic optoelectronic devices, *Science Advances*, *10*, eadk9460.
177. Ding, Y., Xiong, S., Sun, L., Wang, Y., Zhou, Y., Li, Y., Peng, J., Fukuda, K., Someya, T. and Liu, R. (2024). Metal nanowire-based transparent electrode for flexible and stretchable optoelectronic devices, *Chemical Society Reviews*, *53*, 7784–7827.
178. Zhao, J., Lo, L.-W., Yu, Z. and Wang, C. (2023). Handwriting of perovskite optoelectronic devices on diverse substrates, *Nature Physics*, *17*, 964–971.

179. Lee, S., Kim, J., Kwon, H., Son, D., Kim, I.S. and Kang, J. (2023). Photoactive materials and devices for energy-efficient soft wearable optoelectronic systems, *Nano Energy*, *110*, 108379.
180. Liu, Y., Wang, T., Sun, Z., Liu, C., Han, L., Liu, R., Liu, Y. and Zhou, Y. (2023). Nature-inspired pressure interactive alternating current electroluminescent for orientation indication, visual sensing, and remote monitoring, *Advanced Materials Technologies*, *8*, 2301289.
181. Cho, S., Ha, J.H., Ahn, J., Han, H., Jeong, Y., Jeon, S., Hwang, S., Choi, J., Oh, Y.S. and Kim, D. (2024). Wireless, battery-free, optoelectronic diagnostic sensor integrated colorimetric dressing for advanced wound care, *Advanced Functional Materials*, *34*, 2316196.
182. Nam, M., Chang, J., Kim, H., Son, Y.H., Jeon, Y., Kwon, J.H. and Choi, K.C. (2024). Highly reliable and stretchable OLEDs based on facile patterning method: Toward stretchable organic optoelectronic devices, *NPJ Flexible Electronics*, *8*, 17.
183. Liu, X., Li, D., Wang, Y., Yang, D. and Pi, X. (2023). Flexible optoelectronic synaptic transistors for neuromorphic visual systems, *Advanced Materials Letters*, *1*, 031501.
184. Ye, Y., Deng, Q., Wu, J., Zhong, C., Ma, H., Shi, Y., Li, D., Tang, R., Tang, Y. and Jian, J. (2024). Electrostatic force-assisted transfer of flexible silicon photodetector focal plane arrays for image sensors, *ACS Applied Materials & Interfaces*, *30*, 39572–39579.
185. Hua, Q. and Shen, G. (2024). Low-dimensional nanostructures for monolithic 3D-integrated flexible and stretchable electronics, *Chemical Society Reviews*, *53*, 1316–1353.
186. Song, Y.M., Xie, Y., Malyarchuk, V., Xiao, J., Jung, I., Choi, K.-J., Liu, Z., Park, H., Lu, C. and Kim, R.-H. (2013). Digital cameras with designs inspired by the arthropod eye, *Nature*, *497*, 95–99.
187. Larson, C., Peele, B., Li, S., Robinson, S., Totaro, M., Beccai, L., Mazzolai, B. and Shepherd, R. (2016). Highly stretchable electroluminescent skin for optical signaling and tactile sensing, *Science*, *351*, 1071–1074.
188. He, J., Wei, R., Ma, X., Wu, W., Pan, X., Sun, J., Tang, J., Xu, Z., Wang, C. and Pan, C. Contactless user-interactive sensing display for human-human and human-machine interactions, *Advanced Materials*, *36*, 2401931.
189. Wang, C., Hwang, D., Yu, Z., Takei, K., Park, J., Chen, T., Ma, B. and Javey, A. (2013). User-interactive electronic skin for instantaneous pressure visualization, *Nature Materials*, *12*, 899–904.
190. Kim, J., Shim, H.J., Yang, J., Choi, M.K., Kim, D.C., Kim, J., Hyeon, T. and Kim, D.H. (2017). Ultrathin quantum dot display integrated with wearable electronics, *Advanced Materials*, *29*, 1700217.

Index

A
active layer, 5, 7–9, 26, 67, 97, 107–108, 213, 280
artificial intelligence, 14, 270, 330–331

B
biosensors, 2, 4, 62, 170, 193–194, 196–200, 202, 205, 207, 337

C
carbon nanotubes, 2, 21, 67, 160, 217

G
gas sensors, 2, 4, 62, 84, 209, 211, 216, 218–219, 223, 321

H
healthcare monitoring, 4, 16–17, 81, 322, 348

I
Internet of Things, 5, 16, 186, 269–270, 319

M
magnetoresistive sensors, 43
microstructure designs, 11–12

P
pressure sensors, 2, 4, 11–12, 23, 26, 60, 63, 75–76, 82, 95, 98–99, 144, 146, 153, 165, 167, 176, 295, 299, 316, 337, 352

R
resistive sensors, 26, 107, 209, 278
robotics, 1–2, 4, 8, 40, 91, 99, 150, 153, 157, 226, 233, 351

S
stretchable sensors, 2, 156–157

T
temperature sensors, 2, 4, 70, 82, 208, 316, 321, 350, 355

W
wearable devices, 1–2, 8, 16, 25, 40, 73–74, 91, 93, 103, 167, 176, 180, 218, 270–271, 330
wireless communication, 3, 319, 337

www.ingramcontent.com/pod-product-compliance
Lightning Source LLC
Chambersburg PA
CBHW050353090625
27790CB00004B/25